Infrared Fibers and Their Applications

Also Available from SPIE Press Exclusively Published by New Age International Pvt. Ltd.

Fourier-Transform Spectroscopy Instrumentation Engineering
Vidi Saptari
ISBN: 978-81-224-2823-0, Original Edition Price: $ 48.00, Indian Edition Price: Rs. 395.00

Tissue Optics : Light Scattering Methods and *Instruments for Medical Diagnosis,* 2nd Ed.
Valery Tuchin
ISBN: 978-81-224-2829-2, Original Edition Price: $ 99.00, Indian Edition Price: Rs. 995.00

Local Approximation Techniques in Signal and Image Processing
Vladimir Katkovnik, Karen Egiazarian, Jaakko Astola
ISBN: 978-81-224-2830-8, Original Edition Price: $ 99.00, Indian Edition Price: Rs. 695.00

Logic-based Nonlinear Image Processing
Stephen Marshall
ISBN: 978-81-224-2825-4, Original Edition Price: $ 51.00, Indian Edition Price: Rs. 395.00

Introduction to the Optical Transfer Function
Charles S. Williams, Orville A. Becklund
ISBN: 978-81-224-2831-5, Original Edition Price: $ 78.00, Indian Edition Price: Rs. 695.00

Digital and Analog Fiber Optic Communications for CATV and FTTx Applications
Avigdor Brillant
ISBN: 978-81-224-2832-2, Original Edition Price: $ 115.00, Indian Edition Price: Rs. 999.00

Infrared Fibers and Their Applications
James A. Harrington
ISBN: 978-81-224-2827-8, Original Edition Price: $ 78.00, Indian Edition Price: Rs. 495.00

Optoelectronics of Solar Cells
Greg P. Smestad
ISBN: 978-81-224-2826-1, Original Edition Price: $ 46.00, Indian Edition Price: Rs. 395.00

MOEMS: Micro-Opto-Electro-Mechanical Systems
Manouchehr E. Motamedi
ISBN: 978-81-224-2833-9, Original Edition Price: $ 113.00, Indian Edition Price: Rs. 795.00

Introduction to Image Stabilization
Scott W. Teare, Sergio R. Restaino
ISBN: 978-81-224-2822-3, Original Edition Price: $ 51.00, Indian Edition Price: Rs. 395.00

Fundamentals of Polarimetric Remote Sensing
John R. Schott
ISBN: 978-81-224-2834-6, Original Edition Price: $ 68.00, Indian Edition Price: Rs. 395.00

Fundamentals of Infrared Detector Materials
Michael A. Kinch
ISBN: 978-81-224-2824-7, Original Edition Price: $ 61.00, Indian Edition Price: Rs. 395.00

Infrared Fibers and Their Applications

James A. Harrington

SPIE
PRESS
Bellingham, Washington USA

NEW AGE

First Indian Sub-Continent Edition : 2010
by New Age International (P) Ltd., Publishers

Branches:

- 36, Malikarjuna Temple Street, Opp. ICWA, Basavanagudi, **Bangalore**. ✆ (080) 26677815
- 26, Damodaran Street, T. Nagar, **Chennai**. ✆ (044) 24353401
- CC-39/1016, Carrier Station Road, Ernakulam South, **Cochin.** ✆ (0484) 4051303
- Hemsen Complex, Mohd. Shah Road, Paltan Bazar, Near Starline Hotel, **Guwahati**. ✆ (0361) 2543669
- No. 105, 1st Floor, Madhiray Kaveri Tower, 3-2-19, Azam Jahi Road, Nimboliadda, **Hyderabad**. ✆ (040) 24652456
- RDB Chambers (Formerly Lotus Cinema) 106A,1st Floor, S.N. Banerjee Road, **Kolkata**. ✆ (033) 22275247
- 16-A, Jopling Road, **Lucknow**. ✆ (0522) 2209578
- 142C, Victor House, Ground Floor, N.M. Joshi Marg, Lower Parel, **Mumbai**. ✆ (022) 24927869
- 22, Golden House, Daryaganj, **New Delhi**. ✆ (011) 23262370, 23262368

Printed in India at Anand Sons, Delhi.
C-09-11-4219

ISBN: 978-81-224-2827-8
Original Edition Price: $ 78.00
Indian Edition Price: Rs. 495.00

PUBLISHING FOR ONE WORLD
NEW AGE INTERNATIONAL (P) LIMITED, PUBLISHERS
4835/24, Ansari Road, Daryaganj, New Delhi-110002
Visit us at **www.newagepublishers.com**

Contents

Preface

This book is an outgrowth of more than 25 years working in the field of infrared fiber optics. I was first introduced to the new and rather arcane area of IR fibers when I was working at Hughes Research Laboratories (HRL) in Malibu, CA. Back in 1977 Hughes had large research programs aimed at the development of highly transparent IR laser window materials, initially for CO_2 lasers operating at 10.6 μm, but later for other IR gas lasers operating between 3 and 12 μm. During this period of active laser window research at HRL and many other laboratories in the U.S., some excellent optical materials were fabricated with very low loss. Specifically, some of the best materials developed included the alkali halides (polycrystalline NaCl and KBr), the alkaline earth fluorides (CaF_2), two-six compounds (ZnSe and ZnS), and sapphire. At about the same time, the U.S. Army was interested in building a surveillance satellite that employed short, <25- cm-long IR transmissive fibers to relay information from the exterior of the satellite to an interior HgCdTe detector. At the time of this army program the requirements placed on the IR fibers were daunting: develop fibers capable of transmitting wavelengths from 2 to as long as 20 μm! There was no fiber in existence in the late 1970s that could come even close to this transmission range. To accomplish this goal, the research group at HRL decided to exploit the technology developed in the pursuit of low-loss IR window materials and somehow fabricate these materials into fiber optics. The first group of materials studied was the chalcogenide glasses, in particular the arsenic sulfide and selenide glasses. While these glasses had, in fact, been made into fiber some years earlier, they turned out to be unsuitable for the army application. Not only did they not transmit sufficiently far into the infrared, but they were also too brittle. Therefore, HRL researchers developed a new technique of hot extrusion to extrude polycrystalline KRS-5 (TlBrI) into fiber. The KRS-5 fibers did, in fact, transmit well beyond 20 μm, but they were ultimately abandoned as they exhibited a nasty aging problem, i.e., loss of transmission in time; they were also toxic.

At essentially the same time as the polycrystalline fibers were being developed at HRL, fluoride glass was invented and drawn into fiber at the University of Rennes by Lucas and Poulain, while Garmire and her group at University of Southern California were constructing simple and efficient hollow rectangular waveguides made from aluminum strips. The impetus for this spurt of IR fiber activity in the late 1970s and early 1980s was the universal desire to have a flexible and reliable fiber optic analogous to silica glass fibers for non-telecommunication applications such as the delivery of CO_2 laser power for laser surgery and IR fiber sensors. But the holy grail was the possibility of reducing the loss in solid-core IR fibers to the theoretical limit, which for many IR materials was as much as 10^3 times lower than the intrinsic loss of 0.2 dB/km for silica fibers. The idea of having a fiber with losses in the range of 10^{-3} dB/km was very enticing. For example, one could envision thousand-kilometer long, repeaterless fiber links for use in undersea communication systems for connecting submarines to land-based stations. Unfortunately, while the theoretical foundations for the predicted intrinsic losses for many IR crystalline and glassy materials are still sound, no IR fiber has ever been produced to date that has a loss less than silica fiber at 1.55 μm. In fact, the real advantage today for IR fiber optics is their application in short-haul applications, which generally require fibers less than 10 m in length.

Over the many years of my involvement with IR fiber optics, I have found that most people are interested in these special fibers because of their obvious need for the transmission of IR light greater than about 2 μm. Their primary concern is determining which IR fiber is best for a particular application. What I have tried to do in this book is to keep this concern paramount in the discussions of the various types of IR fibers. Simply put, this is not a book that will, for example, exhaustively detail the bulk properties of every fluoride glass made to date or any other IR optical material; rather, the emphasis is on the salient properties of the IR fiber materials that are today the best in their class. To this end the reader will find that the discussions surrounding the properties of the materials are always aimed at a better understanding of the IR fibers made from these materials rather than just a listing of all the bulk material properties. I would expect that the reader will first turn to the chapters on the individual IR fibers, Chapters 4 through 7, to find the most important characteristics of a particular IR fiber, and then refer to Chapters 8 through 11 for details on the many applications of these fibers. In this way this book should serve as a ready reference guide for those wishing to learn about the candidate fibers for their particular application.

This book is intended to be a complete reference for the field of IR fiber optics. I have included a historical recounting of the development of IR fiber optics as well as a summary of the best fibers in each of the three categories of IR fibers: crystalline, glass, and hollow waveguides. The theoretical aspects of solid-core fibers and hollow waveguides are covered in Chapters 2 and 3, respectively. Because the optical principles of solid and hollow-core waveguides are somewhat different, I chose to separate the theory into these separate chapters. The next four chapters detail the four main types of IR fibers available today: fluoride glass, chalcogenide glass, crystalline, and hollow waveguides. Each of these chapters is designed to familiarize the reader not only with the properties of each fiber but also the limitations and shortcomings of each fiber type. In fact, the optical and mechanical properties of all IR fibers often fall far short of those for conventional silica fiber. However, this does not mean that there is not a place for an IR fiber; rather, it suggests that the design considerations of any system employing IR fibers should take into account the lower transparency and reduced strength of these special fibers.

The final four chapters cover a wide range of IR fiber applications, including fiber optic chemical and temperature sensors, laser power delivery, fiber amplifiers, and new and emerging applications. The last chapter includes the most recent IR fiber structures based on hollow 1D, photonic bandgap fibers. Photonic bandgap fibers are emerging as a new type of fiber that has the promise of one day exhibiting ultralow loss across a broad IR wavelength range.

Throughout the book I have tried to keep the units consistent for all of the IR fiber types. In particular, I have attempted to keep the units for loss, usually dB/m, and wavelength, usually microns, the same for all of the optical data. I have referenced theoretical sections and equations in the chapters on the individual fibers and their applications. This is an advantage over an edited text where cross referencing between chapters is difficult. Finally, the references at the end of each chapter include the latest developments and, in particular, the recent applications for these fibers. In fact, what is often of most interest is the myriad of new applications for these fibers. It is my hope that readers of this book will find suitable information and encouragement to implement these special fibers in new and novel applications. Many of the IR fibers discussed in the book are now commercially available so that many of the earlier drawbacks to their use have been overcome by better packaging and better technical assistance from the manufacturer. My best recommendation to all those who have asked me about the best IR fiber for a particular application is to first select the best IR fiber candidates based on their optical and mechanical properties and then carefully test and evaluate the candidate fibers, keeping in mind that IR fibers are not equal to silica fibers.

There are many people with whom I have had contact over my years working in this field. The IR fiber community is rather small in contrast to the enormous efforts in manpower and dollars poured into the development of silica fiber optics. Many of my colleagues have been in the field for as long as I have and they are now also enjoying renewed interest and growth in this field. I want to thank all of my IR fiber colleagues who helped me better understand IR optical materials, fiber fabrication, and applications. I would especially like to thank the following: Slava Artjushenko, Larry DeShazer, Martin Drexhage, Jerry Fitzgibbon, A. Ray Hilton, Sr., Abraham Katzir, Jacques Lucas, Mitsunobu Miyagi, Yuji Matsuura, Jas Sanghera, George Sigel, and Danh Tran. I am forever indebted to the many graduate and undergraduate students who have worked tirelessly with me to develop an ever better IR fiber. Thanks to the staff at SPIE Press, including my editor Sharon Streams and Rick Hermann. The writing of this book has been a labor of love supported throughout by my loving wife Jeanne and daughter Julie. Finally, I found time away from my favorite subject of IR fibers to engage in ballroom dancing—a nonoptical form of relaxation and fun.

James A. Harrington

Chapter 1

Introduction to Infrared Fiber Optics

1.0 Historical overview

Infrared (IR) optical fibers may be defined as fiber optics that transmit radiation with wavelengths greater than approximately 2 μm. The first IR fibers were fabricated in the mid-1960s from a rather special class of IR transparent glasses called chalcogenide glasses.[1] It had been known for some time that blending chalcogen elements such as arsenic and sulfur can form a dark red-colored glass that is transparent well beyond 2 μm. Being an excellent glass former, arsenic trisulfide (As_2S_3) was a logical choice to be drawn into crude fiber using a simple fiber drawing apparatus. Kapany et al.[2] reported on these first IR fibers in 1965, but the losses were very high. In fact, the losses were an untenable 10 dB/m over the IR spectrum from 2 to 8 μm. A loss of 10 dB/m means that a mere 1-m length of As_2S_3 fiber would transmit only 10% of the incident light apart from reflection losses from the fiber end faces. The reflection losses would amount to an additional 31%, as this glass has a high refractive index of about 2.3. On top of these problems, their As_2S_3 fiber was quite brittle. During the mid-1970s, the interest in developing an efficient and reliable IR fiber for short-haul applications increased, partly in response to the need for a fiber to link broadband, long-wavelength radiation to remote photodetectors in military sensor applications. In 1975, Hughes Research Laboratories (HRL) researchers in Malibu, CA, began to search for an IR fiber that could be used in a satellite application to link IR radiation in the 3- to 5-μm, 8- to 12-μm, and longer wavelength bands, which was incident on a surveillance satellite, to an interior IR detector array. The basic fiber/detector system as envisioned by HRL is shown schematically in Fig. 1.1. The idea was simple: Take an array of IR fibers and position them so that they would each intercept a large field of view, and then transmit that signal through an IR fiber for a distance of no more than about 50 cm to a mosaic IR detector array. This particular U.S. Army-funded program, called Mosaic Infrared Sensor Technology (MIST), would ultimately require a fiber array consisting of several hundred fibers, each interfaced with one detector element in the detector array. Initially the investigators at HRL experimented with the chalcogenide fibers, but they found the loss too high and the fibers too brittle for their application. Another unfortunate feature of chalcogenide fibers is that they are made from toxic materials. Some people working with them

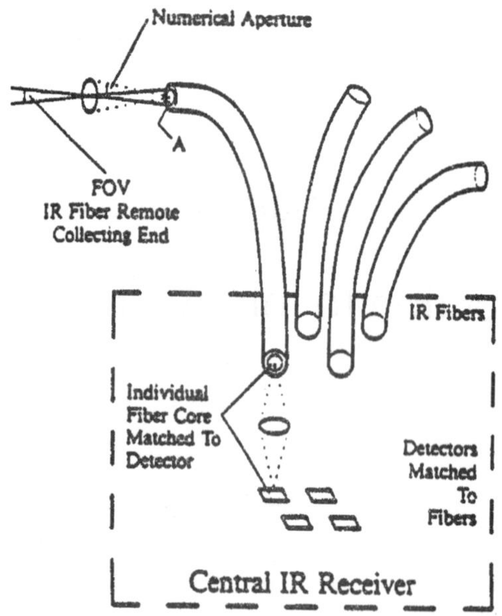

Figure 1.1 Early concept of IR fibers used in a laser threat warning system. A small number of IR fibers access a wide field-of-view (FOV) and transmit the signal to a remote photodetector array.

accidentally poked their fingers with the sharp fiber ends, and this led to some tissue darkening in the areas of broken skin, which they attributed to the death of tissue (necrosis). Needless to say, this aspect contributed to the search for another IR fiber better suited to the satellite application. The IR fiber that grew out of this program was the first polycrystalline fiber made from thallium halides, or KRS-5, as it was known.[3] Oddly enough, KRS-5 fibers were also toxic due to the presence of thallium compounds. Nevertheless, KRS-5 fibers had the advantage of being very flexible and transmitting wavelengths even beyond 20 μm.[4]

At this same time another application emerged that was to help drive IR fiber technology for many years. This application involved the ever-increasing need for a flexible fiber delivery system for transmitting CO_2 laser radiation in surgical applications. Surgical lasers began to be developed only a few years after the invention of the laser in 1960. There were basically two surgical lasers in the early years of their development: the Nd:YAG laser operating at 1.06 μm, and the CO_2 laser operating at 10.6 μm. While each laser could be used for cutting, ablating, and coagulating tissue, the Nd:YAG laser had a distinct advantage over the CO_2 laser in that the 1.06-μm radiation could be delivered by a conventional albeit large-core silica fiber optic. Surgeons quickly became accustomed to the tactile feel and flexibility of a silica fiber; this is especially important when the fiber optic is introduced into the body least invasively through an endoscope into the body. The problem was that there was no counterpart to the silica fiber for the CO_2 laser medical lasers. Instead of a fiber delivery system for 10.6 μm, these lasers employed cumbersome articulated arms. Articulated arms, like the one shown attached to a CO_2 laser in

Fig. 1.2, are made of tubes and mirrors that can move in any direction. It is obvious from Fig. 1.2, however, that these arms are meant only for the delivery of laser power outside the body, as they are too big to be used otherwise. This fact coupled with the easy misalignment of the early arms and their high cost argued strongly for a better, i.e., fiber optic, solution to the problem of delivering of IR surgical laser power. When HRL workers announced the invention of the first polycrystalline KRS-5 fibers in 1976, it was immediately realized that they could also deliver CO_2 laser power. In fact, some quite optimistic scientists at HRL said that KRS-5 fibers could deliver up to 100 W of CO_2 laser power, well before this had ever been tried. It would be some years later before this goal was reached, but the mere ability to deliver this important surgical laser wavelength set many laboratories around the world on the quest for a good 10.6-μm transmitting fiber. After almost 30 years of research on IR fibers, the need for a CO_2 laser fiber similar to silica fiber still remains a goal today. The difference today is that there are many more IR fiber candidates for this wavelength including the popular hollow waveguides. In point of fact, a close analog to silica fiber operating at 10 μm will never be attained because there is no IR material at this wavelength that has optical and mechanical properties equal to that of silica.

There was yet one other motivation for developing IR fibers beside their use in sensor and laser power delivery applications. As will be elucidated in some detail in Chapter 2, many if not most of the IR materials transmitting beyond 2 μm have a theoretical loss that is much less than silica. In fact, when the KRS-5 fiber was developed by HRL it was realized that the intrinsic or fundamental losses for this crystalline material could be as low as 10^{-3} dB/km at approximately 6 μm. This is the minimum theoretical loss for this material, but there are hundreds of other IR optical materials including glasses that have similar low losses. The key issue here

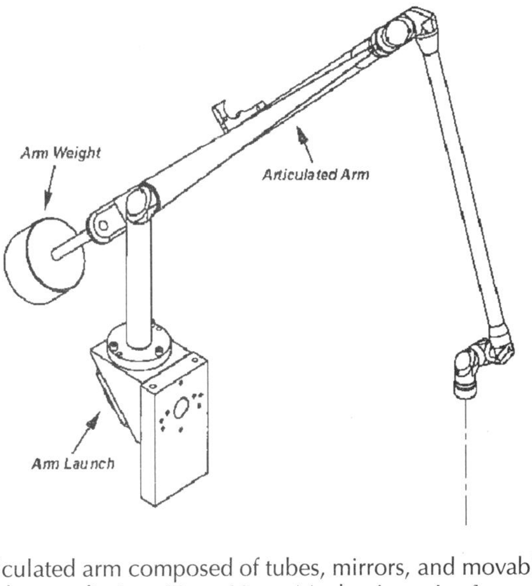

Figure 1.2 Articulated arm composed of tubes, mirrors, and movable joints for the delivery of CO_2 laser radiation. [From Haser Mechanisms, Inc.]

is that silica fibers have a theoretical minimum attenuation of about 0.14 dB/km at 1.55 μm, or about 100 times higher than KRS-5 or many other IR materials. Therefore, if a fiber could be made with a loss this low, then it would be possible to construct telecommunications links thousands of kilometers in length without repeaters. In particular, the U.S. government funding agency DARPA started a program in the late 1970s called the Clear Day Program to fund development of ultralow-loss fibers for undersea applications. The idea was that a submarine could be in constant contact with its base station by paying out an ultralow-loss fiber as it traveled under water. Because the fiber had such a low loss, "On a clear day you can see forever." Unfortunately, no IR fiber was ever developed with a loss lower than silica, much less near its intrinsic loss. This is still a long-term goal, but at least it remains a theoretical possibility even though there are enormous challenges to overcome before it becomes a reality.

Interestingly enough, other types of IR fibers began appearing at about the same time as KRS-5 fiber. These other fibers included the heavy metal fluoride glass (HMFG), other polycrystalline fibers like the silver halides, and hollow rectangular waveguides. While none of these fibers had physical properties even approaching that of conventional silica fibers, they were, nevertheless, useful in lengths of less than 2 to 3 m for a variety of IR sensor and power delivery applications.[1]

1.1 Types of IR fibers

Infrared fiber optics may logically be divided into three broad categories: glass, crystalline, and hollow waveguides. These categories may be further subdivided based on the fiber material or structure or both, as shown in Fig. 1.3. An example of a typical fiber for each subdivision is given in the last row of Fig. 1.3. Another way of viewing IR fibers that is more convenient from the viewpoint of laser power delivery applications, is to ask for the best IR fibers for use at the two most popular IR laser wavelengths: the 10-μm CO_2 laser and the 3-μm Er:YAG laser. The chart in Fig. 1.4 is a guide to the most useful fibers available today for delivering these two laser energies.

Over the past 30 years many IR fibers have been made in an effort to fabricate a fiber optic with properties as close to silica as possible, but only a relatively small number have survived. A good source of general information on these various IR fiber types may be found in the few other books on IR fibers and in review

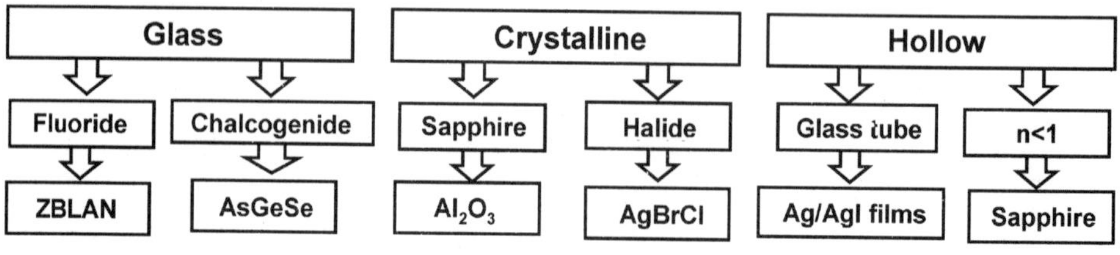

Figure 1.3 Major categories of IR fiber optics and an example of each fiber type.

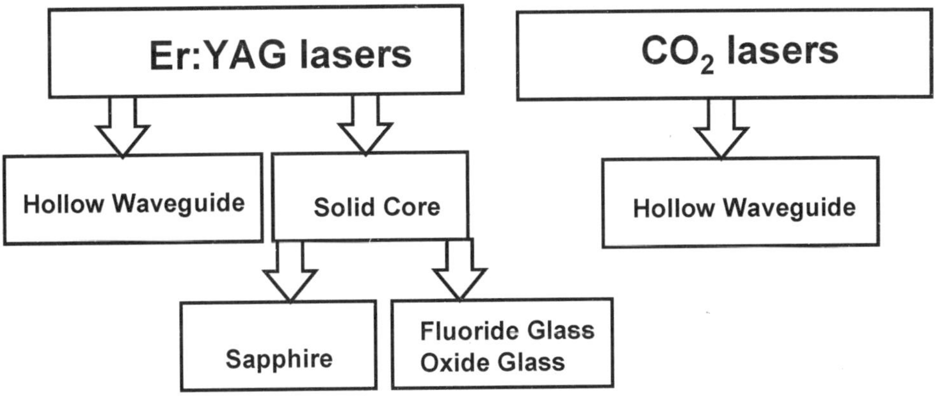

Figure 1.4 IR fibers most commonly used for the delivery of two popular IR lasers: the CO_2 and Er:YAG laser.

articles.[5–8] In this treatise only the best, most viable and, in most cases, commercially available IR fibers are treated in detail, although mention is made of a few other attempts to make a viable IR fiber. One of the most important points to glean from a study of IR fiber optics is that because their physical properties are inferior to silica fibers, they are primarily limited to non-telecommunication, short-haul applications that require only a few meters of fiber rather than the kilometer lengths common to telecommunication applications. The primary reason that their use is limited to short lengths is that the losses for most of the fibers are a few decibels per meter (dB/m) rather than less than 1 dB/km, which is common for silica fibers. An exception is fluoride glass fiber, which can have losses as low as a few dB/km. In addition, IR fibers are much weaker than silica fiber and, therefore, more fragile. These deleterious features have slowed the acceptance of IR fibers and restricted their use to applications in chemical sensing, thermometry, and laser power delivery.

1.2 General properties of IR fibers

The obvious key property of IR fibers is their ability to transmit wavelengths longer than most oxide glass fibers. In some cases the transmittance of the fiber can extend well beyond 20 μm, but there are few applications requiring the transmission of IR radiation longer than about 12 μm. A summary of the spectral loss for five of the six subcategories of fibers listed in Fig. 1.3 is shown in the composite data[1] in Fig. 1.5. From the data it is clear that there is a wide variation in range of transmission for the different IR fibers, and that the loss of most of the IR fibers is quite high compared with silica fibers. Remembering that 1 dB/m is a bulk loss of about 20% per meter, it is again evident that this high loss will restrict applications to meter-long lengths. In fact, the losses for all of these fibers except the hollow waveguide should be much lower. The reason that they are not is that the fibers contain impurities and imperfections, which give rise to a large extrinsic absorption and scattering. Some of these extrinsic absorption bands are evident in

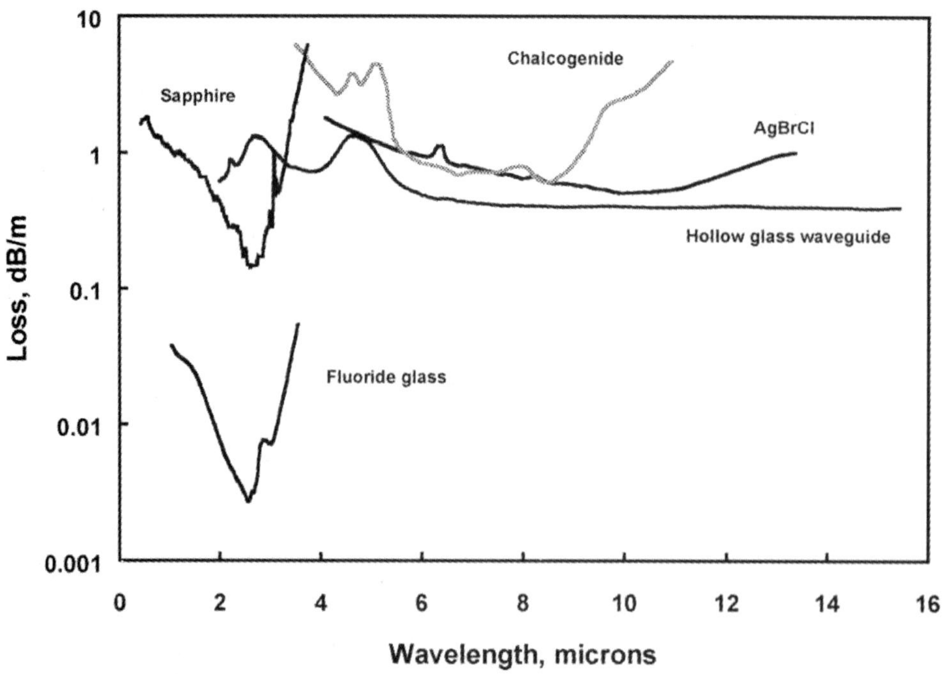

Figure 1.5 Composite loss spectra for some common IR fiber optics: ZBLAN fluoride glass, SC sapphire, chalcogenide glass, PC AgBrCl, and hollow glass waveguide.[1] [Reprinted with permission from The McGraw-Hill companies.]

Fig. 1.5. The absorptions shown for the hollow waveguide are not due to impurities; rather they are due to interference effects resulting from the thin-film coatings used to make the guides. One of the most important research areas of IR fibers is the study of the sources for these extrinsic absorptions and methods to eliminate them. Details of a myriad of extrinsic absorption mechanisms in IR fibers are a major part of each chapter on glass, crystalline, and hollow IR waveguides.

Some of the important optical and mechanical properties of IR fibers are listed in Table 1.1. For comparison, the properties of silica fibers are also listed. The data in the Table 1.1 and in Fig. 1.5 reveal that, compared with silica, IR fibers usually have higher loss, larger refractive indices and dn/dT, lower melting or softening points, and greater thermal expansion. For example, chalcogenide and polycrystalline Ag-halide fibers have refractive indices greater than 2. This means that the reflection or Fresnel loss exceeds 20% for two fiber ends. The higher dn/dT and low melting or softening point leads to thermal lensing and, as a result, low laser-induced damage thresholds for the solid-core IR fibers. Finally, many IR fibers do not have a proper cladding analogous to conventionally clad oxide glass fibers. Nevertheless, core-only IR fibers such as sapphire and chalcogenide fibers can still be useful. This is because their refractive indices are sufficiently high so that there is less evanescent wave energy outside the core. As long as the unclad fiber does not come in contact with an absorbing medium, the fiber can operate reasonably well, as there will be very little leakage of light from the core to the surrounding air.

Table 1.1 Selected physical properties of key IR fibers compared to conventional silica fiber.

		Glass		Crystal		Hollow
		HMFG	Chalcogenide	PC	SC	Hollow Silica
Property	Silica	ZBLAN	AsGeSeTe	AgBrCl	Sapphire	Waveguide
Glass transition or melting point, °C	1175	265	245	412	2030	150 (useable T)
Thermal conductivity, W/m °C	1.38	0.628	0.2	1.1	36	1.38
Thermal expansion coefficient, 10^{-6} °C^{-1}	0.55	17.2	15	30	5	0.55
Young's modulus, GPa	70.0	58.3	21.5	0.14	430	70.0
Density, g/cm^3	2.20	4.33	4.88	6.39	3.97	2.20
Refractive index (λ, μm)	1.455 (0.70)	1.499 (0.589)	2.9 (10.6)	2.2 (10.6)	1.71 (3.0)	NA
dn/dT, 10^{-5} °C^{-1} (λ, μm)	+1.2 (1.06)	−1.5 (1.06)	+10 (10.6)	−1.5 (10.6)	+1.4 (1.06)	NA
Fiber transmission, range, μm	0.24–2.0	0.25–4.0	4–11	3–16	0.5–3.1	0.9–25
Loss* at 2.94 μm, dB/m	~ 800	0.08	5	3	0.4	0.5
Loss* at 10.6 μm, dB/m	NA	NA	2	0.5	NA	0.4

*Typical measured loss, NA = not applicable.

1.3 General applications of IR fibers

The motivation to develop a viable IR fiber stems from many proposed applications. A summary of the most important current and future applications and the associated candidate IR fiber that will best meet the need is given in Table 1.2. There are several noteworthy trends seen in this table. The first is that hollow waveguides are an ideal candidate for laser-power delivery at all IR laser wavelengths. The air core of these special fibers or waveguides gives an inherent advantage over solid-core fibers because IR materials used in solid-core fibers have laser damage thresholds that are frequently very low. The air-core waveguides are capable of delivering close to 3000 W of cw CO_2 laser power, far in excess of any IR solid-core fiber. However, solid-core IR fibers are ideal evanescent-wave sensors for monitoring chemical processes in the sensitive fingerprint region of the infrared spectrum. In these applications the fiber core is surrounded by the chemical or biological agent, and some portion of the light is coupled out of the core into the surrounding medium. This type of chemical sensor is potentially very sensitive and selective. Chalcogenide and silver halide fibers are particularly good for this application, as they are quite inert and their high refractive index means that only a small portion of the light is out-coupled from the core into the absorbing medium. A good fiber for gas sensing is the hollow waveguide, as the core of this fiber can

Table 1.2 Examples of IR fiber candidates for various sensor and power delivery applications.

Application	Comments	Suitable IR fibers
Fiber optic chemical sensors	Evanescent wave principle—liquids	AgBrCl, sapphire, chalcogenide, HMFG
Fiber optic chemical sensors	Hollow core waveguides—gases	Hollow glass waveguides
Radiometry	Blackbody radiation, temperature measurements	Hollow glass waveguides, AgBrCl, chalcogenide, sapphire
Er:YAG laser power delivery	3-μm transmitting fibers with high damage threshold	Hollow glass waveguides, sapphire, germanate glass
CO_2 laser power delivery	10-μm transmitting fibers with high damage threshold	Hollow glass waveguides
Thermal imaging	Coherent bundles	HMFG, chalcogenide
Fiber amplifiers and lasers	Doped IR glass fibers	HMFG, chalcogenide

be filled with gas so that light propagating through the waveguide is partially absorbed by the gas. Temperature measurements using long-wavelength transmissive fibers like the silver halides or hollow waveguides are possible over a large temperature range. Normally, blackbody radiation from a source is transmitted through the fiber and the temperature determined by calibration to a blackbody of known temperature. Since blackbody radiation from room temperature objects peaks near 10 μm IR, fibers are excellent candidates for use in measuring temperatures below 50°C. A host of different IR fiber applications with many examples are given in Chapters 8, 9, and 10.

References

1. J. A. Harrington, "Infrared Fibers," in *Handbook of Optics: Fiber and Integrated Optics*, Vol. 4, M. Bass, J. M. Enoch, E. W. Van Stryland, and W. L. Wolfe, Eds., McGraw-Hill, New York (2000).
2. N. S. Kapany and R. J. Simms, "Recent developments of infrared fiber optics," *Infrared Phys.*, Vol. 5, pp. 69–75 (1965).
3. D. A. Pinnow, A. L. Gentile, A. G. Standlee, A. J. Timper, and L. M. Hobrock, "Polycrystalline fiber optical waveguides for infrared transmission," *Appl. Phys. Lett.*, Vol. 33, pp. 28–29 (1978).
4. A. L. Gentile, M. Braunstein, D. A. Pinnow, J. A. Harrington, D. M. Henderson, J. Myer, R. C. Pastor, R. R. Turk, and L. M. Hobrock, "Infrared fiber optical materials," in *Fiber Optics, Advances in Research and Development*, B. Bendow and S. S. Mitra, Eds., Plenum, New York (1979).
5. T. Katsuyama and H. Matsumura, *Infrared Optical Fibers*, Adam Hilger, Bristol, UK (1989).

6. I. D. Aggarwal and G. Lu, *Fluoride Glass Fiber Optics*, Academic Press, New York (1991).
7. J. S. Sanghera and I. D. Aggarwal, *Infrared Fiber Optics*, CRC Press, Boca Raton, FL (1998).
8. P. France, M. G. Drexhage, J. M. Parker, M. W. Moore, S. F. Carter, and J. V. Wright, *Fluoride Glass Optical Fibres*, Blackie and Son, London (1990).

Chapter 2

Theoretical Foundations of Infrared Fiber Optic Transmission: Solid-core Fibers

2.0 Introduction

The basic principle for guiding light by total internal reflection (TIR) in a dielectric media was first demonstrated around 1840 by D. Colladon. Colladon showed very simply that sunlight could be guided by a stream of water flowing out through a hole in a water-filled vessel. Later he applied these ideas to lighting water fountains. In fact, his ideas were responsible for the spectacular lighted and colored fountains at the Universal Exposition in Paris in 1889. In an excellent account on the history of fiber optics, J. Hecht, in the *City of Light—The Story of Fiber Optics*,[1] discusses the role of Colladon and that of Tyndall, who repeated Colladon's experiments in 1853. Because Tyndall was a well-known scientist at the time, he is often wrongly given credit for the first observation of light guiding in water jets.

The first glass fiber optic was proposed in 1926 by C.W. Hansell.[1] Hansell was a prolific inventor who was trying to transmit an image of an instrument dial to a remote location for viewing around corners and curves. In fact, up until the mid-1950s, the driving force behind fiber optic transmission was the need for a flexible image bundle for industrial and medical applications. In the latter case, it was the wish of several teams of surgeons and engineers to have flexible endoscopes for seeing inside the body. Until then only crude arrangements of rigid tubes and mirrors, the forerunners of today's modern articulated arms used to transmit CO_2 laser power, were used for visualization within the body. Also, in the early 1950s, van Heel, O'Brien, and others came onto the idea of a clad fiber as a means of more effectively trapping and confining the light within a single glass fiber. This was important because crosstalk present in the core-only, coherent image bundles distorted the image. These investigators realized that it was necessary to have a lower refractive-index material, i.e., a cladding layer, surround the glass core for TIR. The first cladding materials were oils; most oils have a lower refractive index than glass and they were readily accessible.

Much of the early work on glass-cladding fiber optics took place at American Optical in the late 1950s and early 1960s. At American Optical, W. Hicks fabricated glass-clad fibers and E. Snitzer recognized that very small core fibers would transmit only a single mode of light. In 1964, C. Kao made the startling prediction that silica glass fiber with losses as low as 20 dB/km could be made and used for telecommunication applications.[2] At the time this seemed like an impossible goal, but today nearly intrinsic losses of 0.14 dB/km at 1.55 μm have been realized. As will be seen, however, losses for IR fibers are well above the intrinsic loss for silica fibers.

2.1 Ray optic theory of fiber transmission

2.1.1 Reflection and refraction at a boundary

The confinement of light in a fiber optic waveguide may be analyzed most simply in terms of the reflection and refraction of light at a boundary between two dielectric materials. A dielectric-core fiber is composed of a core with a refractive index, n_1, surrounded by a lower refractive index clad, n_2. For simplicity, consider first a planar structure composed of two dielectric media, as shown in Fig. 2.1. The incoming ray in the core material is incident at the boundary between the core (material 1) and cladding (material 2), where it is either (1) totally reflected back into the core or (2) reflected back into the core and also partially refracted into the cladding. In the first case, shown in Fig. 2.1(a), there is TIR. This is the basic principle of operation of guiding light through a fiber optic. In the second instance shown in Fig. 2.1(b), light is lost from the core as a result of light propagating in the cladding. Light energy that propagates in the cladding is called evanescent wave energy. For the case when θ_1 is less than the critical angle, Fig. 2.1(b), the refracted or evanescent wave energy will be greatly attenuated and normally this situation is undesirable. However, in some instances it is useful to outcouple light into the cladding layer for fiber optic sensor applications. Fiber sensors based on evanescent coupling into a "cladding" layer, for example, an absorbing liquid or intimately bonded solid, may be used as a chemical sensor. This subject is discussed in greater detail in Chapter 8.

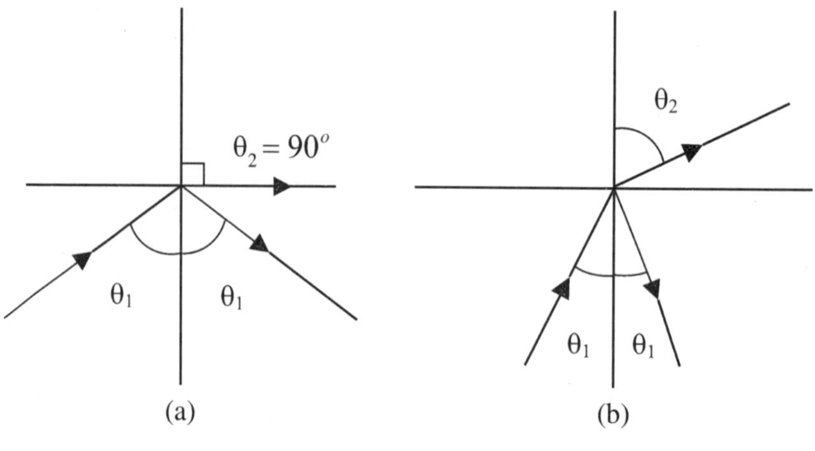

Figure 2.1 (a) Total internal reflection (TIR) necessary for fiber optic transmission. In this case θ_1 is the critical angle. (b) Rays incident at angles less than the critical angle. For this case some light will be reflected and some refracted into medium 2.

The actual amount of light that is reflected and refracted from a dielectric boundary layer is determined by solving Maxwell's equations in terms of the components of the incident, reflected, and refracted electric and magnetic fields. The result of solving Maxwell's equations is a set of four Fresnel equations that give the coefficients of reflectivity, r, and transmissivity, t, in terms of the refractive indices, n_1 and n_2, and the incident and refracted angles, θ_1 and θ_2, respectively. There are four equations because the polarization state of the light must be taken into account. In Fig. 2.2, the incident, reflected, and refracted components of the electric field vectors, **E**, are shown. The components of **E** lying in the plane of incidence, defined by the incident and reflected rays, are the parallel or p-polarization (p-pol) components; those perpendicular to the plane of incidence are the s-polarization (s-pol) components. Equivalently, these are the transverse magnetic (magnetic field perpendicular to the plane of incidence) or TM (p-pol) and the tranverse electric (electric field perpendicular to the plane of incidence) or TE (s-pol) components.

The four Fresnel equations are

$$r_p \equiv \frac{E_{rp}}{E_{ip}} = \frac{n_2 \cos\theta_1 - n_1 \cos\theta_2}{n_2 \cos\theta_1 + n_1 \cos\theta_2}, \tag{2.1a}$$

$$r_s \equiv \frac{E_{rs}}{E_{is}} = \frac{n_1 \cos\theta_1 - n_2 \cos\theta_2}{n_1 \cos\theta_1 + n_2 \cos\theta_2}, \tag{2.1b}$$

$$t_p \equiv \frac{E_{tp}}{E_{ip}} = \frac{2n_1 \cos\theta_1}{n_2 \cos\theta_1 + n_1 \cos\theta_2} = 1 - r_p, \tag{2.1c}$$

$$t_s \equiv \frac{E_{ts}}{E_{is}} = \frac{2n_1 \cos\theta_1}{n_1 \cos\theta_1 + n_2 \cos\theta_2} = 1 + r_s. \tag{2.1d}$$

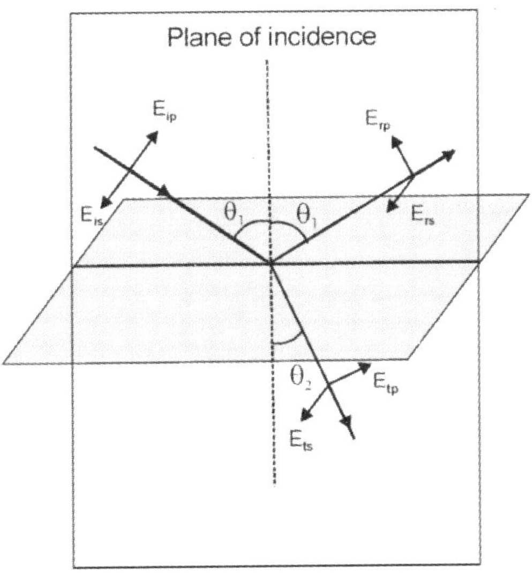

Figure 2.2 Electric field components perpendicular (s-pol) and parallel (p-pol) to the plane of incidence.

These equations are particularly useful in the development of some of the basic ray optic properties of fibers, but they become cumbersome when used to determine the optical properties of fibers for a large number of incident angles. Therefore, these equations will only be used to develop the concept of total internal reflection and a few other basic optical properties needed to better understand fiber optics.

Two further consequences of solving Maxwell's equations are the Law of Reflection, which says that the angle of incidence equals the angle of reflection, and Snell's law, which states that

$$n_1 \sin\theta_1 = n_2 \sin\theta_2. \tag{2.2}$$

The most important application of Snell's law in fiber optics is the case when θ_2 equals 90° and $n_1 > n_2$. This is the case for internal reflection from a boundary in contrast to external reflection when $n_1 < n_2$. The situation for internal reflection is shown in Fig. 2.1(a). Here $\theta_2 = 90°$ and θ_1 equals the critical angle, θ_c, and all of the light is totally reflected back into the core. That is,

$$\sin\theta_c = \frac{n_2}{n_1}\sin 90° = \frac{n_2}{n_1} \text{ or } \theta_c = \sin^{-1}\frac{n_2}{n_1}, \tag{2.3}$$

for $n_1 > n_2$. Therefore, all rays are internally reflected when $\theta_1 > \theta_c$ and there is no refracted light leakage into the cladding region. Normally, for conventional silica fiber optics $n_1 \cong n_2$ which means that θ_c is typically greater than approximately 80°. However, for many of the important IR fibers the cladding index may differ significantly from that of the core index. For example, the first IR polycrystalline fibers were made from KRS-5 (TlBrI), which has a refractive index of 2.37 at 10.6 μm. In addition, these fibers were initially made without a material clad, i.e., core-only, and therefore the critical angle for an air clad ($n_2 = 1$) and a KRS-5 core would be approximately 25°.

There are two other important relationships that are important for fiber optics. One is the Brewster angle. When p-pol or TM polarized light is incident at the Brewster angle, θ_B, then there is no reflected p-pol light. This special case occurs when $\theta_1 + \theta_2 = 90°$ or $\theta_B = 90° - \theta_2$. Using the Fresnel equation for r_p [Eqs. (2.1a)] and setting $r_p = 0$, the Brewster angle is found to be

$$\theta_B = \tan^{-1}\frac{n_2}{n_1}. \tag{2.4}$$

The other important relation is the normal incidence reflectance. In general, the measured reflected intensity or reflectance, R, of light is equal to the square of the reflectivity coefficient. For example, $R_s = r_s^2$. At normal incidence $\theta_1 = \theta_2 = 0$, and again using Eqs. [2.1(a) and (b)] for r_s and r_p, the normal incidence reflectivity is found to be

$$R_s = R_p = r^2 = \left(\frac{n_1 - n_2}{n_1 + n_2}\right)^2. \tag{2.5}$$

The structure of a typical solid-core fiber is shown in Fig. 2.3. The inner core material is surrounded by the lower-index cladding material, followed by a buffer layer for strength and protection against optical and mechanical degradation. While this structure is commonplace for silica glass fibers, it is only recently that some of the IR fiber optics have been fabricated with all three layers. As mentioned above, the polycrystalline fibers were extruded for many years without a cladding or tight binding buffer coating. Only recently have they been made with a proper cladding layer; however, they are generally still fabricated without a proper cladding layer.

The ray-optic approach for the guiding of light in a fiber is shown for the simplest meriodinal rays in Fig. 2.4. Meriodinal rays are those that pass through the core or optic axis. In Fig. 2.4(a), a large-core, multimode (MM), step index (SI) fiber is shown. This is the most common IR fiber made today. For this fiber the core radius, a, is much larger than the wavelength, λ, and, in general, a is greater than approximately 100 μm. The cladding radius, b, may also be rather large compared again to silica glass fibers. Figure 2.4(a) shows two rays, one for the case of TIR, or guided light, and the other for refracted, or unguided light. The other important fiber geometry is the single-mode (SM) fiber shown in Fig. 2.4(b). For this

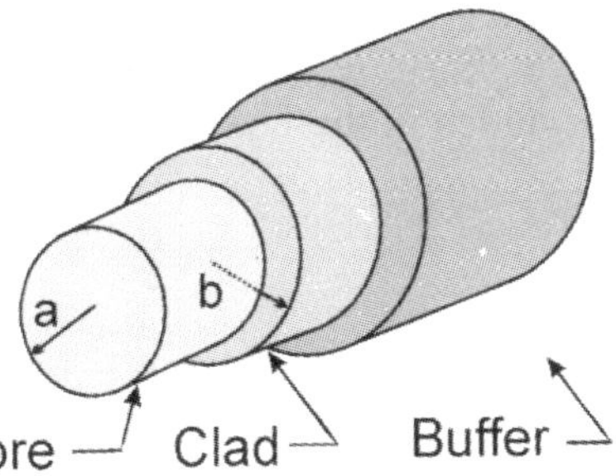

Figure 2.3 Structure of a typical core/clad/jacketed MM fiber optic. Multimode cores are considerably larger than a single-mode core.

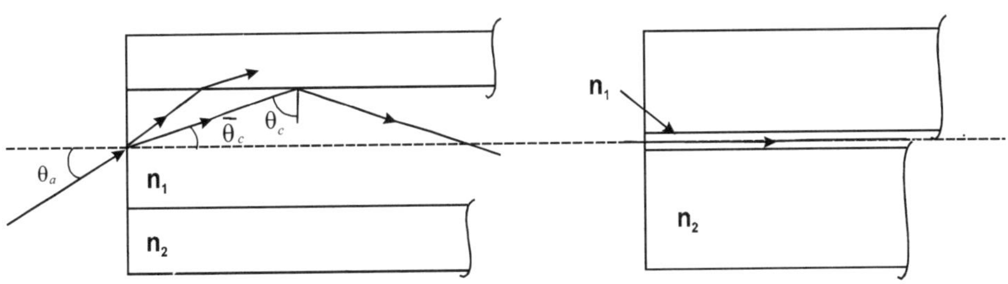

(a) (b)

Figure 2.4 (a) MM fiber showing guided ($\theta < \theta_C$) and unguided rays ($\theta > \theta_C$) and (b) SM fiber with small core compared to MM fiber.

fiber $a \ll b$ and the core diameter $2a$ is usually only several times the wavelength. An SM fiber propagates the ray with the shortest pathlength in the fiber. This fiber also has an SI profile.

The refractive index profile of the MM, SM, and a graded index (GI) fiber are illustrated in Fig. 2.5. As indicated above, the MM, Fig. 2.5(a), and SM, Fig. 2.5(b), fibers have an abrupt change in index at $r = a$. In contrast, the GI fiber, Fig. 2.5(c), is structured so that there is a gradual, usually parabolic, increase in n from $r = a$ to $r = 0$. The GI fiber is a multimode fiber but it guides rays more efficiently than the SI MM fibers. This is a result of gently bending light rays that are traveling near the critical angle back toward the core; these rays do not have as great a pathlength as they would in the SI MM structure. However, the GI fiber structure is very uncommon for IR fibers, and therefore the MM and SM fiber structures will be emphasized here. Furthermore, except for emerging heavy-metal fluoride glass fiber amplifiers, most IR fibers today are multimode with a step-index profile.

2.1.2 Numerical aperture and acceptance angle

The numerical aperture, NA, is one of the most important basic fiber parameters. The NA of a fiber is related to the critical angle in the fiber and, therefore, to those rays which are propagated by only TIR. The NA is easily related through Snell's law to the acceptance angle, θ_a. The acceptance angle defines a cone of angles within which all rays are guided in the fiber. Figure 2.4(a) shows θ_a for the MM fiber.

The NA of a fiber is defined as

$$\mathrm{NA} \equiv \sqrt{n_1^2 - n_2^2}. \tag{2.6}$$

It may also be conveniently written in terms of the half angle of the acceptance cone, θ_a. Referring to Fig. 2.4(a) and using Snell's law,

$$n_o \sin\theta_a = n_1 \sin(90^\circ - \theta_c) = n_1 \cos\theta_c = n_1(1 - \sin^2\theta_c)^{1/2},$$

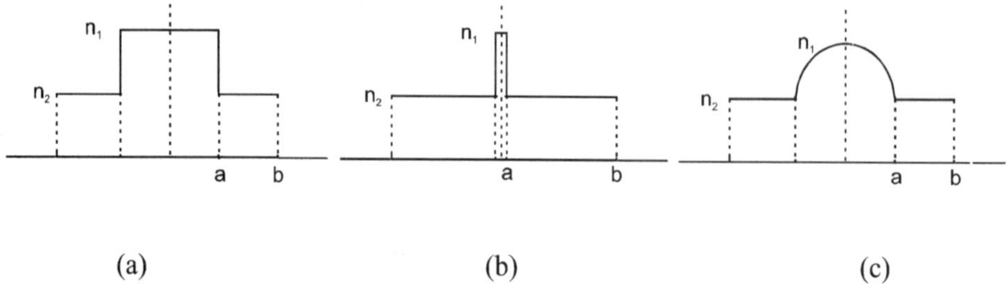

Figure 2.5 Refractive index profile for core/clad fibers for (a) MM, (b) SI, and (c) GI fibers.

or, using Eq. (2.2),

$$n_o \sin\theta_a = n_1\left[1-\left(\frac{n_2}{n_1}\right)^2\right]^{1/2} = \sqrt{n_1^2-n_2^2} = \text{NA}, \tag{2.7}$$

where n_o is the refractive index of the medium surrounding the fiber. Normally this medium is air so $n_o = 1$. There are, however, important instances when $n_o \neq 1$, such as when fibers are immersed in water for some laser surgical procedures. In this case n_o (H_2O) = 1.33, and sin θ_a is smaller by a factor of n_o. That is, the output cone angle decreases when the fiber is immersed in water.

When $n_1 \cong n_2$, then NA can be conveniently written as

$$\text{NA} = \sqrt{n_1^2-n_2^2} = \sqrt{(n_1-n_2)(n_1+n_2)} \cong \sqrt{(n_1-n_2)(2n_1)} = n_1\sqrt{2\Delta}, \tag{2.8}$$

where the relative refractive index difference, Δ, is given by

$$\Delta = \frac{n_1-n_2}{n_1}. \tag{2.9}$$

The approximation given in Eq. (2.8) is commonly used in typical telecommunication fibers where it is advantageous to have $n_1 \cong n_2$. But for many IR fibers and their applications, it is often the case that n_1 is not at all close to n_2, and then the full definition of the NA given in Eq. (2.6) must be used.

One method used to determine a practical value for the NA of an SI MM fiber is simply to measure the angular dependence of the accepted light energy. For silica fibers the angular dependence of the accepted energy typically looks like the data shown in Fig. 2.6. That is, there is not an abrupt decrease in energy at $r = a$; rather,

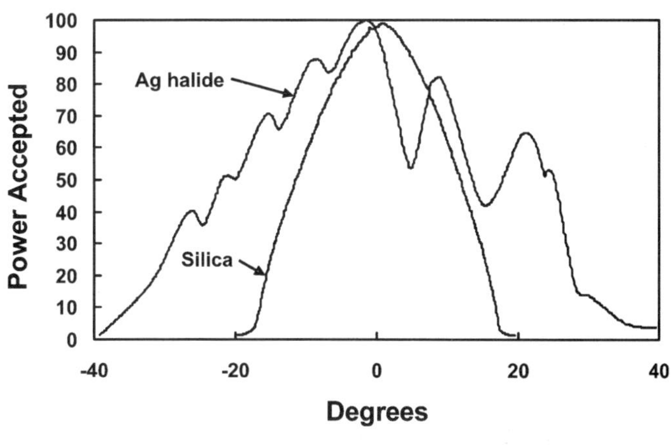

Figure 2.6 Numerical aperture for conventional core/clad silica fiber compared to unclad Ag-halide fiber. The source of the structure in the Ag-halide fiber is unknown.[3] [Reprinted with permission.]

there is a more gradual falloff in the transmitted power. The NA of the fiber is practically defined by the Electronics Industry Association as the sine of the angle at which the power has fallen to 5% of the peak value. This value agrees well with that calculated from Eq. (2.8) for silica fibers.

The situation is not so simple when one measures the NA of some IR fibers. For example, PC fibers often have no clad, so the NA is usually calculated to be very high. In the case of AgClBr fibers, the core-only fiber with $n_1 = 2.15$ would give a calculated NA equal to $\sqrt{2.15^2 - 1}$ or, since the NA cannot be greater than 1, the NA for unclad AgClBr fiber equals 1. Based on this, it would appear any ray incident would be internally reflected. However, when a similar measurement of the angular dependence of the accepted power is made for this PC fiber, the data shown in Fig. 2.6 is obtained.[3] It can be clearly seen from the data that the NA is not even close to 1. The fact that the NA is not 1 is due in part to the lossy nature of the high-angle rays or equivalently higher-order modes that are launched into the fiber but are quickly damped out. In practice, therefore, it is necessary to define an effective NA, NA_{eff}, for those IR fibers with a less-than- ideal NA profile, i.e., an index profile similar to that of silica fiber. The NA_{eff} is defined to be the half angle at which the power falls to one-half the maximum or the 3-dB point. From the data for the AgBrCl fiber in Fig. 2.6, this would give an NA_{eff} of approximately 0.35.

The above discussion pertains to solid-core fibers with or without a proper cladding. For the case of hollow-core fibers, the situation is clearly different because, in most cases, there is not a conventional cladding material surrounding the air core. Therefore, the NA of an air-core fiber is best defined in practical terms as it is for the AgBrCl fiber discussed above. In fact, the NA profile of many hollow waveguides is similar in appearance to the data in Fig. 2.6 for the silver halide fiber. The difference here is that the NA depends more strongly on the modal properties of the hollow-core waveguides. In general, the NA_{eff} is very small for a hollow waveguide. This subject is discussed further in Chapter 7.

2.1.3 Modes of a fiber

The ray-optic approach explains many of the most basic properties governing the guiding of light in a fiber. A simple Snell's law analysis leads to the concept of the numerical aperture of a fiber. Rays incident within the NA or acceptance angle of the fiber are guided in the fiber by TIR. That is, the propagation angles in a fiber must be less than the complement of the critical angle, $\overline{\theta}_c = (90° - \theta_c)$, as shown in Fig. 2.4(a). From this simple analysis it would seem that no light would propagate in the cladding when the propagation angle is less than θ_c. In fact this is not the case. Sometimes a significant portion of the light can propagate in the clad region. In addition, a ray-optic approach would also lead one to believe that any propagation angle is possible as long as the angle is less than θ_c.

A more accurate picture of the propagation of light in a fiber is obtained through a careful analysis using Maxwell's equations and treating light as a wave rather than a ray. The wave-optic approach leads naturally to the concept of the modes of a fiber. In a wave-optic approach, it is found that light propagates only at discrete angles and that some energy does indeed propagate in the cladding of the fiber. The fact that light propagates only at specific angles results from the phase shift that takes place on reflection from the core/clad boundary. This effect can be most easily understood by an analogy with planar mirror waveguides. Consider plane waves reflecting from perfect mirror surfaces as shown in Fig. 2.7. The plane waves travel in the z direction and the electric and magnetic fields of the plane wave lie in the xy plane, which is perpendicular to the direction of propagation. This is called a transverse electromagnetic wave or TEM wave. Assume that perfectly reflecting mirrors form the top and bottom boundaries of the guide that is filled with air, as shown in Fig. 2.7. On reflection from one mirror there is a phase shift of π for light waves polarized either with the electric field parallel transverse magnetic (TM) or perpendicular transverse electric (TE) to the plane of incidence (xz plane). For a plane wave to reproduce itself upon making two reflections, one from the top mirror and one from the bottom mirror, requires that the wave be in phase after a total 2π phase shift. This in turn means that only certain plane waves traveling at discrete angles $\pm\theta_m$ will reproduce the original plane wave. This is the principle of self consistency and the θ_m are called the bounce angles.

A similar analysis applies in the case of cylindrical fiber optics and dielectric interfaces. However, the phase shift is not merely π/bounce as it is for a perfect mirror; rather it is much more complicated. The subscript m on the bounce angle indicates the mode number and, for the hollow mirror guides, $m = 1, 2, 3, \ldots$. Therefore, a mode may be simply thought of as relating to the bounce angle of a ray propagating through a fiber. The smallest bounce angle occurs for $m = 1$ in a mirror guide; this is referred to as the lowest-order mode. Higher-order modes for large m always have a greater bounce angle. In terms of the TE or TM, the mode designation is properly written as TE_m or TM_m, respectively. The mode parameter m is related to the transverse distribution of the electric field. Further discussion of the modal properties of hollow planar waveguides and their field distributions are given in Chapter 3.

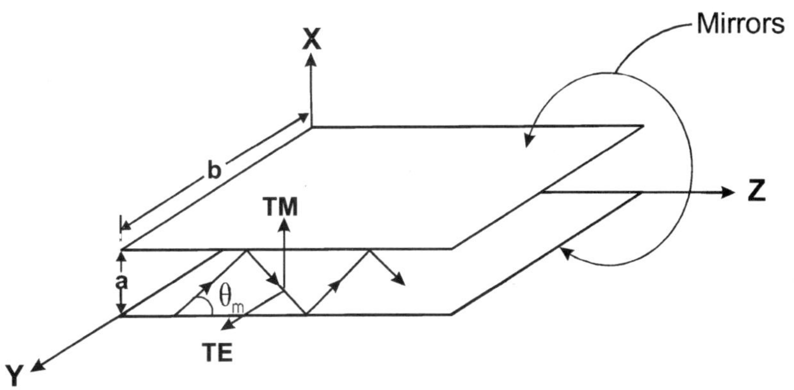

Figure 2.7 Hollow planar waveguide composed of two mirrors usually coated to improve reflectivity.

The situation for a fiber optic is more complex than for a mirror waveguide. This is because of the cylindrical geometry of a fiber. Therefore, instead of identifying the mode by a single mode parameter m, it is necessary to use two mode parameters, l and m. As for a planar waveguide, m designates the radial or transverse dependence of the electric field whereas l designates the azimuthal dependence. The transverse modes for a fiber are then TE_{lm} and TM_{lm} and these modes correspond to the meriodinal rays. There is also a hybrid or mixed mode designated by HE_{lm} and EH_{lm}. These hybrid fiber modes have a small component of the electric field in the z direction or along the fiber or optic axis and correspond to skew rays (a skew ray is a ray that does not cross the fiber axis; rather it travels in a corkscrew or helical path down the fiber). A further simplification of fiber optic modes is possible when the core and clad indices are nearly equal as they are in all telecommunication and many other applications. When $n_1 \cong n_2$ or $\Delta \ll 1$, the weakly guiding approximation holds. In this approximation light rays are nearly paraxial; i.e., they travel approximately parallel to the fiber axis. The electromagnetic waves are essentially TEM waves. Thus, the longitudinal components of the electric and magnetic fields are much less than the transverse components. In this approximation, a new linearly polarized or LP mode can be formed from a linear combination of the exact TE, TM, EH, and HE modes. The LP mode is a mode with polarizations in the x and y (transverse) directions and it is an excellent approximation to the exact solutions of the wave equation. The corresponding mode designation for an LP mode is LP_{lm}.

An important parameter related to the modal properties of a fiber is the fiber or V parameter. The V parameter governs the number of modes that can propagate as well as certain aspects of the propagation of light in a fiber. It is given by

$$V = \frac{2\pi a}{\lambda_o}\text{NA} \cong \frac{2\pi a n_1\sqrt{2\Delta}}{\lambda_o}, \tag{2.10}$$

where λ_o is the free space wavelength. For small NA and ratios of a/λ_o, V is also small. An important case is that for SM fibers. For SM fibers only the lowest order LP_{01} mode propagates in the fiber. This occurs only when $V \leq 2.405$. When $V > 2.405$, more thanone mode can propagate. The number of modes that can propagate in an MM fiber is given by

$$M \approx \frac{4V^2}{\pi^2}. \tag{2.11}$$

Equation (2.11) is an approximation that gives the number of modes for large V-number fibers. As an example, consider a core-only (NA = 1), MM AgCl PC fiber with an $n_1 \cong 2.0$ and a typical large core size of $2a = 600$ μm operating at the CO_2 laser wavelength of 10.6 μm. This fiber would have a V-parameter equal to 177.8 and $M > 12{,}800$! Clearly this fiber, which is typical of many IR fibers, is very multimode with the potential of thousands of propagating modes.

2.2 Refractive indices of optical materials

The refractive index is one of the most important optical parameters characterizing a dielectric material. Historically, it has been studied for over two centuries and its importance in optics from the earliest optical instruments to the understanding of a variety of electrooptic effects has been long recognized. In the simplest terms, the refractive index is the dimensionless ratio of the velocity of light in vacuum to that in the dielectric medium, or, $n = c/v$. That is, the velocity of light is less in a medium compared with its velocity in vacuum. This allows a simple measurement of n by measuring the degree of bending of light (i.e., the angle of refraction) in a medium, or

$$n = \frac{\sin\theta_{vac}}{\sin\theta_{medium}}.$$

In most cases, $n \geq 1$. However, n can be less than 1 in the region of anomalous dispersion. This $n < 1$ behavior has been exploited to make a most interesting hollow waveguide in which the core index of air ($n \cong 1$) is greater than the surrounding $n < 1$ "cladding" material.[4,5]

The refractive index of IR fiber materials is reasonably well characterized for many wavelengths of interest in the 2- to 10-μm region. In general, however, the indices have not been as precisely measured as they have for the oxide fiber materials such as silica. One reason for this is that IR fibers are not used in long- haul telecommunications. For telecom applications, it is essential to know the refractive indices of core and cladding materials to many significant figures to accurately determine the dispersive properties of the fibers. Since the IR fiber applications are short haul and usually involve MM fibers, the relative index difference between core and clad is reasonably large. Thus, it is not essential to have the same precision in n for IR fiber materials as for oxide materials.

2.2.1 The optical constants n and k

The common or measured refractive index n is the real part of the complex index of refraction, n^*, where $n^* = n - ik$ and k is the extinction coefficient. Together, the optical constants n and k form the basis of many of the key optical properties of materials. In fact, they are not constant at all; rather they are functions of wavelength. In the transparent regime, which is clearly of importance for most IR fibers, n decreases slowly as the wavelength increases. This is the region of normal dispersion. This behavior is shown for a typical dielectric material, NaCl, in the dispersion data shown in Fig. 2.8.[6] As the wavelength increases beyond the region of normal dispersion, the index is seen to vary strongly. This is the region of anomalous dispersion or the Reststrahl region where n most often decreases to values less than 1 over some small region of the spectrum. The extinction coefficient is related to the absorption coefficient, α [cm^{-1}], in the material or specifically, $\alpha = 4\pi k/\lambda$. In the transparent regime, the absorption is naturally very low and k is small. However, in the region of anomalous dispersion k becomes large. The result

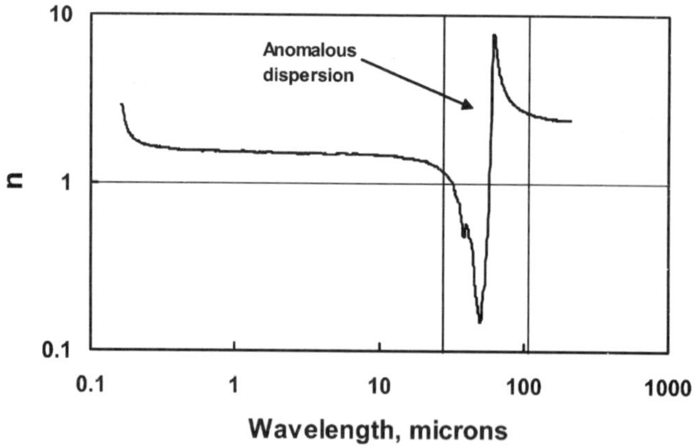

Figure 2.8 Normal and anomalous dispersion for single-crystal NaCl over the visible and infrared spectral regions. [Reprinted from Ref. 6, Copyright 1985, with permission from Elsevier.]

is that the material can become opaque and/or highly reflective in this region. This effect has led to a host of important applications for dielectric materials including Reststrahl filters and mirrors. It is also an important feature of the $n < 1$ hollow waveguides discussed in Chapter 7.

The variation of n with wavelength or the dispersion of a material depends on a variety of microscopic mechanisms arising from the interaction of light and matter. These mechanisms and how they contribute to n are covered in many optics textbooks. In brief summary, the electric field, E, of a light or electromagnetic (EM) wave applied to a dielectric material distorts the charge surrounding the atoms, ions, or molecules. This leads to the creation of electric dipoles and a polarization of the material that is proportional to E. In turn, this gives rise to the concept of the polarizability, α_{pol}, of the material, which may be interpreted as the ability of the E field to polarize or create dipoles in a solid. The refractive index may be written in terms of the polarizability as

$$n = \sqrt{1 + \frac{N\alpha_{pol}}{\varepsilon_o}} = \sqrt{\kappa}, \tag{2.12}$$

where N is the dipole density; ε_o is permittivity of free space; and κ is the dielectric constant. From Eq. (2.12) it is seen that the exact nature of the polarizability determines the value of both n and κ. In general, α_{pol} is strongly frequency dependent and, therefore, one must be careful when using Maxwell's relation, $n = \sqrt{\kappa}$, to be sure that the dielectric constant is measured at the same frequency as the refractive index. As the frequency increases, the index decreases because low-frequency polarization mechanisms, such as those resulting from permanent or polar molecules and/or from orientation polarization, cannot follow high-frequency fields. In the IR and the UV/visible optical spectrum, the significant mechanisms contributing to the total α_{pol} are the polarizabilities due to ionic and electronic motion, respectively. At the highest frequencies α_{pol} will approach zero and the index will approach 1.

The polarizability of the IR fiber materials of interest depends on many factors. It depends on atomic and molecular bonding with covalently bonded materials having a higher index than ionically bonded materials. This is a result of covalent bonds being more polarizable than ionic bonds. Both polarizability and n also depend on the density of the material, with liquids and gases having generally lower indices than solids. Finally, the index varies with temperature but, most important, it varies with wavelength.

The optical constants have been well studied since the early 1900s. The best resource for n and k as functions of wavelength is the three-volume set of handbooks containing the optical constants of solids compiled by E. D. Palik.[6] These handbooks patch together optical data taken from the literature that cover the far UV to far IR region of the spectrum. The data includes the regions of normal and anomalous dispersion. In addition, Palik includes the parameters to an empirical or Sellmeier fit to n as a function of wavelength. The extinction coefficient is also given in the tables but these values are not as complete as they are for n, as k is often too small to reliably measure in the highly transparent regime.

The number of dielectric materials that have been used for the fabrication of IR fibers is rather small. Table 2.1 lists the more common IR materials and their refractive indices at two important IR laser wavelengths. From Table 2.1 it may be seen that, while some materials like the fluoride glasses and the alkali halide materials have values of n close to that of the common oxide glasses used for conventional fiber optics, other materials have values of n considerably higher. The chalcogenide glasses, for example, have very high indices. This can be a problem in some applications because Fresnel reflections, as seen in Eq. (2.5), can approach 50% for two fiber ends.

Table 2.1 Refractive index for some important IR transmissive materials used in the fabrication of IR fibers.

Category	Material	n (3 μm)	n (10.6 μm)
Crystalline	AgCl	2.00	1.98
	AgBr	2.27	2.25
	AgBrCl	—	2.15
	KRS-5	2.38	2.37
	Al_2O_3	1.71	NA
Glass–HMFG	ZBLAN*	1.50	NA
Glass–Chalcogenide	As_2S_3	2.415	NA
	GeAsSe (Amtir 1)	2.52	2.49
	As_2Se_3	2.799	2.775
Semiconductor	Ge	4.05	4.00
	Si	3.43	NA
	ZnSe	2.44	2.41
	ZnS	2.26	2.19

*Exact value depends on doping for either core or clad glass.

2.3 Loss mechanisms in IR fibers

The overarching property that distinguishes IR fibers from conventional silica fibers is their higher loss. Over the past 25 years of IR fiber development, the loss for these special fibers continues to be not only higher than theoretically predicted, but often considerably greater than the minimum loss of ~0.14 dB/km for silica fiber. This aspect alone has hindered the acceptance of these fibers in many applications. Only recently has the fiber community been able to adapt to IR fiber's higher losses and greater fragility. This has resulted in IR fiber applications that commonly involve links no greater than approximately 10 m. There are many reasons why IR fibers have a generally higher loss than silica fibers. However, before describing the extrinsic loss mechanisms that ultimately limit the fiber loss, it is of interest to first look at the intrinsic or fundamental loss mechanisms in solid-core IR fibers.

2.3.1 Intrinsic loss mechanisms

The loss for a fiber optic is specified in terms of the absorption coefficient, α. In general,

$$\alpha = \frac{10}{l}\log\frac{1}{T} = \frac{10}{l}\log\frac{P_{in}}{P_{out}}, \tag{2.13}$$

where T is the transmission; P is the power; and l is the length of the fiber. An accurate measurement of α is often very important and, therefore, it is usually necessary to make a cut-back measurement to eliminate coupling, higher-order mode, and other measurement losses in determining α. This may be done by first measuring the loss for the entire length of fiber. This determines the value of P_{out}. Then the fiber is cut, leaving a short length of fiber for the cut-fiber measurement. This determines the value of P_{in}. The length of the fiber is the difference between the long and short length of fiber. The units for α are either dB/km or dB/m. The latter unit is more appropriate for IR fibers because IR fibers are typically meters in length.

The total absorption coefficient is the sum of the intrinsic and extrinsic absorption mechanisms, i.e., $\alpha = \alpha_{int} + \alpha_{ext}$. The intrinsic losses are those attributable to various inherent absorption mechanisms of pure transparent solids. As such intrinsic losses cannot be lowered below some fundamental value. For silica fiber this loss is 0.14 dB/km at 1.55 μm. Extrinsic losses, however, are due to impurity absorption and scattering from defects. A primary goal of IR fiber research continues to be the reduction of α_{ext} to the point where intrinsic losses dominate. This holy grail of IR fiber optic research remains elusive, but progress has been made to the point where IR fibers are still useful in spite of their high extrinsic loss.

The fundamental loss mechanisms in IR fiber materials would suggest that IR fibers hold the solution to all fiber applications, including telecommunications. This is because the projected losses for these materials can be as much as 1000 times less than the loss for silica fiber. In practice, however, no IR fiber has ever been fabricated with a loss as low as the minimal loss for silica and it is doubtful that one ever will.

There are three intrinsic loss mechanisms for any dielectric material: UV absorption, scattering, and IR absorption. These three mechanisms combine to give a total intrinsic loss that may be written schematically as

$$\alpha_{\text{int}} = Ae^{\frac{a}{\lambda}} + \frac{B}{\lambda^4} + Ce^{-\frac{c}{\lambda}}. \tag{2.14}$$

The first term describes the exponentially increasing absorption in the UV- or short-wavelength regime as the wavelength decreases. This is the familiar Urbach tail discovered many years ago first by F. Urbach in 1953.[7] It results from electronic or exitonic absorption, which becomes important when the energy of the photons approaches the electronic bandgap energy. The second term represents the scattering losses. In general, scattering results from Rayleigh (elastic) and Brillouin (inelastic) scattering. Rayleigh scattering dominates in glassy materials. Here it arises as a result of light scattering from density or compositional fluctuations in the refractive index that occur naturally in disordered amorphous materials. For crystalline fibers that have perfect order (i.e., in a perfect single-crystal fiber), there is no Rayleigh scattering and only Brillouin scattering remains. This is one reason why single-crystal fibers can, in principle, have a very low theoretical loss. The final term in Eq. (2.14) describes the exponential increase in attenuation with increasing wavelength that characterizes infrared or multiphonon absorption. Strong IR or multiphonon absorption results from the coupling of light (photons) to lattice vibrations (phonons). Multiphonon absorption becomes significant as n decreases rapidly in the region of anomalous dispersion (see Fig. 2.8). In the Reststrahl region, the material becomes opaque and then transparent again on the long wavelength side of the Reststrahl peak where the photon-phonon interaction becomes weak.

A wavelength dependence of the intrinsic losses for a typical transparent solid is shown in Fig. 2.9.[8] In general, the Urbach or UV absorption edge is rather sharp, as might be expected for an absorption that depends on the electronic bandgap energies. The fact that this edge is not more abrupt is a result of a series of closely spaced energy levels in the conduction and valence bands that give rise to a broader spectrum of excitonic transitions than if there were merely one electronic bandgap energy. Nevertheless, the Urbach tail is sufficiently sharp in most materials and does not contribute significantly to the overall intrinsic absorption in the visible or IR region. The exception is the chalcogenide glasses, in which there is a much broader UV edge. In these glasses, there is generally a weak absorption tail that extends to much longer wavelengths. This is in addition to the usual Urbach tail. It is a result of electronic transitions from intrinsic defect states lying within the bandgap. The overall electronic absorption, therefore, is the sum of the usual Urbach and weak absorption tails. This absorption is discussed further in Chapter 5 on chalcogenide glass fibers.

The minimum loss for a transparent solid occurs at a wavelength, λ_{min}, where the scattering loss equals the loss due to IR absorption. As may be seen in Fig. 2.9, the scattering loss and multiphonon edge form a so-called V-curve; the minimum in the V-curve at λ_{min} gives the minimum loss α_{min}.[9,10] If the electronic edge absorption term in Eq. (2.14) is neglected in favor of the scattering and multiphonon absorption terms (i.e., set $A = 0$), then it is possible to minimize the resultant equation to solve for λ_{min} and α_{min} in terms of B, C, and c. Lines[11] has done this and, using experimental values of B, C, and c taken from the literature, he has computed λ_{min} and α_{min}. Some values for λ_{min} and α_{min} along with the scattering constant, B, for glasses are given in Table 2.2.[12] For example, the value of α_{min} for silica is 0.14 dB/km at approximately 1.55 μm. As mentioned earlier, this value has almost been achieved experimentally so that silica fiber is virtually intrinsic at the 1.55-μm telecom band.

There are, however, materials that have significantly lower values of α_{min} than silica. From the limited data in Table 2.2 it can be seen that some glassy materials, and as will be seen later, many crystalline materials have losses that are often a factor of more than 100 times less than the minimum loss for silica. This can be seen in the intrinsic data for selected IR glassy materials shown in Fig. 2.10.[12] In

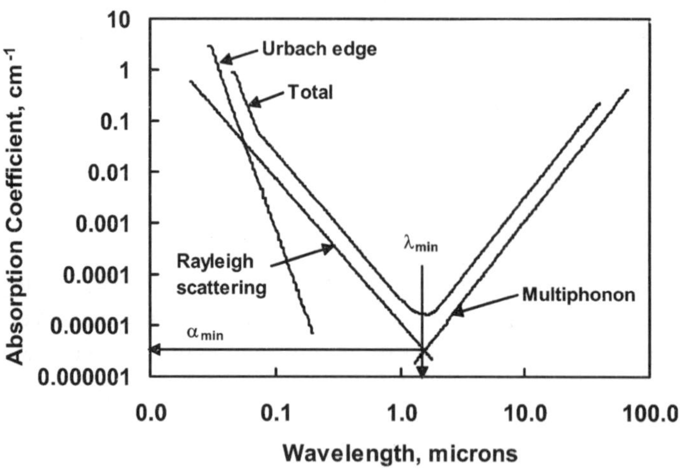

Figure 2.9 Intrinsic loss mechanisms in low-loss fiber optic materials. [Reprinted with permission of The American Ceramic Society, www.ceramics.org. Copyright 1981. All rights reserved.]

Table 2.2 Some representative values of the minimum intrinsic absorption in different glasses.

Material	B, μm^4 dB/km	α_{min}, dB/km	λ_{min}, μm
SiO_2	0.47	0.14	1.55
GeO_2	0.79	0.06	2.0
ZBLAN	—	0.013–0.022	2.5
(Heavy metal fluoride glass)			
$ZnCl_2$	1.10	0.0034	4.6

Fig. 2.10, only the calculated scattering loss and extrapolated multiphonon absorption are plotted, as these are the only significant loss mechanisms contributing to the minimum loss. The reason that the predicted losses are so low for these nonoxide glassy materials in contrast with the oxide glasses is twofold. First the heavier mass of the metallic and halide ions produces a lower fundamental resonance frequency and, therefore, the IR absorption edge is shifted to longer wavelengths. This is predicted by the Szigeti relation, which relates the fundamental resonance frequency, ω_0 to the reduced mass, μ, of the ions. Specifically, $\omega_0 \propto \sqrt{k/\mu}$, where k is the force constant. By way of an example, a comparison of ω_0 between silica and a material with heavier masses can be made. Specifically, assuming a diatomic oscillator model with different reduced mass but no change in force constants, the Szigeti relation for SiO_2 and TlBrI indicates that $\omega_o(SiO_2)$ is 2.35 times greater than ω_o(TlBrI). In practice the force constant for these two materials is not equal, but the trend is clear. That is, the shift to longer wavelengths is particularly large when the ionic mass and/or the difference between the two masses is large. Historically, the crystalline material KRS-5 (TlBrI) was observed as early as 1977 to have an intrinsic loss at least 10^3 times less than that for silica.[13] At the time this caused quite a sensation, as many saw the potential of a polycrystalline fiber made from this material as a replacement for silica fiber optics. In fact, optimists considered undersea links made from KRS-5 fiber that would be thousands of kilometers

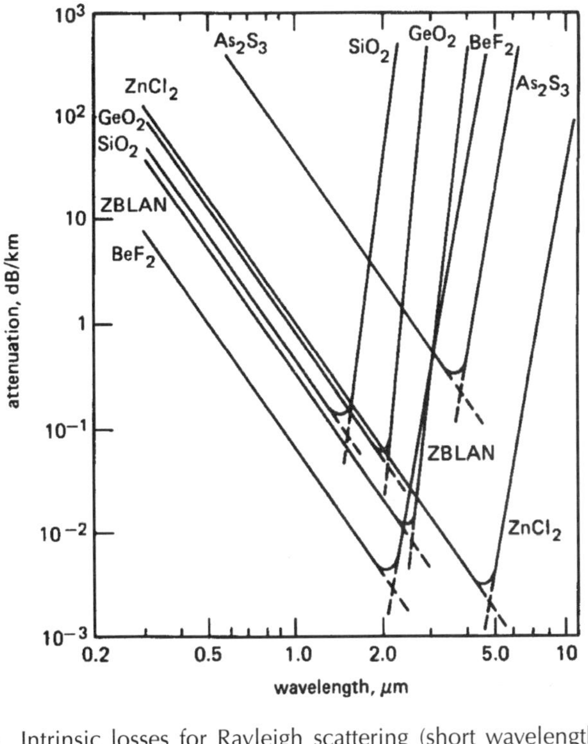

Figure 2.10 Intrinsic losses for Rayleigh scattering (short wavelength) and multiphonon absorption (long wavelength) in selected IR glassy materials. [Reprinted with permission from Ref. 12, Copyright CRC Press, Boca Raton, Florida.]

in length without repeaters. Unfortunately, the promise of such ultralow-loss fibers never materialized, and today it remains only a distant dream for these and related IR fiber materials.

The second reason that the intrinsic loss for IR materials is so low is that their scattering losses are sometimes less than those for silica. Pinnow et al.[14] have given the intrinsic scattering loss for glasses as

$$\alpha_{scat} = \frac{8\pi^3}{3}\frac{1}{\lambda^4}(n^8P_{12}^2)(k_BT\beta_T), \tag{2.15}$$

where P_{12}^2 is the appropriate Pockel's coefficient; k_B is Boltzmann's constant; T is the absolute temperature or the glass transition temperature in the case of glass; and β_T is the isothermal compressibility. Equation (2.15) shows the expected λ^{-4} dependence characteristic of Rayleigh and Brillouin scattering and a very strong dependence on n. Many investigators have used this relationship to calculate the ultimate scattering loss for both glassy and crystalline materials and then used these losses to obtain the scattering loss dependence for construction of the V-curves.

A good example that illustrates well the wavelength dependence of the intrinsic losses in an IR material is crystalline sapphire (Al_2O_3). In Fig. 2.11, data is given for all three intrinsic loss mechanisms.[15] The UV and IR edge data are extrapolations of UV and IR absorption coefficient data taken from the literature for bulk, single-crystal sapphire. Clearly, the Urbach tail is not only rather sharp for sapphire but also well displaced to shorter wavelengths and not a factor in determining α_{min}. The scattering loss was calculated using Eq. (2.15). The minimum in the V-curve for sapphire gives $\alpha_{min} = 3.4 \times 10^{-3}$ dB/km or 8×10^{-9} cm^{-1} at $\lambda_{min} = 1.78$ μm. While this intrinsic loss is approximately 40 times less than that for silica, the actual measured losses are considerably higher. This is seen in Fig. 2.11, where the

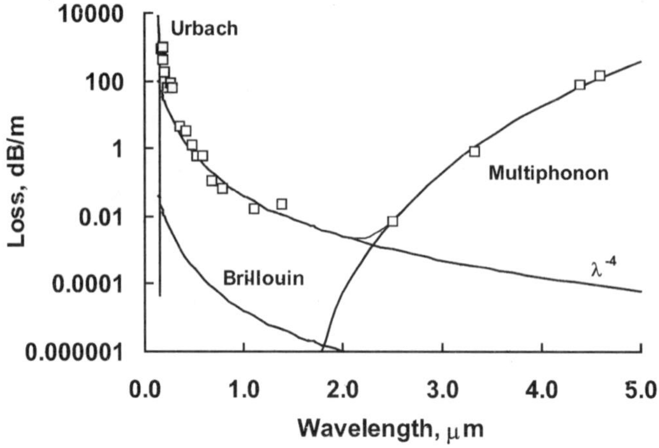

Figure 2.11 Intrinsic and measured losses for bulk sapphire crystals. Brillouin scattering is the minimal intrinsic scattering loss for perfect crystals.[15] [Reprinted with permission.]

experimental data taken from Innocenzi et al.[16] is plotted for high-purity bulk sapphire. Comparison of the measured results with the intrinsic loss, shows that the bulk data differs significantly from the theoretical loss over most of the visible- and near-IR region of the spectrum. Interestingly, a power-law fit varying as λ^{-4} to the data of Innocenzi et al. for wavelengths below ~1.5 μm (solid curve through data points in Fig. 2.11) suggests that Rayleigh scattering from inhomogeneities in the crystal contributes to the measured loss. This is an example of an extrinsic loss mechanism that, in general, dominates the practical losses in IR fibers.

2.3.2 Extrinsic loss mechanisms

Extrinsic losses in fibers or bulk materials arise as a result of imperfections in the IR material. Specifically, they occur as a result of impurity absorption and extrinsic scattering. In general, these losses can, in principle, be eliminated if appropriate steps in the fiber fabrication process are taken. In practice, however, it has become virtually impossible to entirely eliminate these loss mechanisms, and so today they continue to dominate the total loss in IR fiber optics. An example of the dominant nature of these mechanisms is the extrinsic scattering loss seen in sapphire data in Fig. 2.11.

There are numerous extrinsic loss mechanisms that limit the loss in IR fibers. Specific mechanisms for the different IR fiber categories will be elucidated in subsequent chapters. Some of the most common extrinsic mechanisms are listed in Table 2.3 along with an example of an IR fiber where these losses are most likely to occur. Water-related impurities such as OH and H_2O are some of the most common impurities in IR fibers, as they are in the oxide glass fibers. They are seen in virtually all IR fibers. In most cases it is much more difficult to remove these impurities in IR materials, as many are far more sensitive to the absorption and adsorption of water compared with silica. The HMFG fibers such as ZBLAN, for example, are frequently contaminated with water both during the batching of the starting materials and during the fiber drawing. Special conditions, such as dry boxes and inert atmosphere drawing environments, are required to reduce the water

Table 2.3 Major extrinsic loss mechanisms found in IR fibers.

Extrinsic loss	Source	IR fiber
Impurity absorption	OH^-, H_2O	PC, chalcogenide, HMFG
	SeH, SH	Chalcogenide
	Transition metals, e.g., Cu, Fe, Ni	HMFG
	Rare earth ions, e.g., Nd^{3+}, Cr^{3+}	HMFG
	CO^{3-}	PC
Scattering	Bubbles	HMFG
	Crystallites	HMFG
	Grain boundaries	PC
	Core/clad irregularities	HMFG, PC

absorption in the fiber. Other difficult-to-remove contaminants are the transition- and rare-earth-metal impurities. In the chalcogenide glasses, hydrogen present in the batching process bonds very strongly with sulfur and selenium, giving rise to quite intense absorption bands in the 4.0- to 4.3-μm region. Extrinsic scattering results from bubbles, small crystallites, and inhomogeneities in the material. Examples are given in Table 2.3. In addition, IR fibers often suffer from irregularities in the core/clad interface and other drawing defects. These irregularities come about during preform preparation and in fabrication of the fibers. It is difficult to prepare preforms that are as perfect as the oxide glass preforms prepared by vapor deposition techniques. Rod-in-tube methods for IR glass preform fabrication and the extrusion process for PC fibers are less precise and, therefore, IR glass fiber drawing or extruded PC fiber is less perfect than silica fiber.

2.4 Dispersion

The dispersive properties of a fiber control the temporal spreading of a pulse as it travels through the fiber. Dispersion and attenuation are the key optical properties determining the ultimate performance of an SM fiber used in telecommunication applications. Since IR fibers are not generally used in telecommunications, dispersion is not as important a property as it is for silica fibers. Nevertheless, because fluoride and chalcogenide glass fibers are potentially useful in short-haul telecommunication systems and as active fiber amplifiers, it is worthwhile reviewing the fundamental mechanisms giving rise to pulse dispersion in fibers.

Pulse spreading or pulse dispersion results from both waveguide and material properties of a fiber. The most important source of dispersion in MM fibers is intermodal or just modal dispersion. Modal dispersion occurs only in MM fibers as a result of differences in the group velocities of the different modes. This is easily understood from a ray-optic viewpoint. Referring to Fig. 2.4(a), rays launched at highly oblique angles near θ_c will have a longer path than those traveling near the optic axis. That is, higher-order modes take a longer time to travel a fiber of length L compared with the lowest-order modes, whose angle with respect to the fiber axis is essentially zero. The difference in transit times for the longest and shortest times is given by

$$\Delta t = t_{\text{max}} - t_{\text{min}} = \frac{n_1 L}{c}\left(\frac{n_1}{n_2} - 1\right) \cong \frac{n_1 L}{c}\Delta, \tag{2.16}$$

where Δ is the relative index difference defined in Eq. (2.9). For an SI, MM fiber with a core index of 1.5 and a reasonable $\Delta = 0.01$ for a MM fiber, the difference in transit times would be 50 ns/km. This is quite large in comparison with the dispersion in a SM fiber. Therefore, MM fibers are limited to short-haul applications.

Dispersive effects are most important for SM fibers. The important dispersion mechanisms are

- material dispersion;
- waveguide dispersion;

- nonlinear effects;
- polarization mode dispersion.

Of these, the two most important are material and waveguide dispersion, these, when taken together, are referred to as chromatic dispersion. It is possible in a silica fiber optic telecommunication system to have nearly zero chromatic dispersion through a judicious choice of materials and a correctly configured index profile for the waveguide. But chromatic dispersion is not a common or even necessary consideration for an SM IR fiber. Rather, the most important quantity for IR fibers is material dispersion, as it has the greatest impact on the pulse dispersion that might occur in, for example, IR fiber amplifiers.

2.4.1 Material dispersion

All dielectric materials are dispersive. This means that the refractive index varies with wavelength, i.e., $n = f(\lambda)$. This may be seen in the typical dispersion curve for a dielectric material (NaCl) shown in Fig. 2.8. There are several ways to measure dispersion in transparent materials. A simple measure is the Abbe number, V_D. The Abbe number is obtained by measuring the index at several key wavelengths, or

$$V_D = \frac{n_D - 1}{n_F - n_C}, \tag{2.17}$$

where n_F, n_D, and n_C, are the refractive indices at three standard visible wavelengths: $\lambda_F = 486.1$ nm (blue), $\lambda_D = 589.2$ nm (yellow), and $\lambda_C = 656.3$ nm (red). Abbe numbers are normally measured and tabulated for optical glasses that are transparent in the visible region. The most dispersive glasses are the heavier flint glasses with V_D ranging from 30 to 40. Less dispersive glasses like the crown glasses have higher Abbe numbers. Silica has a $V_D = 68$. Fluoride glasses have quite low dispersion, with V_D ranging from 70 to 80.

Another measure of dispersion is the derivative $dn/d\lambda$. The derivative may be most easily obtained from dispersion data like those shown in Fig. 2.8 once a curve fit to the data has been made to determine n as a function of wavelength. It is quite common to fit the dispersion data using a Sellmeier power-law fit. The Sellmeier equation is of the form

$$n^2 = 1 + \sum_i \frac{A_i \lambda^2}{\lambda^2 - \lambda_{oi}^2}, \tag{2.18}$$

where A_i and λ_{oi} are the Sellmeier constants: λ_{oi} is a vacuum wavelength related to natural vibrational frequencies $\lambda_{oi}\nu_{oi} = c$. In general, a very good fit is obtained in the region of normal dispersion, with $i = 3$. The Sellmeier constants for many of the common IR materials are given in the Handbook of Optical Constants series by Palik.[6]

A pulse of light will travel with a group velocity v_g. This pulse is a wavepacket composed of a spectrum of different frequencies (or wavelengths) centered at a frequency ν_o with a width $\Delta\nu \ll \nu_o$. The group velocity is given by

$$v_g = \frac{c}{n - \lambda_o \dfrac{dn}{d\lambda_o}}. \tag{2.19}$$

The pulse itself broadens as a result of the individual components traveling with a different velocity. This leads to an overall temporal spreading of the pulse as it moves through the material. The temporal width σ_τ (in ps) of a pulse after traveling a length L (in km) in a fiber is given by

$$\sigma_\tau = |D_\lambda|\sigma_\lambda L, \tag{2.20}$$

where σ_λ is the initial source width (in nm); and D_λ is the dispersion coefficient for material dispersion (in ps/km-nm). D_λ is given by

$$D_\lambda = -\frac{\lambda_o}{c}\frac{d^2n}{d\lambda^2}. \tag{2.21}$$

The units for the dispersion coefficient are practical units more relevant to the long lengths of fibers used in telecommunication applications. In summary, it can be seen from Eqs. (2.19) and (2.21) that material dispersion affects pulse propagation though both the first derivative (group velocity) and the second derivative (dispersion coefficient).

As mentioned above, $dn/d\lambda$ has a minimum value at the point of inflection in the dispersion curve. At the wavelength where $dn/d\lambda$ is a minimum, $d^2n/d\lambda^2$ is zero and the group velocity is a maximum. This is the very important zero material dispersion wavelength (ZMD). In Fig. 2.12 data is shown for the dispersion coefficient D_λ for a typical fluoride glass (ZBLAN) and, for comparison, silica.[17] The zero dispersion wavelengths occur at 1.724 μm and 1.312 μm for ZBLAN and silica, respectively. The zero dispersion wavelengths are always less than λ_{min}, where the attenuation is least (see Table 2.2). For silica the minimum in attenuation occurs at 1.55 μm, and for ZBLAN at 2.45 μm. In modern telecommunication systems, both λ_{min} and λ_{ZMD} and the respective dispersion coefficients are very important in the design of both low-loss and low- dispersion systems.

2.4.2 Waveguide and chromatic dispersion

Even if there is no material dispersion, there is still dispersion in an SM fiber. The most important additional dispersive mechanism is waveguide dispersion. Waveguide dispersion is present in both MM and SM fibers, but it is only significant in SM fibers. It occurs because the amount of light energy traveling in the

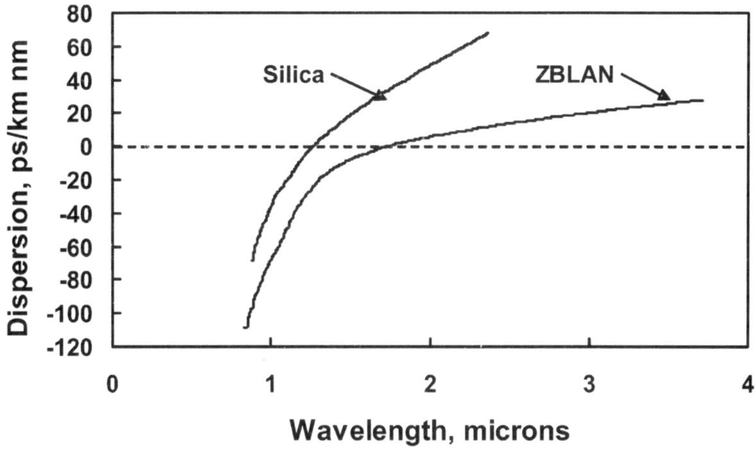

Figure 2.12 Material dispersion, D_λ , for conventional silica fiber compared to that of the fluoride glass, ZBLAN. [Reprinted from Ref. 17, Copyright 1995, with permission from Elsevier.]

core and in the cladding of a waveguide depends on the ratio a/λ_o. If the wavelength is altered, there will be a different distribution of light in the given core/clad structure. This results in different phase velocities (equal to c/n) in the core and in the clad, which in turn means that the group velocity of the mode is altered. To calculate the pulse spreading that results from waveguide dispersion, an equation equivalent to Eq. (2.20) is used, but only the dispersion coefficient for waveguide dispersion, D_W, is given by

$$D_W = -\frac{1}{2\pi c}V^2\frac{\partial^2\beta}{\partial V^2}, \tag{2.22}$$

where β is the propagation constant. D_w and β are complicated functions; further details, including graphical solutions for D_w, are available in most advanced books on fiber optics.

The combined effect of material and waveguide dispersion produces the major contribution to the total dispersion in SM fibers. The sum of material and waveguide dispersion is called chromatic dispersion, and it is controlled by the chromatic dispersion coefficient,

$$D(\lambda) = D_\lambda(\lambda) + D_W(\lambda), \tag{2.23}$$

and the pulse spreading due to chromatic dispersion is

$$\sigma_\tau = |D|\sigma_\lambda L. \tag{2.24}$$

As mentioned earlier, pulse dispersion is not as important in IR fibers as it is in conventional silica for telecommunications. Modern SM telecommunication fibers have been developed with index profiles, a/λ_o ratios, and selected compositions tailored to reduce $D(\lambda)$ to as low a value as possible over a relatively broad wavelength range. This has not been done for IR fibers.

2.5 Mechanical properties of fibers

The mechanical behavior of fibers is determined by several factors. Most important is the strength or the stress required to produce failure. Flaws and defects in the fiber determine the ultimate strength. While these flaws and defects can be introduced at any stage in the life of the fiber, they are usually introduced during handling or fiber drawing, or they reside in the preform prior to drawing. The fiber flaws are distributed randomly throughout the fiber and they vary in intensity. This naturally leads to the interpretation of fiber strength in terms of a "weakest link" theory, which is a statistical analysis based on the largest expected flaw in the fiber. The strength is also time dependent because the fiber is exposed to moisture and other environmental changes. That is, the strength frequently degrades with time.

The strength of a fiber can be determined in either tension or bending. Tensile testing involves gripping the ends of a certain length of fiber (called the gauge length) and then stretching the fiber until it breaks. The failure stress and/or strain at failure are obtained in a straightforward manner. An advantage of this method is that relatively long gauge lengths can be used. As a result, the failure stress measured in tension will be lower than the values obtained in bending, where smaller pieces of fiber are used. This is because there are bigger defects in a longer length of fiber. Bend testing involves using a bent loop of fiber constrained between two polished or grooved faceplates. A schematic of a fiber bend tester is shown in Fig. 2.13.[18] The advantage of bend testing over tensile testing is that there are no problems associated with gripping the fiber. This is especially important because gripping coated fibers can be difficult. That is, the fiber can often fail at the grip and this can lead to erroneous tensile strengths. Another advantage of bend testing is that only a small length ($\approx$3 cm) is needed; this can be important when long lengths of fiber are not available for testing.

The maximum stress σ_{max} can be obtained from the geometry of the bend. The result is

$$\sigma_{max} = 1.198E\frac{2r}{D-d}, \tag{2.25}$$

where E is Young's modulus, D is the plate separation as shown in Fig. 2.13, d is the overall fiber diameter including coating, and r is the radius of the fiber.[19] The mean bending strength of a fiber can be determined by measuring σ_{max} for at least 10 specimens cut from the original fiber. Each specimen will have different values for σ_{max} because each short length of fiber has a different flaw population and, therefore, different fracture strengths. The mean strength can then be determined by making a statistical analysis of the strength data.

A statistical analysis of the failure data[19] allows one to obtain the probability of failure for the fibers. In general, the weakest-link theory, in which the fibers break

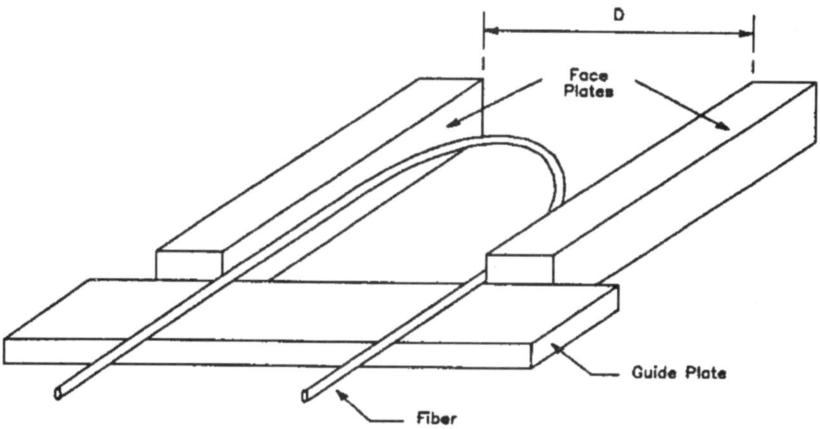

Figure 2.13 Schematic representation of two-point bending apparatus used to measure bending strength in fibers. [Reprinted with permission of The American Ceramic Society, www.ceramics.org. Copyright 1986.]

at statistically random points of greatest weakness, is used to obtain the cumulative probability of failure, $F(\sigma)$:

$$F(\sigma) = 1 - \exp\left[-\int_A f(\sigma)dA\right], \tag{2.26}$$

where $f(\sigma)$ is the probability per unit area that failure has occurred from a stress σ. The most commonly used form for $f(\sigma)$ is the Weibull distribution, or

$$f(\sigma) = \left(\frac{\sigma}{\sigma_0}\right)^m, \tag{2.27}$$

where m is the Weibull modulus, which is a measure of the scatter in the data or the flaw size distribution. Typically, m varies from a low of 5 for weak fiber to 100 or more for pristine silica. In terms of strength, this means that the strain (stress) to failure ranges from 0.1% (~50 MPa) to 20% (~14 GPa) for pristine silica. Substituting Eq. (2.27) into Eq. (2.26) gives the cumulative probability of failure,

$$F(\sigma) = 1 - \exp\left[\left(\frac{\sigma}{\sigma_0}\right)^m\right]. \tag{2.28}$$

In the presence of stress and certain chemical species such as water, the size of flaws existing in the glass can grow to a size that causes failure. This means that the breaking strength of the fiber also depends on the velocity that the faceplates (see Fig. 2.13) move together. A slow faceplate velocity will yield a low value of the breaking strength because there is more time for the flaw to grow and, thus, the fiber to fail.

Strength data for fibers is most often displayed in a Weibull plot. This is a plot of the cumulative probability of failure, Eq. (2.28), as a function of failure stress that has been plotted on logarithmic axes to yield a straight line. A typical Weibull plot for silica glass fiber[18] is given in Fig. 2.14. From a practical viewpoint, this data may be simply interpreted to give an indication of a safe bending strength limit for a particular fiber. For example, if it is desired to use the fiber with a less-than-5% probability of failure in bending, then the fiber should not be stressed beyond ~5.6 GPa. Conversely, knowing or assuming the safe limit of failure strength, one can obtain a minimum-bending radius using Eq. (2.25). More simply, a very useful approximate expression obtained from Eq. (2.25) may be used, i.e.,

$$\sigma_{\max} = E\frac{2r}{D}. \tag{2.29}$$

A fundamental and important aspect of the Weibull data concerns the physical interpretation of the Weibull exponent m, which is the slope obtained from the Weibull plot. Specifically, m may be obtained from the Weibull data in Fig. 2.14 by taking from the graph two values of the fractional failure probability, F_1 and F_2, two corresponding stress values, σ_1 and σ_2, and the equation

$$m = \frac{\left\{\dfrac{\ln\ln[1/1 - F_1]}{\ln\ln[1/1 - F_2]}\right\}}{\ln\left(\dfrac{\sigma_1}{\sigma_2}\right)}. \tag{2.30}$$

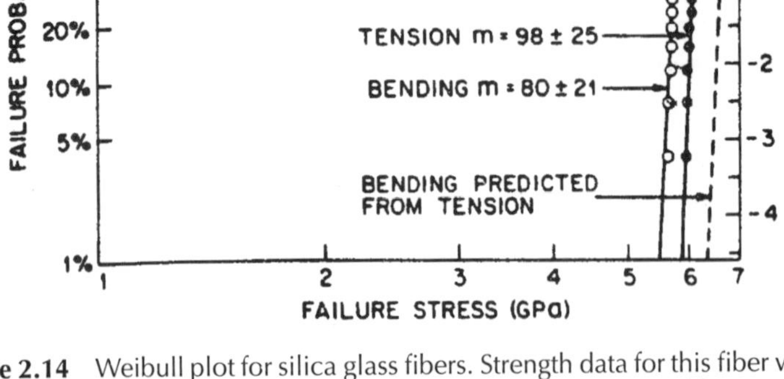

Figure 2.14 Weibull plot for silica glass fibers. Strength data for this fiber was taken using the two-point bending technique shown in Figure 2.13 and by standard tensile testing. The reason that the tensile strength is greater than the bending strength for this particular case is that the gauge length of the specimens for tensile testing was rather short. [Reprinted with permission of The American Ceramic Society, www.ceramics.org. Copyright 1986.]

That is, the slope may be found directly from the data in Fig. 2.14 if one uses the top abscissa and right ordinate axes. In general, a high m value indicates that there is a narrow distribution of flaw sizes in the fiber. Therefore, coatings providing good mechanical protection should result in large values of m. It is also possible to obtain a bimodal Weibull plot. In this case, there are two slopes and two different m values. The initial slope is usually broad and it has a small Weibull modulus compared to the high strength slope that is shown in Fig. 2.14. Such a two-slope plot indicates a bimodal distribution of flaw sizes. These fibers are often too weak or brittle for practical applications. Examples of this behavior will be found in Chapters 4 and 5 for the fluoride and chalcogenide glasses.

References

1. J. Hecht, *City of Light*, Oxford University Press, New York (1999).
2. C. Kao and G. A. Hockham, "Dielectric-fiber surface waveguides for optical frequencies," *Proc. IEEE*, Vol. 133, pp. 1151–1155 (1966).
3. V. G. Artjushenko, L. N. Butvina, V. V. Vojtsekhovsky, E. M. Dianov, and J. G. Kolesnikov, "Mechanisms of optical losses in polycrystalline KRS-5 fibers," *J. Lightwave Technol.*, Vol. LT-4, pp. 461–465 (1986).
4. T. Hidaka, T. Morikawa, and J. Shimada, "Hollow-core oxide-glass cladding optical fibers for middle-infrared region," *J. Appl. Phys.*, Vol. 52, pp. 4467–4471 (1981).
5. J. A. Harrington and C. C. Gregory, "Hollow sapphire fibers for the delivery of CO_2 laser energy," *Opt. Lett.*, Vol. 15, pp. 541–543 (1990).
6. E. D. Palik, *Handbook of Optical Constants of Solids*, Academic Press, New York (1985).
7. F. Urbach, *Phys. Rev.*, Vol. 92, pp. 1324 (1953).
8. M. G. Drexhage, K. P. Quinlan, C. T. Moynihan, and M. Saleh-Boulos, "Fluoride glasses for visible to mid-IR guided wave optics," in *Advances in Ceramics: Physics of Fiber Optics*, Vol. 2, B. Bendow and S. S. Mitra, Eds., pp. 57–73 (1981).
9. S. Takahashi, "Optical properties of fluoride glasses," *J. Non-Cryst. Solids*, Vol. 95/96, pp. 95–106 (1987).
10. J. A. Harrington, *Selected Papers on Infrared Fiber Optics*, SPIE Press, Bellingham, WA (1990).
11. M. Lines, "Scattering losses in optic fiber materials. II. Numerical estimates," *J. Appl. Phys.*, Vol. 55, pp. 4058–4063 (1984).
12. P. France, M. G. Drexhage, J. M. Parker, M. W. Moore, S. F. Carter, and J. V. Wright, *Fluoride Glass Optical Fibres*, Blackie and Son, London (1990).
13. D. A. Pinnow, A. L. Gentile, A. G. Standlee, A. J. Timper, and L. M. Hobrock, "Polycrystalline fiber optical waveguides for infrared transmission," *Appl. Phys. Lett.*, Vol. 33, pp. 28–29 (1978).
14. D. A. Pinnow, T. Rich, F. W. Ostermayer, and M. DiDomenico, "Fundamental optical attenuation limits in the liquid and glassy state with application to fiber optical waveguide materials," *Appl. Phys. Lett.*, Vol. 22, pp. 527–529 (1973).

15. R. Nubling and J. A. Harrington, "Optical properties of single-crystal sapphire fibers," *Appl. Opt.*, Vol. 36, pp. 5934–5940 (1997).
16. M. E. Innocenzi, R. T. Swimm, M. Bass, R. H. French, A. B. Vilaverde, and M. R. Kokta, "Room-temperature optical absorption in undoped α-Al_2O_3," *J. Appl. Phys.*, Vol. 67, pp. 7542–7546 (1990).
17. F. Gan, "Optical properties of fluoride glasses: a review," *J. Non-Cryst. Solids*, Vol. 184, pp. 9–20 (1995).
18. M. J. Matthewson, C. R. Kurkjian, and S. T. Gulati, "Strength measurement of optical fibers by bending," *J. Am. Cer. Soc.*, Vol. 69, pp. 815–821 (1986).
19. C. R. Kurkjian, J. T. Krause, and M. J. Matthewson, "Strength and fatigue of silica optical fibers," *J. Lightwave Techn.*, Vol. 7, pp. 1360–1370 (1989).

Chapter 3

Theoretical Foundations of Infrared Fiber Optic Transmission: Hollow-core Fibers

3.0 Introduction

Hollow waveguides present an attractive alternative to other solid-core IR fibers. The key features of hollow guides are their ability to transmit wavelengths well beyond 20 μm, their inherent advantage of having an air core for high-power laser delivery, and their relatively simple structure and potential low cost. Initially these waveguides were developed for medical and industrial applications involving the delivery of CO_2 laser radiation, but more recently they have been used to transmit incoherent light for broadband spectroscopic and radiometric applications.[1] In general, hollow waveguides enjoy the advantages of high laser power thresholds, low insertion loss, no end reflection, ruggedness, and small beam divergence. Their potential disadvantages, however, include an additional loss on bending and a small NA. Nevertheless, they are today one of the best alternatives for both chemical and temperature sensing as well as for power delivery in IR laser surgery or in industrial laser delivery systems, with losses as low as 0.1 dB/m and transmitted cw laser powers as high as 2.7 kW.[2]

Hollow-core waveguides may be grouped into two categories: (1) those whose inner core materials have refractive indices greater than 1 (leaky guides), and (2) those whose inner wall material has a refractive index less than 1 (attenuated total reflectance, or ATR guides).[1] Leaky, or $n > 1$, guides have metallic and dielectric films deposited on the inside of metallic, plastic, or glass tubing. ATR guides are composed of dielectric materials with refractive indices less than 1 in the wavelength region of interest. Therefore, $n < 1$ guides are fiberlike in that the core index ($n \approx 1$) is greater than the clad index. Hollow sapphire fibers operating at 10.6 μm ($n = 0.67$) are an example of this class of hollow guide.[3,4] The structure and key parameters for circular cross-section hollow guides are shown in Fig. 3.1. In general, hollow structures with $n > 1$ have been made from metal, plastic, and glass tubes while the $n < 1$ or ATR guides are made of sapphire or some special $n < 1$ oxide glass.

The optical principles governing light transmission in hollow waveguides differ somewhat from those for solid-core fibers. A theoretical analysis of the transmis-

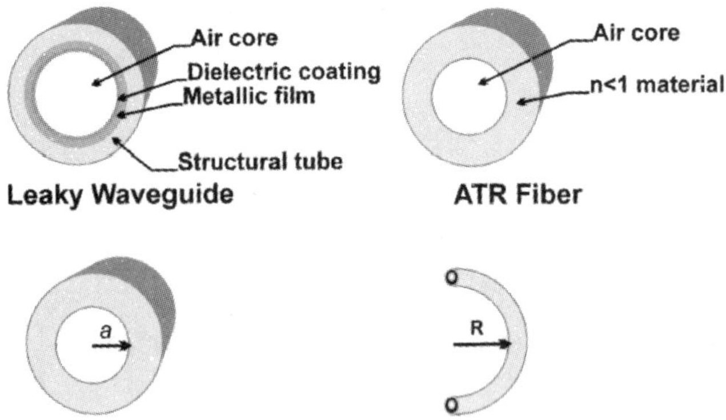

Figure 3.1 Structure of two different hollow waveguides: the leaky waveguide with inner metallic and dielectric coatings acting as an enhanced reflector, and the attenuated total reflectance (ATR) waveguide. The ATR hollow guides are fiberlike in that they work on the principle that the inner wall material has $n < 1$ and the air core has $n \cong 1$. [Copyright 2000. From Ref. 5. Reproduced by permission of Taylor & Francis, Inc., http://www.routledge-ny.com.]

sive properties for a hollow guide, based on both a ray and wave optic approach, will be given in the next two sections. In summary, the most important differences between hollow- and solid-core fibers are that the loss (absorption coefficient) for hollow waveguides depends on bore size and bending radius. Specifically,

$$\alpha \sim \frac{1}{a^3} \tag{3.1}$$

and

$$\alpha \sim \frac{1}{R}, \tag{3.2}$$

where a is the bore radius and R the radius of curvature (see Fig. 3.1). Equations (3.1) and (3.2) are valid for most launch conditions, bore sizes, and bending radii, and they are a characteristic property for all hollow guides with only one exception. This exception is a new class of omnidirectional or photonic-bandgap hollow guides in which the reflected light does not depend on the angle of incidence; i.e., it is omnidirectional.[6] For these guides, which are just now under active development, there would be no dependence on R as given by Eq. (3.2). It is important to note that Eqs. (3.1) and (3.2) that govern hollow guides do not have a counterpart in solid-core fibers. That is, these same relationships do not hold for the more conventional solid-core IR fibers.

An interesting consequence of the $1/a^3$ effect is that the losses for hollow waveguides can be made arbitrarily small merely by increasing the bore size. Practically speaking this usually means an upper limit of approximately a 1000-μm bore diameter for hollow silica guides and ~2000-μm bore diameter for the hollow plastic and metal guides. Nevertheless, it is often very useful in sensor applications, where energy throughput considerations are critical, to go to the larger bore sizes. For some radiometric applications, for example, hollow polycarbonate guides with a 2000-μm bore diameter have proven to have a very low loss. On the other hand, the losses increase dramatically when it is desired to make very flexible waveguides. Recall from Eq. (2.29) that the bending radius is directly proportional to the fiber radius, so small bending radii necessarily mean small bore sizes and, unfortunately, high loss.

The $1/R$ effect indicates that there is an additional loss that must be added to the straight loss for hollow guides. This feature would seem to offset many of the advantages of hollow guides. In practice, however, this additional loss is often insignificant for many applications. This is because most applications do not require that the fiber be sharply bent or that a significant portion of the fiber be under bend. In Chapter 7, experimentally determined bending losses will be given for a variety of hollow guides, and it will be seen that these losses are least for the lowest-loss guides. From this data it is possible to determine the total loss for a particular hollow waveguide merely by adding the bending loss contribution for each section of bent fiber to the straight loss.

3.1 Rectangular metal guides

Guiding EM radiation in rectangular or square cross section waveguides has been a very well studied subject for many years. The earliest work, and that covered in most textbooks on EM theory, involves the transmission of microwaves in hollow-metallic cavities. Microwaves have longer wavelengths than optical waves, and the wavelength of typical microwaves is usually close to the dimensions of the waveguide cavity. This is not true for hollow guides operating at optical frequencies. Instead, the optical wavelengths are much smaller than the waveguide or cavity dimensions. The first and truly seminal paper to thoroughly investigate the losses in hollow circular waveguides at optical wavelengths is that by Marcatili and Schmeltzer (MS) published in 1964.[7] The primary objective of the paper was to study the possibility of using a hollow pipe as an ultralow-loss conduit in place of conventional SM silica fiber for telecommunication applications. The authors concluded that hollow waveguides would not be suitable even though the losses could be very low because of the $1/a^3$ effect. The primary reason for this was the loss on bending that they surmised would ultimately overwhelm the small straight loss. In any case this paper had considerable importance beyond that for just telecommunication purposes. Their theory is the basis for losses in waveguide lasers such as the slab CO_2 lasers and, of course, for the current subject of flexible IR transmissive waveguides.

The study of flexible hollow guides for the delivery of IR energy began in earnest in the mid-1970s with the theoretical and experimental work of Garmire and her group at University of Southern California at Los Angeles.[8] Much of the work by Garmire was aimed at developing a waveguide for transmitting high power CO_2 laser energy at 10.6 μm. The first rectangular waveguides were simply constructed of two parallel aluminum metal plates separated by metal shim stock spacers, as shown in Fig. 3.2. The straight loss for these waveguides was determined

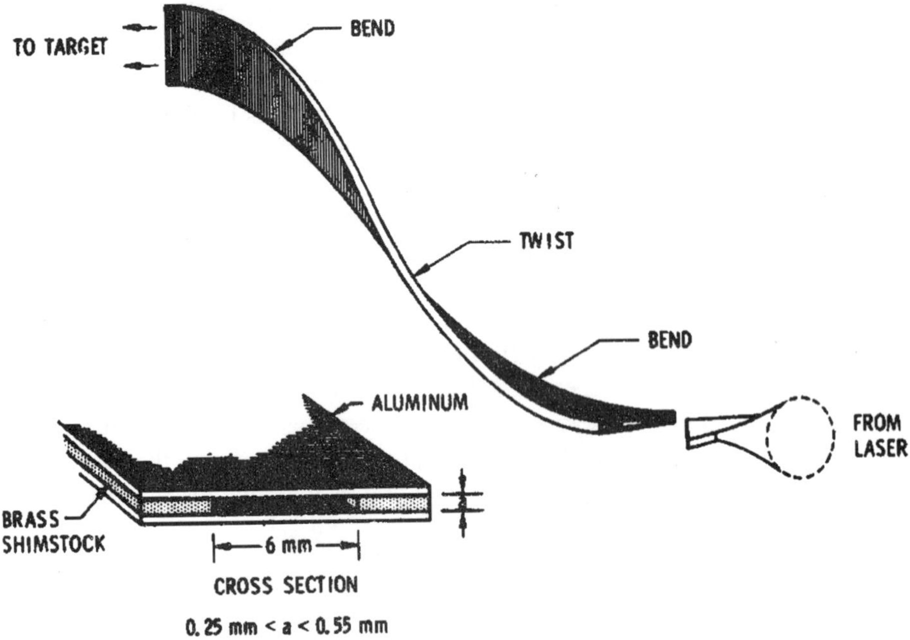

Figure 3.2 Rectangular waveguide structure of Garmire et al., using ordinary aluminum strips and shim stock to form the waveguide.[8] [Copyright 2003 IEEE. Reprinted with permission.]

to be only a few percent per meter, and even with a 1-m bending radius the losses would be less than 5% per meter. Garmire was also able to transmit over 200 W of CO_2 laser power using the waveguide shown in Fig. 3.2. This was a remarkable result at the time, as this power could never be delivered by a solid-core IR fiber. This result clearly delivered on the promise of hollow guides for laser power transmission. As may be seen from Fig. 3.2, however, the rectangular guide suffers from the fact that its dimensions are rather large in comparison to circular fibers, and because the guide cannot be bent in any arbitrary plane.

The losses for rectangular waveguides comprising two mirror surfaces can be readily calculated using a ray-optic approach. The geometry of the hollow planar waveguide is that shown in Fig. 2.7. The direction of propagation is along the z direction and the TE_m and TM_m modes have their electric field vectors pointing in the y and x directions, respectively. That is, the electric field vector for TE_m mode is parallel to the long dimension or to the side of length b. The spacer width or the side of length a is such that $b \gg a$. This ensures that essentially all of the light is reflected from the mirror plates and not from the spacers. For purposes of obtaining the essential loss parameters, it is sufficient to consider only the TE_m modes. As discussed in Section 2.1.3, only rays propagating at discrete bounce angles, θ_m, can propagate in the guide. The mode parameter $m = 1, 2, 3, \ldots$. The lowest-order modes usually have very small bounce angles and, as will be shown, the lowest loss. In practice this means that essentially only the lowest-order modes are transmitted and that light rays propagate at near-grazing-incidence angles. By applying the principle of self-consistency it can be shown that

$$\sin\theta_m \cong \theta_m = m\frac{\lambda}{2a}. \tag{3.3}$$

Assuming that the ray makes $N(\theta)$ bounces per unit length and that the loss per bounce is $A(\theta)$, then the total straight waveguide loss is $\alpha = N(\theta)A(\theta)$. From the geometry shown in Fig. 2.7 and for small bounce angles, $\sin\theta_m \cong \tan\theta_m \cong \theta_m$,

$$N(\theta) = \frac{\tan\theta_m}{a} = \frac{m\lambda}{2a^2}. \tag{3.4}$$

The absorption coefficient is, therefore,

$$\alpha = \frac{m\lambda A(\theta)}{2a^2}. \tag{3.5}$$

The loss per bounce or $A(\theta)$ may be determined using the s-pol or TE Fresnel equation, Eq. (2.1b), Eq. (2.5), and the fact that

$$A^{TE}(\theta_m) = 1 - R_s = 1 - |r_s|^2. \tag{3.6}$$

It is important to note that θ_m is the complement of the angle of reflection given in the Fresnel equations and that the refractive index in the Fresnel equations must now be formally replaced with the complex refractive index, $n^* = n - ik$ (see Sec. 2.2.1). The complex index is used because it is necessary to include the k value in the determination of the loss coefficient. Substituting Eqs. (3.6) and (2.1b) into (3.5) yields

$$\alpha_m^{TE}(\theta_m) = \frac{m^2\lambda^2}{a^3}\,\mathrm{Re}\left(\frac{1}{n^*}\right), \tag{3.7}$$

where it has been assumed that $|n^*|^2 \gg 1$. This is true for essentially all metals in the IR region. A similar analysis for the TM modes yields

$$\alpha_m^{TM}(\theta_m) = \frac{m^2\lambda^2}{a^3}\,\mathrm{Re}(n^*). \tag{3.8}$$

Equations (3.7) and (3.8) illustrate several important properties common to all hollow guides irrespective of their cross-sectional profile. The loss is seen to vary as $1/a^3$, where a is the mirror spacing for the rectangular waveguide and, as will be seen later, a similar relation with a being the bore radius for a circular waveguide. The loss also depends on the square of the wavelength and the mode parameter, m, as well as a dependence on the material's optical properties through the complex refractive index. The lowest-loss modes are the TE_m modes. Additionally, metals with large values of n^* and waveguides propagating low-order modes, i.e., small m values operating at short wavelengths, exhibit the lowest loss. It is also the case that most guides have values of a that are typically much larger than the wavelength of light. Normally a ranges from 300 to 1000 μm, as may be seen from Fig. 3.2. In principle, this means that hollow rectangular waveguides can transmit many modes or, equivalently, that they are highly overmoded. In practice, however, higher-order modes do not propagate very far in hollow guides. This is because the losses for higher-order modes increase as m^2, and, therefore, the losses for the high-order modes can be quite large. The net result is that only the lowest-order modes propagate because they have the lowest loss. In fact, when a is less than approximately 250 μm, it is often observed that only the lowest-order TE_1 mode will propagate.

The rectangular waveguide illustrated in Fig. 3.2 was made by Garmire et al.[8] using different mirror materials. Most of the guides were fabricated using an ordinary smooth aluminum sheet for the reflecting surface. While aluminum is not a particularly good reflector at 10.6 μm, it did make a respectable guide for the CO_2 laser. For an aluminum guide with a spacing of 0.5 mm ($a = 500$ μm) operating at 10.6 μm with n^* (10.6 μm) $= 20 - i59$, the straight loss for the TE_1 mode calculated using Eq. (3.7) is 4.6×10^{-5} cm^{-1} or 0.02 dB/m. This would be the loss for an ideal, brightly polished aluminum surface. The aluminum that Garmire used was commercial grade material. The waveguide that they made from this material had a measured loss approximately 16 times higher than that calculated above for ideal aluminum.

The bending losses for the rectangular, as well as for the circular cross section guides are not easy to calculate. This is in part due to the exact path that the ray or mode takes as it travels through the guide. For guides that are bent only slightly, MS have shown that the bending loss varies as $1/R^2$. From a ray-optic viewpoint, slight bending refers a radius of curvature for which the rays bounce back and forth between the two mirrors, as shown in Fig. 2.7. When the guide is bent to the point that rays bounce against only the outer wall, then the bending loss varies as $1/R$. Modes traveling in this bending regime are known as whispering gallery modes. A sketch for the whispering gallery mode is shown in Fig. 3.3. From a wave-optic viewpoint, the intensity profile of, for example, the lowest-order TE_1 mode, moves from the center closer to the outer wall. Further details of bending losses are given in the next section on circular waveguides.

3.2 Circular metal guides

The rectangular waveguides fabricated by Garmire et al.[8] provided important information on the propagation of light in hollow structures, but otherwise they proved to be impractical. This is because they are rather large in their overall dimensions (~0.1 cm × 1 cm) and their flexibility is limited in some directions. Their use seemed to be relegated largely to the delivery of high-energy laser power both in the IR and later in the UV range of excimer lasers. Nevertheless, this technology never caught on. This is largely due to the concomitant development of circular

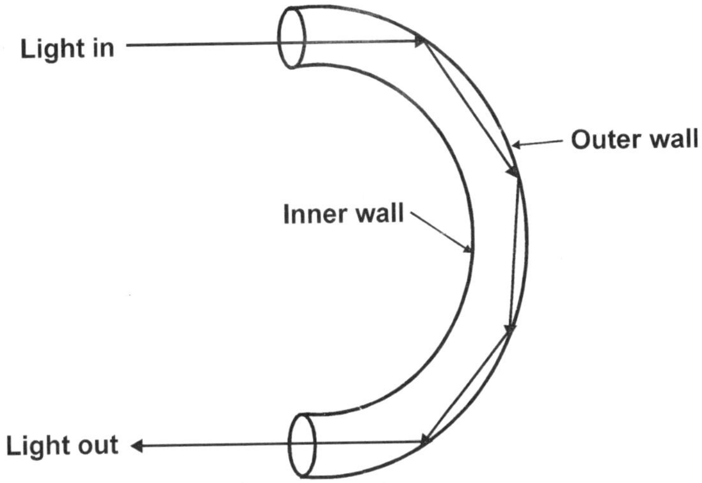

Figure 3.3 Geometry of bent rectangular waveguide illustrating the propagation of a whispering gallery mode.[8] [Copyright 2003 IEEE. Reprinted with permission.]

cross section guides with a much smaller waveguide cross section and capable of delivering very large laser powers. Today, circular cross-section guides are the waveguide of choice for essentially all hollow guide applications.

The theory describing light transmission in circular guides is again that due to MS.[7] Their theory, however, is restricted to only one wall material with complex refractive index n^*. With only one inner wall material the losses are rather high, especially for small bore sizes. The way around this dilemma is to overcoat the metallic inner wall with a transparent dielectric material to enhance the reflectivity; that is, to take advantage of the increased reflectivity that results when dielectric layers of the appropriate thickness are applied to any mirror surface. This is what is commonly done to enhance the reflectivity of, for example, laser mirrors. High-reflectance mirrors are made by applying alternating high and low refractive index materials to form a multilayer dielectric stack that will greatly improve the reflectivity.

3.2.1 Hollow waveguide theory: MS

The wave optic theory of MS involves solving Maxwell's equations for an air-core guide in cylindrical coordinates: r, z, and θ. The geometry of a circular waveguide is shown in Fig. 3.4. The MS analysis gives a solution for the electric and magnetic fields for the TE_{lm}, TM_{lm}, HE_{lm}, and EH_{lm} modes. In cylindrical coordinates, the l parameter is the number of periods in the azimuthal or θ direction. (Note that MS use the index n instead of l). The m parameter is the number of maximums and minimums for each component in the radial direction. In their analysis, MS simplify the solution of Maxwell's equations by making two assumptions. One is

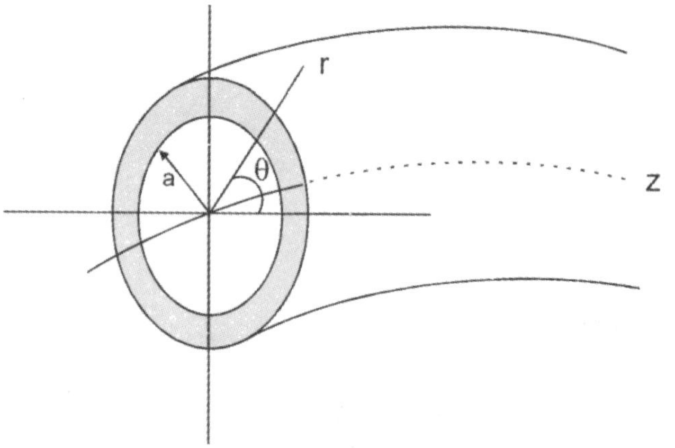

Figure 3.4 Cylindrical coordinate system used to describe hollow waveguides.

that the bore size is considerably larger than the wavelength of light or

$$k \cdot a = \frac{2\pi a}{\lambda} \gg |n^*| \cdot u_{lm}, \tag{3.9}$$

where k is the free space wavevector, $k = 2\pi/\lambda$ and u_{lm} is the mth root of the Bessel function $J_{l-1}(x)$, that is, when $J_{l-1}(u_{nm}) = 0$. They also assume that the propagation constant, β_{lm}, is approximately equal to the wavevector k. This is a good approximation for the lowest-order modes present in hollow guides as they travel at near grazing incidence, i.e., when they are nearly paraxial. MS derive an expression for the complex propagation constant, γ_{lm}, as

$$\gamma_{lm} \approx \left[1 - \frac{1}{2}\left(\frac{u_{lm}\lambda}{2\pi a}\right)^2 \left(1 - i\frac{\nu_l \lambda}{\pi a}\right)\right], \tag{3.10}$$

where ν_l is a mode-dependent material parameter specified below. The term γ_{lm} can be separated into its real (β_{lm}) and imaginary (α_{lm}) parts,

$$\beta_{lm} = \mathrm{Re}(\gamma)\frac{2\pi}{\lambda}\left\{1 - \frac{1}{2}\left(\frac{u_{lm}\lambda}{2\pi a}\right)^2 \cdot \left[1 + \mathrm{Im}\left(\frac{\nu_l \lambda}{\pi a}\right)\right]\right\} \tag{3.11}$$

and

$$\alpha_{lm} = \mathrm{Im}(\gamma) = \left(\frac{u_{lm}}{2\pi}\right)^2 \frac{\lambda^2}{a^3} \mathrm{Re}(\nu_l). \tag{3.12}$$

The $\mathrm{Re}(\nu_l)$ term in Eq. (3.12) depends on the mode and the complex refractive index n^* in a manner similar to the discussion surrounding the development of

Eqs. (3.7) and (3.8) for planar guides. The explicit dependence on n^* for the different modes is given by

$$\alpha_{lm} = \left(\frac{u_{lm}}{2\pi}\right)^2 \frac{\lambda^2}{a^3} \mathrm{Re} \left\{ \begin{array}{ll} \dfrac{1}{\sqrt{n^{*2}-1}}, & \text{for } \mathrm{TE}_{0m} \text{ modes } (l=0) \\ \dfrac{n^{*2}}{\sqrt{n^{*2}-1}}, & \text{for } \mathrm{TM}_{0m} \text{ modes } (l=0) \\ \dfrac{n^{*2}+1}{2\sqrt{n^{*2}-1}}, & \text{for } \mathrm{HE}_{1m} \text{ modes } (l \neq 0) \end{array} \right\}. \tag{3.13}$$

These results by MS embodied in Eq. (3.13) are used almost exclusively to calculate the loss for straight hollow waveguides made from an inner wall material with refractive index n^*. The attenuation coefficient α_{lm} in Eq. (3.13) is the absorption coefficient determined from Maxwell's field equations. Normally we measure the power loss $P_{\text{in}}/P_{\text{out}}$ and the corresponding power loss attenuation coefficient α as given by Eq. (2.13). Since the power or intensity is proportional to the square of the electric field, $\alpha = 2\alpha_{lm}$. It is interesting to note that Eq. (3.13) form the starting point not only for calculating losses in flexible hollow guides, but also for determining losses in hollow waveguide laser cavities. From these equations it is observed that α_{lm} depends on (λ^2/a^3) the loss increases as the square of the mode parameter u_{lm}, and the lowest-loss modes are the HE_{lm} modes when $n^* < 2.02$. The values for u_{lm} are given in Table 3.1. The electric field lines for the two lowest-order modes are shown in Fig. 3.5. The TE_{11} mode is a circular electric mode and the HE_{11} mode is a hybrid mode. The TE_{11} mode has the lowest-loss mode when $n^* > 2.02$.

The dependence on u_{lm} is similar to the dependence on m for the planar waveguides discussed in Section 3.1 and given in Eqs. (3.7) and (3.8). The difference is that m is a simple integer while the u_{lm} are zeros of Bessel functions. The main point here is that the losses for the higher-order modes increase quite rapidly, as may be seen from the values for u_{lm} given in Table 3.1. Since $a \gg \lambda$ for the circular guides, these guides are also very overmoded. But just as for the planar guides, the loss for the higher-order modes becomes so high that only the lowest-order modes tend to propagate. For example, data for 250-μm-bore hollow glass waveguides given in Chapter 7 will show that effectively only the lowest-order HE_{11} mode propagates in either straight or bent waveguides. As the bore

Table 3.1 Values for the mode parameter u_{lm} for some of the lowest-order modes.

	m = 1	2	3	4
l = 1	2.405	5.52	8.654	11.796
2	3.832	7.016	10.173	13.324
3	5.136	8.417	11.62	14.796
4	6.380	9.761	13.015	16.223

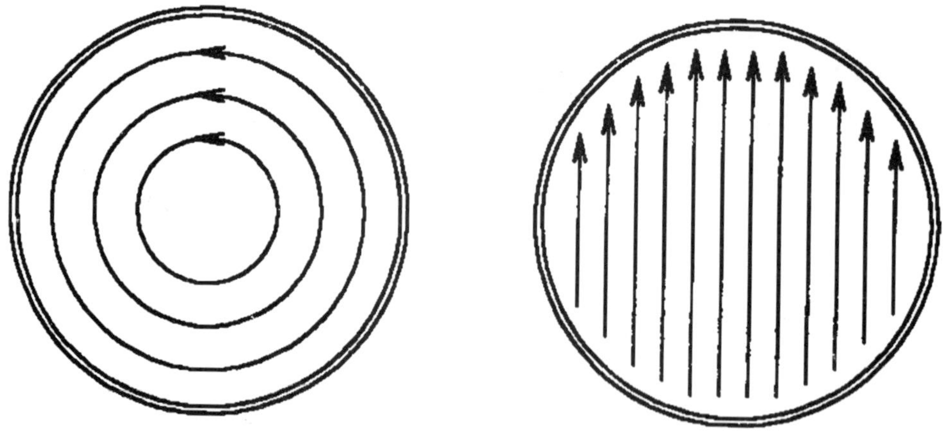

Figure 3.5 The TE_{01} (left) and HE_{11} (right) modes in a cylindrical dielectric hollow waveguide.

size increases more modes propagate, because higher-order mode loss calculated from Eq. (3.13) is not great as it is when $2a < 250$ μm.

The bending loss for the hollow guides calculated by MS involves a somewhat complex expression. For their calculation, MS assume that the bore size is small, normally less than approximately 100 μm, and that the bending radii are large; i.e., no whispering gallery modes. This assumption may be expressed mathematically as

$$\frac{2a}{R}\left(\frac{2\pi a}{u_{lm}\lambda}\right)^2 \ll 1. \tag{3.14}$$

Under this assumption MS determine that the loss due to bending, α_R, varies as

$$\alpha_R \propto \frac{a^3}{R^2}. \tag{3.15}$$

The total loss is, therefore,

$$\alpha_{tot} = \alpha_\infty + \alpha_R, \tag{3.16}$$

where α_∞ is the straight loss given by Eq. (3.13). MS also found that the loss depends on the polarization of the electric field with respect to the plane of the bend. The lowest bending losses occur for polarizations that are perpendicular to the plane of bending or s-pol (TE). This can also be seen through considerations of the Fresnel equations.

The range of bending radii for which Eq. (3.15) is valid lie in the range of 5 to 15 m or greater. Thus, Eq. (3.15) is not practical, as most applications of fibers require bending to smaller radii. Furthermore, bending data taken by Garmire et al.[8] did not show a $1/R^2$ dependence; rather they found that the bending losses were much less than that predicted by MS. It was left to Miyagi and his colleagues[9] to determine that the losses for most bending radii of interest varied as $1/R$ rather than $1/R^2$. This is discussed further in the next section.

3.2.2 Hollow waveguide theory: Miyagi

The losses in MS-type waveguides that incorporate only one wall material with complex refractive index n^* can be quite large. To reduce the loss it is necessary to somehow deposit thin dielectric films over the metallic layer to basically enhance the reflectivity over the metal alone. Miyagi and Kawakami[10–12] first proposed such multilayer structures in the early 1980s. The basic idea is to deposit alternating layers of high (n_H) and low (n_L), index transparent, dielectric materials. Each layer of the high-index dielectric normally has the same thickness, which is frequently equal to an optical thickness of $\lambda/4$. Similarly, the low-index dielectric layers usually have equal thickness, normally equal to $\lambda/4$. The physical thickness, d, of each material is $d = \lambda/4n$, so that the actual thickness of the two materials is not equal. Such a dielectric stack can greatly enhance the reflectivity. This is a well-documented, commonplace approach used for many years to make high-reflectance mirrors. The same general idea holds for hollow waveguides. A dielectric stack, as shown in Fig. 3.6, is deposited on the inside of a hollow guide. In practice this has not been so easy to do, because of the long lengths and small bore sizes of hollow guides and the need to have the coatings on the inside surface. Today, most of the hollow guides made employ only a single dielectric layer deposited on a high-

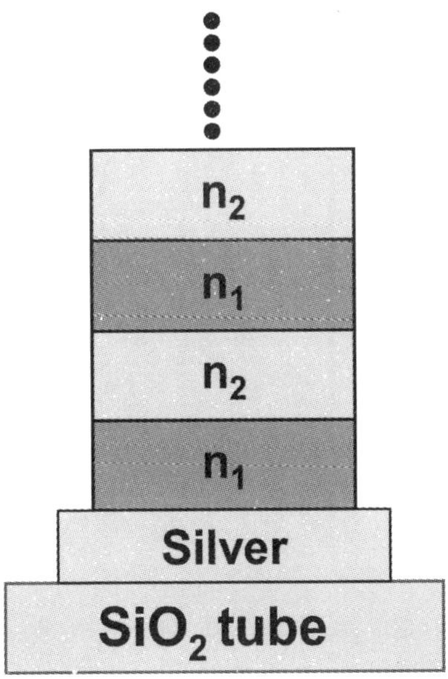

Figure 3.6 Mulilayer dielectric stack deposited on a silver film inside a silica tube. Multilayer structures made up of alternating layers (n_1/n_2) of two different dielectric materials have inherently lower loss than the single dielectric film over silver.

reflectance metal such as silver. Perhaps somewhat surprisingly, a single dielectric layer can still substantially reduce guide loss.

The theory of Miyagi and Kawakami gives a result similar to that of MS given in Eq. (3.13), but with additional terms reflecting the properties of the dielectric layers used in the stack. In its simplest terms, the attenuation coefficient for a straight waveguide with an innermost metallic layer with optical constants n and k is given by

$$\alpha_{lm} = \left(\frac{u_{lm}}{2\pi}\right)^2 \frac{\lambda^2}{a^3} \left(\frac{n}{n^2+k^2}\right)_{metal} \cdot F_{film}, \tag{3.17}$$

where F_{film} is a term which accounts for the loss due to the dielectric film(s).[11,12] The explicit nature of F_{film} is complicated, especially when two materials and multilayers are used. To illustrate the effect of dielectric films deposited on metallic layers, consider the easiest case of one dielectric film deposited on silver. The thickness of the dielectric film chosen for minimum loss depends on the wavelength of operation (for example, the laser wavelength) or on a range of wavelengths. Specifically, Miyagi and Kawakami showed that the minimum loss occurs at the optimum wavelength of λ_{opt} when the film thickness d is calculated from

$$d = \frac{\lambda_{opt}}{2\pi\sqrt{n_d^2-1}} \tan^{-1}\left[\frac{n_d}{(n_d^2-1)^{\frac{1}{4}}}\right], \tag{3.18}$$

where n_d is the refractive index of the dielectric film.[12] Further assuming that there is no absorption in the dielectric film, Miyagi and Kawakami derived the following expression for the losses for the three modes as

$$\alpha_{lm} = \left(\frac{u_{lm}}{2\pi}\right)^2 \frac{\lambda^2}{a^3} \left(\frac{n}{n^2+k^2}\right) \left\{ \begin{array}{ll} \left(1+\dfrac{n_d^2}{\sqrt{n_d^2-1}}\right), & \text{TE}_{0m} \text{ modes} \\ \dfrac{n_d^2}{\sqrt{n_d^2-1}}\left(1+\dfrac{n_d^2}{\sqrt{n_d^2-1}}\right), & \text{TM}_{0m} \text{ modes} \\ \dfrac{1}{2}\left(1+\dfrac{n_d^2}{\sqrt{n_d^2-1}}\right), & \text{HE}_{1m} \text{ modes} \end{array} \right\}. \tag{3.19}$$

The last expressions in braces are the exact values for F_{film} given in Eq. (3.17).[12]

As an illustration of the ability of a single dielectric layer to reduce the loss over a metallic-only hollow guide, Eq. (3.19) will be used to calculate the loss in an Ag/AgI hollow glass waveguide. The calculated losses for the HE_{11} mode propagating in a hollow glass waveguide with a bore size of 700 μm are shown in

Fig. 3.7 for both Ag/AgI and Ag-only films. There are several key features of this spectral data. First, note the peaks and valleys in the spectrum. These are not due to any absorption in the films; rather this is the typical interference phenomenon that one obtains from a thin film structure. For this calculation the design wavelength was $\lambda_{opt} = 10.6$ μm, which resulted in an optimum film thickness, according to Eq. (3.18), of $d = 0.85$ μm. Second, the losses for the Ag-only film guide are a factor of ~100 higher than for the Ag/AgI film guide. While this is the easiest thin film structure that can be deposited, it illustrates well the importance of having a dielectric film(s) incorporated in the waveguide structure. The addition of more dielectric layers further decreases the loss. Examples of these more complicated structures can be found in Chapter 7.

Miyagi also considered the bending losses in hollow guides. In contrast to the work of MS, Miyagi dwelt on the strongly bent guides or guides for which the bending radius was less than some critical radius, R_c. Below this radius, light would be guided in the whispering gallery mode as shown in Fig. 3.3. Miyagi[13] obtained the critical radius as

$$R_c = \frac{(2\pi/\lambda)^2 a^3}{[3/8(m-1/4)\pi^2][1+(2m-1/2)^{-2/3}]^3}, \tag{3.20}$$

where m is the mode number. When $R < R_c$, Miyagi determined that the bending losses were proportional to $1/R$. The exact expressions for the bending loss can be found elsewhere. To summarize, the bending losses are found to exhibit a $1/R$ behavior for $R < R_c$, but they vary as $1/R^2$ when $R > R_c$. The $1/R$ dependence has been well established experimentally by many researchers.

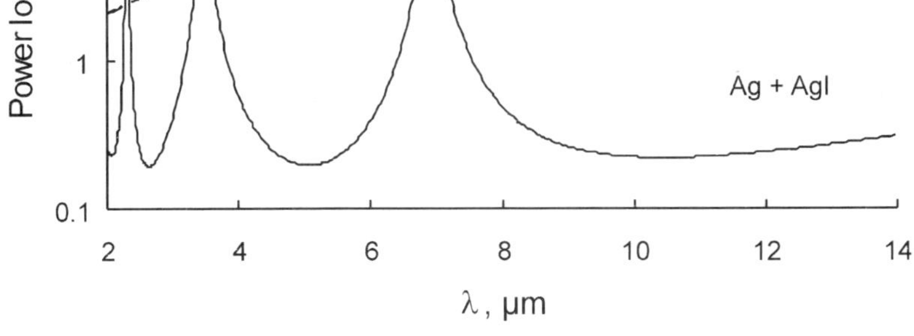

Figure 3.7 Reduction in theoretical power loss of the HE_{11} mode via the addition of a silver iodide layer to a silver metallic hollow waveguide with a 700-μm bore, straight hollow waveguide.

The group of researchers at Tohoku University has been extremely active over many years, both in their analysis and in their fabrication of hollow waveguides. Much of their work is covered in Chapter 7, but it should be mentioned here that they have also analyzed their circular waveguides for incoherent light transmission as well as making a comparison of the wave- and ray-optic approaches for calculating losses in hollow guides. The loss characteristics for incoherent light take on a somewhat different look than the losses simply calculated for some low-order mode using Eqs. (3.13) or (3.19). This is because the launch of incoherent light from, for example, a spectrometer is typically accomplished using low f/# optics. This results in the excitation of many higher-order modes in the guides. Matsuura et al. have computed the losses for incoherent light for straight[14] waveguides using ray-optic theory, while Saito et al., also using a geometric optic approach, have studied bent[15] hollow guides. An excellent review and comparison of both wave- and ray-optical methods for the computation of waveguide losses is given by Saito et al.[16] Miyagi and Karasawa[17] compute the losses for square and circular cross section waveguides for CO_2 laser transmission and show that it is possible to achieve losses for square cross-section guides as low as for circular cross section guides.

Finally, a number of investigators have studied other loss mechanisms in hollow guides, such as the effects of surface roughness on loss and tapered waveguides. Sagi-Dolev et al.[18] and Croitoru et al.[19] have used a ray-optic model to calculate losses for smooth as well as rough surfaces. Danilov et al.[20] and Rabii et al.[21] have also studied the effects of surface roughness on waveguide losses. Tapered waveguides have been studied by Zhou et al.[22] and by Gibson and Harrington.[23]

3.3 Omnidirectional waveguides

Omnidirectional hollow waveguides are a new type of photonic bandgap (PBG) fiber optic, which was first proposed several years ago by Y. Fink et al. at MIT.[6,24] Photonic bandgap or photonic crystal fibers have been studied since about 1996, beginning with the work of Knight et al.[25,26] at the University of Bath. Their work, and also collaborative work with Corning,[27] involved silica holey fibers based on a microstructure that comprised a series of holes in a silica glass fiber matrix. A picture of such a 2D microstructure fiber is shown in Fig. 3.8.[28] The basic idea behind

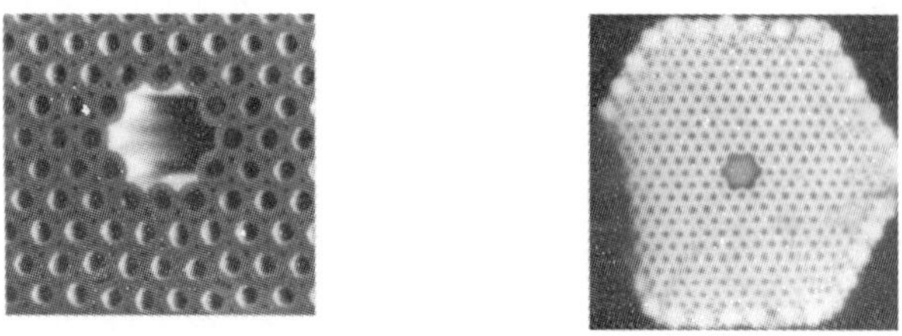

Figure 3.8 Microstructured or holey fiber formed in pure silica glass fiber. The air holes are uniformly spaced, and the center defect may be a hole, as shown, or a solid core.[28] [Reprinted with permission.]

a PBG fiber is that under the right conditions, a gap can exist in the photon band similar to the gap that can occur in electronic structures between the valence and conduction bands. The gap in the photon spectrum occurs as a result of large refractive index changes between the core glass and the surrounding holes. In effect, the holes serve to define the photonic bandgap and to form a defect state within the photon bandgap. The defect region is usually an axially symmetric, small-bore hollow core as shown in Fig. 3.8, but it may also be a solid core. Light within a discrete wavelength band is largely confined to this defect region. To date, all holey microstructure PBG fibers have been made from oxide glasses and, therefore, they do not transmit IR wavelengths. The omnidirectional waveguide described by Fink et al. does not rely on oxide glass, so it does transmit IR wavelengths. The omnidirectional guide concept builds on the basic idea of a photonic bandgap fiber, but instead of a microstructure or holey all glass fiber, they describe a 1D structure that is composed of alternating cylindrical layers of high/low-index dielectric materials surrounding a hollow core, as shown in Fig. 3.9. The dark and light regions surrounding the hollow core are thin dielectric films with a large difference between the high- (n_H) and low- (n_L) index layers. When the index contrast is sufficiently high there exists a gap in the photon band. Ideally the index contrast ratio, n_H/n_L, should be approximately two. When a sufficient number of high/low-index pairs are deposited, then this hollow structure can have a substantial photonic bandgap region. Within this bandgap region, the waveguide has the remarkable property of reflecting light equally for all angles of incidence. This so-called omnidirectional waveguide, in contrast to waveguides without a bandgap like the Ag/AgI structure, will not exhibit a $1/R$ loss on bending because the losses do not depend on the angle the rays strike the surface of the guide. A more detailed treatment of omnidirectional waveguides is given in Sec. 11.1.2, and the photonic bandgap structure is shown in Fig. 11.4.

To achieve a true omnidirectional waveguide, it is necessary to have a high-index contrast between the high- and low-index dielectric films. The higher the index contrast, the greater will be the width of the photonic bandgap. Additionally, fewer high/low dielectric pairs are necessary to achieve low loss. That is, with

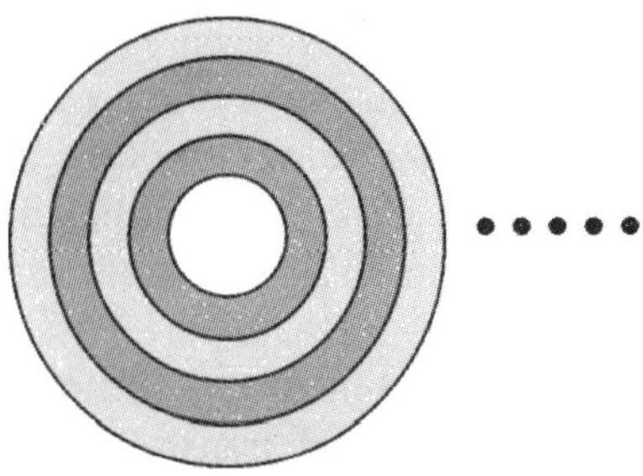

Figure 3.9 Omnidirectional waveguide composed of low- (light gray)/high- (dark gray) index pairs of thin-film dielectric coatings surrounding an air core. The index contrast should be high to achieve a large photonic bandgap and high reflectance with as few layers as possible.

high-index contrast between the dielectric layers, there will be greater reflectivity of the TM or p-polarized light in a relatively few (3 to 4 dielectric pairs) layers. To demonstrate the principles and feasibility of such an omnidirectional structure, Fink used a combined total of nine layers of high-index Te ($n = 4.2$, film thickness 0.8 μm) and low-index polystyrene ($n = 1.6$, film thickness 1.65 μm) surrounding a hollow core.[6] A general photonic band structure for alternating layers of high- (2.8) and low- (1.55) index films is shown in Fig. 11.4. The omnidirectional region is bounded by the light lines which, for simplicity, may be thought of as light incident at grazing incidence. The cross- hatched regions in Fig. 11.4 are the photonic bandgap regions for the fundamental and harmonic overtones. Fink has shown experimentally that the hollow structure composed of Te/polystyrene film did exhibit omnidirectional behavior for TE and TM light rays.[6] They also deposited the same coatings on the inside of a tube. This early work involved a somewhat laborious coating procedure in that they had to sputter Te and then liquid-deposit polystyrene over a hollow mandrel. Then the mandrel was etched away, leaving the thin film structure that was protected on the outside with heat-shrink tubing.[24] The coated hollow waveguide did also exhibit omnidirectional behavior. This is a very promising technology that could lead to a much improved hollow waveguide, both from the viewpoint of lower loss and no additional loss on bending.

References

1. J. A. Harrington, "A review of IR transmitting, hollow waveguides," *Fiber and Integrated Optics*, Vol. 19, pp. 211–227 (2000).
2. A. Hongo, K. Morosawa, K. Matsumoto, T. Shiota, and T. Hashimoto, "Transmission of kilowatt-class CO_2 laser light through dielectric-coated metallic hollow waveguides for material processing," *Appl. Opt.*, Vol. 31, pp. 5114–5120 (1992).
3. C. C. Gregory and J. A. Harrington, "Attenuation, modal, polarization properties of $n < 1$, hollow dielectric waveguides," *Appl. Opt.*, Vol. 32, pp. 5302–5309 (1993).
4. J. A. Harrington and C. C. Gregory, "Hollow sapphire fibers for the delivery of CO_2 laser energy," *Opt. Lett.*, Vol. 15, pp. 541–543 (1990).
5. J. Sanghera, L. B. Shaw, L. Busse, V. Nguyen, P. Pureza, B. Cole, B. B. Harbison, and I. Aggarwal, "Development and infrared applications of chalcogenide glass optical fibers," *Fiber and Integrated Optics*, Vol. 19, pp. 251–274 (2000).
6. Y. Fink, J. N. Winn, S. Fan, C. Chen, J. Michel, J. Joannopoulos, and E. Thomas, "A dielectric omnidirectional reflector," *Science*, Vol. 282, pp. 1679–1682 (1998).
7. E. A. J. Marcatili and R. A. Schmeltzer, "Hollow metallic and dielectric waveguides for long distance optical transmission and lasers," *Bell Syst. Tech. J.*, Vol. 43, pp. 1783–1809 (1964).
8. E. Garmire, T. McMahon, and M. Bass, "Flexible infrared waveguides for high-power transmission," *IEEE J. Quantum Electron.*, Vol. QE-16, pp. 23–32 (1980).

9. M. Miyagi, K. Harada, and S. Kawakami, "Wave propagation and attenuation in the general class of circular hollow waveguides with uniform curvature," *IEEE Transactions on Microwave Theory and Techniques*, Vol. MTT-32, pp. 513–521 (1984).
10. M. Miyagi and S. Kawakami, "Losses and phase constant changes caused by bends in the general class of hollow waveguides for the infrared," *Appl. Opt.*, Vol. 20, pp. 4221–4226 (1981).
11. M. Miyagi, A. Hongo, and S. Kawakami, "Transmission characteristics of dielectric-coated metallic waveguide for infrared transmission: slab waveguide model," *IEEE J. Quantum Electron.*, Vol. QE-19, pp. 136–144 (1983).
12. M. Miyagi and S. Kawakami, "Design theory of dielectric-coated circular metallic waveguides for infrared transmission," *J. Lightwave Techn.*, Vol. LT-2, pp. 116–126 (1984).
13. M. Miyagi, "Bending losses in hollow and dielectric tube leaky waveguides," *Appl. Opt.*, Vol. 20, pp. 1221–1229 (1981).
14. Y. Matsuura, M. Saito, M. Miyagi, and A. Hongo, "Loss characteristics of circular hollow waveguides for incoherent infrared light," *J. Opt. Soc. Am. A*, Vol. 6, pp. 423–427 (1989).
15. M. Saito, Y. Matsuura, M. Kawamura, and M. Miyagi, "Bending losses of incoherent light in circular hollow waveguides," *J. Opt. Soc. Am. A*, Vol. 7, pp. 2063–2068 (1990).
16. M. Saito, S. Sato, and M. Miyagi, "Loss characteristics of infrared hollow waveguides in multimode transmission," *J. Opt. Soc. Am. A*, Vol. 10, pp. 277–282 (1993).
17. M. Miyagi and S. Karasawa, "A comparative study of rectangular and circular dielectric-coated metallic waveguides for CO_2 laser light: theory," *Opt. Commun.*, Vol. 68, pp. 18–20 (1988).
18. A. M. Sagi-Dolev, J. Dror, A. Inberg, J. R. Ferncz, and N. Croitoru, "The influence of roughness on scattering of transmitted radiation through hollow waveguides," in Proc. SPIE, Vol. 2677, pp. 72–80 (1996).
19. N. Croitoru, J. Dror, and I. Gannot, "Characterization of hollow fibers for the transmission of infrared radiation," *Appl. Opt.*, Vol. 29, pp. 1805–1809 (1990).
20. O. B. Danilov, M. I. Zintchenko, Y. A. Rubinov, and E. N. Sosnov, "Transmission losses and mode-selection characteristics of a curved hollow dielectric waveguide with a rough surface," *J. Opt. Soc. Am. B*, Vol. 7, pp. 1785–1790 (1990).
21. C. Rabii, D. Gibson, and J. A. Harrington, "Processing and characterization of silver films used to fabricate hollow glass waveguides," *Appl. Opt.*, Vol. 38, pp. 4486–4493 (1999).
22. J. Y. Zhou, Y. J. Deng, N. Ling, G. Lu, and Z. X. Li, "Laser transmission through a dielectric-coated hollow tapered metallic waveguide," *Opt. Commun.*, Vol. 142, pp. 30–33 (1997).
23. D. Gibson and J. A. Harrington, "Tapered and noncircular hollow glass waveguides," Proc. SPIE, Vol. 3596, pp. 8–13 (1999).

24. Y. Fink, D. J. Ripin, S. Fan, C. Chen, J. D. Joannopoulos, and E. L. Thomas, "Guiding optical light in air using an all-dielectric structure," *J. Lightwave Tech.*, Vol. 17, pp. 2039–2041 (1999).
25. J. C. Knight, T. A. Birks, P. Russell, and D. M. Atkin, "All-silica single-mode optical fiber with photonic crystal cladding," *Opt. Lett.*, p. 1549 (1996).
26. J. C. Knight, T. A. Birks, P. Russell, and D. M. Atkin, "All-silica single-mode optical fiber with photonic crystal cladding: errata," *Opt. Lett.*, Vol. 22, pp. 484–485 (1997).
27. R. F. Cregan, B. J. Mangan, J. C. Knight, T. A. Birks, P. St. Russel, P. J. Roberts, and D. C. Allan, "Single-mode photonic band gap guidance of light in air," *Science*, Vol. 285, pp. 1537–1539 (1999).
28. T. A. Birks, J. C. Knight, and P. St. Russel, "Endlessly single-mode photonic crystal fiber," *Opt. Lett.*, Vol. 22, pp. 961–963 (1997).

Chapter 4

Halide Glass Fiber Optics

4.0 Introduction

Poulain and Lucas discovered heavy metal fluoride glasses (HMFGs) based on zirconium fluoride, also called fluorozirconate fluoride glasses, accidentally in 1975 at the University of Rennes.[1,2] In general, the typical fluoride glass has a glass transition temperature, T_g, four times less than silica, is considerably less stable, and has practical failure strains of only a few percent compared to silica's greater than 5%. While an enormous number of multicomponent fluoride glass compositions have been fabricated, comparably few have been drawn into fiber. This is because the temperature range for fiber drawing is normally too small in most HMFGs to permit fiberization of the glass. The most popular HMFGs for fabrication into fibers are the fluorozirconate and fluoroaluminate glasses, of which the most common are ZBLAN (ZrF_4-BaF_2-LaF_3-AlF_3-NaF) and AlF_3-ZrF_4-BaF_2-CaF_2-YF_3, respectively. The key physical properties that contrast these two glasses are summarized in Table 4.1. An important feature of the fluoroaluminate glass is its higher T_g, which largely accounts for the higher laser damage threshold for the fluoroaluminate glasses compared to ZBLAN at the Er:YAG laser wavelength of 2.94 μm.

The fabrication of HMFG fiber is similar to any glass-fiber drawing technology, except that the preforms are made using some type of melt-forming method rather than by the chemical vapor deposition process common to silica fibers. Specifically, a casting method based on first forming a clad glass tube and then adding the molten core glass is used to form either multimode or single-mode fluorozirconate-fiber preforms.[3] Other preform fabrication techniques include rod-in-tube and crucible techniques.[4] The fluoroaluminate-fiber preforms have been made using an unusual extrusion technique in which core and clad glass plates are extruded into a core/clad preform.[5] All methods, however, involve fabrication from the melt glass rather than from the more pristine technique of vapor deposition used to form SiO_2-based fibers. A melt-glass process, however, creates inherent problems such as the formation of bubbles, core/clad interface irregularities, and small preform sizes. To minimize contamination by moisture or oxygen impurities, which significantly weaken the fiber, fluoride glass fibers are usually drawn in a controlled atmosphere.

Table 4.1 Comparison between fluorozirconate and fluoroaluminate glasses of some key properties that relate to laser power transmission and durability of the two HMFG fibers. Other physical properties are relatively similar.

Property	Fluorozirconate ZBLAN	Fluoroaluminate AlF_3-ZrF_4-BaF_2-CaF_2-YF_3
Glass transition temperature, °C	265	400
Durability	Medium	Excellent
Loss at 2.94 μm, dB/m	0.01	0.1
Er:YAG laser peak output energy, mJ	300	850
Core size	300-μm	500-μm

Either a UV acrylate or Teflon coatings are applied to the fiber. In the case of Teflon, heat-shrink Teflon FEP is generally applied to the glass preform prior to the draw.

The attenuation in HMFG fibers is predicted to be about 10 times less than that for silica fibers.[6] Based on extrapolations of the intrinsic losses resulting from Rayleigh scattering and multiphonon absorption, the minimum in the loss curves or V-curves is projected to be about 0.01 dB/km at 2.55 μm.[7] Recent refinements of the scattering loss have modified this value slightly to be 0.024 dB/km or about 8 times less than that for silica fiber.[8] In practice, however, extrinsic loss mechanisms still dominate fiber loss. The lowest measured loss for a 60-m-long ZBLAN fiber is 0.45 dB/km at 2.3 μm.[9] This loss is dominated by extrinsic loss mechanisms due to scattering (crystallites, oxides, and bubbles) and impurities such as Ho^{3+}, Nd^{3+}, Cu^{2}, and OH^{-}.

The reliability of HMFG fibers depends on protecting the fiber from attack by moisture and on pretreatment of the preform to reduce surface crystallization. In general, the HMFGs are much less durable than oxide glasses. The leach rates for ZBLAN glass ranges between 10^{-3} and 10^{-2} $g/cm^2/day$. This is about five orders of magnitude higher than the leach rate for Pyrex glass. The fluoroaluminate glasses are more durable with leach rates that are more than three times lower than the fluorozirconate glasses. The strength of HMFG fibers is less than that for silica fibers. The largest bending strength measured has been about 1.4 GPa, well below the theoretical value.[10]

4.1 Glass composition and key physical properties

4.1.1 HMFG compositions

HMFGs differ in many ways from their oxide glass counterparts. As the name implies, HMFGs contain only fluoride compounds. More recently, however, oxides have been added to fluoride glasses to form a limited number of oxyfluoride glass compositions.[11,12] In general, fluoride glasses are kept free of oxides, as they tend to weaken the glass and to shorten the IR transmission range. The most significant difference between oxide and fluoride glasses, however, is that metal fluorides do not form a glass as easily as metal oxides. This is in part due to the more ionic nature of the bonds in fluoride glass compared with greater covalent bonding in oxide glasses. Simple oxide glasses like silica form a very stable tetrahedral structure. This structure can form a glass network through a sharing or bridging of oxygen atoms with neighboring silica tetrahedra. In short, fourfold-coordinated oxide glasses like silica form a very stable structure that has been well studied for many, many years. Silica is not only a single component glass former, it can be combined with one (binary), two (ternary), or more oxides to form a wide variety of stable glass compositions with many disparate optical properties.

Fluoride glasses, however, are not fourfold coordinated and they are much less stable than silica. In practice, this means that it is usually necessary to combine three or more fluoride compounds to obtain a stable glass. The only single component fluoride glass former is BeF_2. While BeF_2 glass has been known for more than 75 years, BeF_2 and the fluoroberyllate glasses in general have proven to be of little practical value. This is due to the toxicity of Be and the hygroscopic nature of BeF_2. Excluding the fluoroberyllate glasses, there still have been an enormous number of different multicomponent fluoride glass compositions developed over the past 26 years of fluoride glass research. While literally hundreds of HMFG compositions have been made, today only the fluorozirconate and fluoroaluminate glasses are drawn into fiber.

The two most important HMFGs for fiber optic applications are ZBLAN and the AlF_3-based glasses. Typical compositions for core and cladding glasses based on these two glass systems are given in Table 4.2. It should be noted that the percentage of each component given in Table 4.2 merely represents a typical value and not a particularly unique composition. For example, many other variations and other dopants have been added to the base glass to tailor the refractive index either up or down. An example of this is shown in the index data in Fig. 4.1. The addition of AlF_3 or HfF_4 to ZBLAN lowers the index, making this glass suitable for a clad glass as suggested by the data in Fig. 4.1 and Table 4.2. Conversely, the addition of heavier PbF_2 increases the index, making these glasses suitable for core glass. In any case, it is clear that the refractive index of the fluoride glasses can be tailored to produce a small Δ, as would be necessary for SM fibers or fiber amplifiers, or to produce a large Δ, common for large-core, laser-power delivery fibers. This is an important advantage of IR glass fibers over the crystalline fibers, where it is very difficult to so precisely control the refractive index.

Table 4.2 Glass compositions (mole%) for two of the most common HMFGs used in fiber drawing.

HMFG	Core/clad	ZrF_4	AlF_3	BaF_2	LaF_3	NaF	HfF_4	YF_3	RF_2*
ZBLAN	Core	53	3	20	4	20	—	—	—
	Clad	40	3	18	4	22	13	—	—
AlF_3-based	Core	13	25	—	—	6	—	11	44
	Clad	10	30	—	—	9	—	6	45

*$RF_2 = (MgF_2 + CaF_2 + BaF_2 + SrF_2)$.

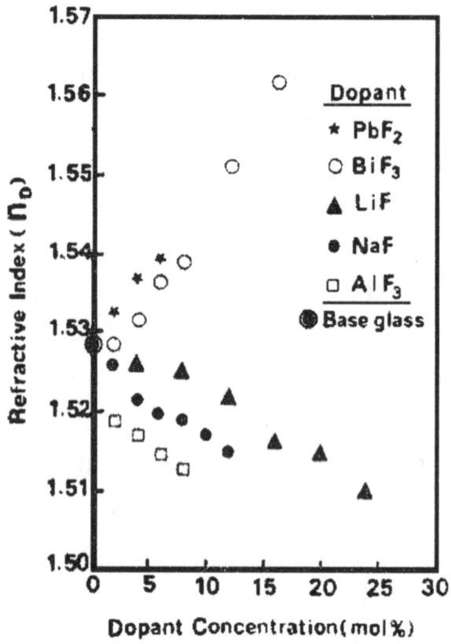

Figure 4.1 Variation in the refractive indices of fluorozirconate (63ZrF$_4$-33BaF$_2$-4GdF$_3$) glass as a function of dopant concentration. [Reprinted from Ref. 18, Copyright 1987, with permission from Elsevier.]

There are many other compositions of fluoride glasses that have been studied, and some of these are suitable for fiber drawing. The reader interested in more detail on HMFGs, especially in their many compositions and myriad properties, is referred to several good books on this subject.[7,10] An interesting variation on the usual ZrF_4- or AlF_3-based glasses are fluoride glasses that have been fabricated for extended IR transmission. That is, it is possible to increase the long wavelength transmission by substituting heavier cations for the lighter ions.[13] In particular, the elimination of Zr from the glass to form, for example, the fluoride glass BaF_2-InF_3-ZnF_2-YbF_3-ThF_4 (BIZYT) extends the transmission range by about 1 μm beyond the cutoff for ZBLAN.[14] BIZYT glasses and similar compositions also containing Ga and/or Mn are reasonably stable, and they have been drawn into fiber. Other glasses of interest are those with high thorium content and those for which some of the fluorine ions have been replaced with other halides (Cl^-, Br^-, and I^-).[15,16]

4.1.2 Key physical properties

In the early development of HMFGs there was an emphasis on finding compositions of metal fluorides that would not only be well-behaved glass formers but that would also have a large difference between the crystallization temperature, T_x, and the glass transition temperature, T_g. This difference is called the working range, and a primary focus of the early research was to obtain $\Delta T = T_x - T_g$ as large as possible. A large value of ΔT means that the glass is more stable against devitrification or crystallization, and that it has a smaller change in viscosity with changing temperature. This latter property is important if the glass is to be easily drawn into fiber. Therefore, one desires a glass with a high ΔT. This requirement eliminates most of the fluoride glass compositions from consideration for fiberization because normally ΔT is too small for fiber drawing. The few fluoride glass compositions that have a large working range normally have a ΔT near 100°C. This is considerably lower than that for silica glass; nevertheless, the working range is sufficiently high that these fluoride glasses can be formed into fiber or cast into optical components. The key thermal data for ZBLAN and the AlF_3-based glasses is given in Table 4.3. From the data in the Table 4.3 it can be seen that both glasses have the requisite ΔT for fiber drawing and that the T_g for the AlF_3-based glasses is substantially higher than that for ZBLAN. The higher T_g is one of the reasons that the high Al-content glasses are more durable, i.e., less sensitive to moisture and more stable than the fluorozirconate glasses. In addition, the higher T_g also results in higher laser damage thresholds for the AlF_3 fibers compared with ZBLAN. Further details on laser damage thresholds are given in Chapter 9.

In addition to having a large ΔT, it is also important to have a viscosity-temperature relationship suitable for fiber drawing. In particular, it is desirable to have a small change in viscosity with changing temperature. In Fig. 4.2 the viscosity-temperature dependence of ZBLAN and, for reference, silica are shown on a normalized temperature scale.[7] The temperature for each glass is normalized to the glass transition temperature, which for silica is 1175°C and for ZBLAN is 259°C. Fiber drawing from a preform normally requires a viscosity near 10^5 poise, whereas crucible drawing can be accomplished as low as 10^3 poise. From Fig. 4.2 it can be seen that the change in viscosity with temperature for ZBLAN is steeper than it is for silica. As a result, silica has a rather large working temperature range over which the viscosity is suitable for fiber drawing. This is why silica can so easily be drawn into fiber. In the fiber drawing process, the neckdown length (the length of the region where the glass is reduced in size from the preform diameter to the fiber diameter) is typically similar to the preform diameter. Since silica preforms are generally greater than ~3 cm in diameter and the viscosity curve is not steep, as shown in Fig. 4.2, they have a large neckdown region. ZBLAN preforms are usually less than 10 mm in diameter; this fact coupled with their steep viscosity profile means that only a narrow hot zone is used. Therefore, the neckdown region is relatively small for ZBLAN compared to silica. Notice also from Fig. 4.2 that the drawing temperature for ZBLAN is nearer T_x, which means that crystallization during fiber drawing can be a problem.

Table 4.3 Key thermal properties of the two most important HMFGs for fiber drawing. Note the significantly higher T_s for the AlF_3-based glass. This is a major reason that this fiber has a higher Er:YAG laser threshold energy compared to ZBLAN fiber.

Property	ZBLAN	AlF_3-based glass*
T_g, °C	259	446
T_x, °C	359	554
$\Delta T = T_x - T_g$, °C	100	108

*AlF_3 (28.7%)-YF_3 (28.7%)-ThF_4 (22.6%)-BaF_2 (20%).

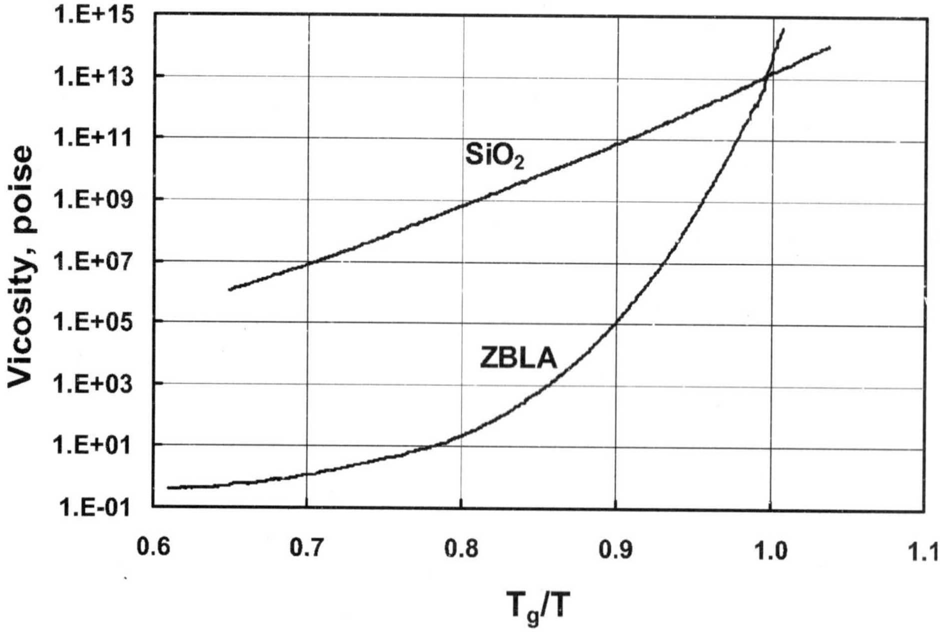

Figure 4.2 Comparison of viscosity-temperature curves for silica and fluorozirconate glass, illustrating the rapid change in viscosity for the fluoride glasses in contrast to the much slower change in viscosity for the oxide glasses. [Reprinted with permission from Ref. 7. Copyright CRC Press, Boca Raton, Florida.]

The thermal expansion coefficient and chemical durability for the HMFGs are generally inferior to those for silica and even other oxide glasses. The thermal expansion coefficient for ZBLAN is more than 30 times greater than that of silica. The fluoroaluminate glasses have slightly lower expansion coefficients, but they are still much higher than that of silica. The chemical durability of the fluoride glasses is significantly less than for oxide glasses. Generally speaking, the most important aspect of the durability of a glass is its ability to resist attack by moisture. The fluorozirconate glasses are attacked by water. Experiments by Simmons[10] on the leach rate ($g_{water}\,cm^{-2}\,d^{-1}$) show that the leach rate for the fluorozirconate glasses is on the order of $10^{-2}\,g_{water}\,cm^{-2}\,d^{-1}$ whereas for Pyrex glass it is less than $10^{-7}\,g_{water}\,cm^{-2}\,d^{-1}$. A much more durable fluoride glass composition is obtained if AlF_3 is substituted for ZrF_4. In practice, this means that the fluoride glass preforms must be kept very dry during and after fabrication. Normally the preforms are enclosed in some sort of low-moisture, low-oxygen environment while they are being drawn. Of course, the fibers need to be coated immediately after drawing to not only protect against degradation from moisture, but also to retain their strength. Fluoride fiber that is left unprotected will be greatly weakened by exposure to moisture.

A summary of the some of the key physical properties of a typical fluoride glass (ZBLAN) contrasted with the properties of silica is given in Table 4.4. It is quite clear from Table 4.4 that there is a stark contrast in properties between

Table 4.4 Some key thermal, mechanical, and optical properties of ZBLAN glass in contrast to SiO_2 glass.

Property	ZBLAN	Silica glass
T_g, °C	259	1175
Specific heat, cal/g °C	0.151	0.179
Thermal conductivity, W/m °C	0.628	1.38
Expansion coefficient	17.2×10^{-6}	0.55×10^{-6}
Density, g/cm^3	4.33	2.20
Knoop hardness, kg/mm^2	225	600
Toughness (K_{1C}), MPa m$^{1/2}$	~0.32	0.72
Young's modulus, GPa	52.7	72.2
Poisson's ratio	0.31	0.17
Refractive index (0.59 μn)	1.499	1.458
Abbe number	76	68
ZMD wavelength, μm	1.6	1.3
dn/dT, °C^{-1}	-14.75×10^{-6}	$+11.9 \times 10^{-6}$
Minimum intrinsic loss, dB/km	0.01 (2.55 μm)	0.14 (1.55 μm)
Lowest measured loss, dB/km	0.45 (2.3 μm)	0.16 (1.55 μm)

fluoride and oxide glasses. Almost every property of ZBLAN is inferior to that of silica. Sometimes this is dramatic, as is the case when comparing T_g (more than four times less for ZBLAN), thermal expansion, hardness (about three times less than silica), and Young's modulus. The optical properties are relatively similar for the two glasses except that the measured loss for silica fiber is less than that for ZBLAN by a factor of about 3, even though the theoretical loss for ZBLAN is about 8 times less than that for SiO_2.

4.2 Optical properties of bulk glass

The optical properties of the fluoride glasses depend on a variety of compositional as well as structural properties of the glass. The factors contributing to the basic losses in IR fiber materials were discussed in Sec. 2.3. In general, the losses are due to intrinsic and extrinsic mechanisms. These are described in general in Secs. 2.3.1 and 2.3.2. As is the case for essentially all of the IR fibers, the losses in the HMFGs are dominated by extrinsic losses. A major emphasis in fluoride glass research has been the study of the extrinsic loss mechanisms and the means to reduce these losses.

4.2.1 Intrinsic loss

The fundamental or theoretical losses at IR wavelengths for fluoride glass are determined by Rayleigh scattering at short wavelengths and by multiphonon absorption at long wavelengths. These losses are shown in the two V-curves for ZBLAN and silica in Fig. 4.3. The minima in the V-curves are listed in Table 4.4. From the

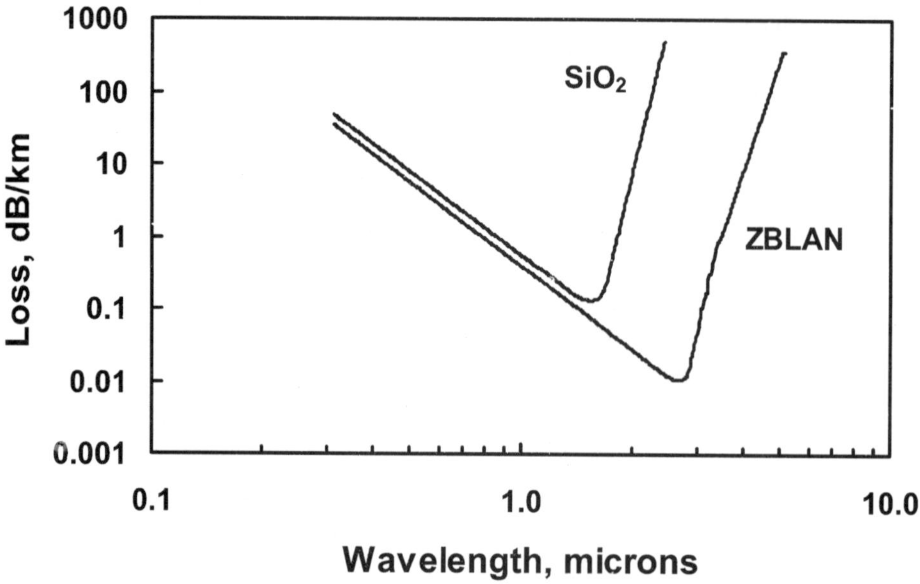

Figure 4.3 Intrinsic losses or V-curves for silica and fluorozirconate glass showing the very low loss theoretically predicted for ZBLAN. After France et al. [Reprinted with permission from Ref. 7. Copyright CRC Press, Boca Raton, Florida.]

ZBLAN V-curve it can be seen that the minimum loss is projected to be approximately 0.01 dB/km at 2.55 μm. This loss is approximately 10 times less than the minimum loss for silica. However, recent refinements of the Rayleigh scattering loss have modified this value slightly to be 0.024 dB/km or about 8 times less than that for silica fiber.[8] The wavelength dependence of the scattering and IR absorption losses shown in Fig. 4.3 follow the general form given in Eq. (2.14).

The position of the IR absorption edge depends in large part on the mass of the cations, as discussed in Sec. 2.3.1. When lighter Al ions (atomic mass 27) are substituted for Zr (atomic mass 91.2), the IR edge shifts to shorter wavelengths. The shift to shorter wavelengths for the AlF_3-based glasses, however, is not so large compared to ZBLAN that applications involving ZBLAN fiber would be comprised by the substitution of a fluorozirconate for a ZBLAN fiber. For example, both ZBLAN and the AlF_3-based fibers function well at the Er:YAG laser wavelength of 2.94 μm, as the loss for both fibers is reasonably close at that wavelength. When heavier ions such as Yb (atomic mass 173) and Th (atomic mass 232) are used in the glass composition, then the IR edge can move to significantly longer wavelengths. This trend can be seen in the composite V-curves taken from F. Gan[17] and shown in Fig. 4.4. Unfortunately, adding heavier ions to shift the edge to longer wavelengths often reduces T_g and ΔT such that the glasses are less stable and harder to draw into fiber. A better alternative for producing a long wavelength transmissive, halide glass fiber is to add heavier halide ions like Br^- and I^-. Furthermore, if the chalcogen elements such as Te, As, and Se are combined with these anions to form the so-called chalcohalide glasses, then the IR edge is greatly extended. Chalcohalide glasses are discussed in Chapter 5. Therefore, in practice,

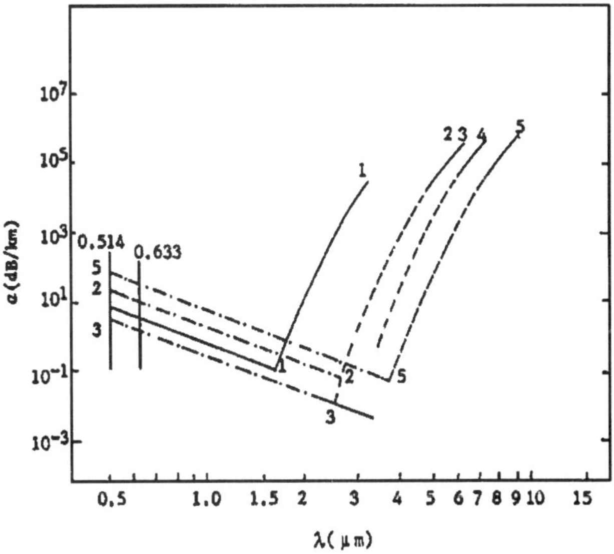

Figure 4.4 Intrinsic losses for different glasses: (1) SiO_2, (2) BaF_2-LaF_3-AlF_3-ZrF_4, (3) LiF-BaF_2-AlF_3-LaF_3-ZrF_4, (4) BaF_2-ZrF_4-ThF_4, (5) BaF_2-ZnF_2-YbF_3-YF_3. The minima of the HMFGs are shifted to longer wavelengths with the addition of heavier cations. [Reprinted from Ref. 17, Copyright 1995, with permission from Elsevier.]

the ZBLAN fibers are still the fluoride fiber of choice for most applications, as they are very stable and easily drawn into fiber.

Intrinsic scattering losses for ZBLAN are somewhat lower than those for silica, as may be seen from Figs. 4.3 and 4.4. Other fluoride glass compositions, however, have higher scattering loss. In general, intrinsic Rayleigh scattering is due to the inherent compositional and density fluctuations in the glass. These fluctuations produce local variations in the refractive index, and these index variations are small compared with the wavelength. In principle, scattering losses can be calculated using Eq. (2.15). Most often, however, the parameters needed to solve for α_{scat} in Eq. (2.15) are not well known for the multicomponent fluoride glasses, so the attenuation due to scattering is measured. The scattering losses for several fluorozirconate glass compositions are given in Fig. 4.5.[18] From the data it is seen that the losses vary as λ^{-4}, as expected for Rayleigh scattering and as predicted by Eq. (2.15).

The intrinsic UV absorption edge for the HMFGs is a result of electronic transitions from the valence to conduction band. This short wavelength absorption edge, or Urbach tail, is represented by an exponential increase in attenuation with decreasing wavelength in Eq. (2.14). Normally one would expect a perfectly sharp or well-defined electronic bandgap, but electron-phonon interactions lead to a weak tail. In addition, amorphous solids possess a distribution of atomic configurations, each with its own energy level. This leads to well-defined but different energy levels for the different configurations. The sum of these different levels gives a small spread in the value of the energy gap. Therefore, the cutoff wavelength depends on the glass and its composition. To a good approximation, the cutoff wavelength for

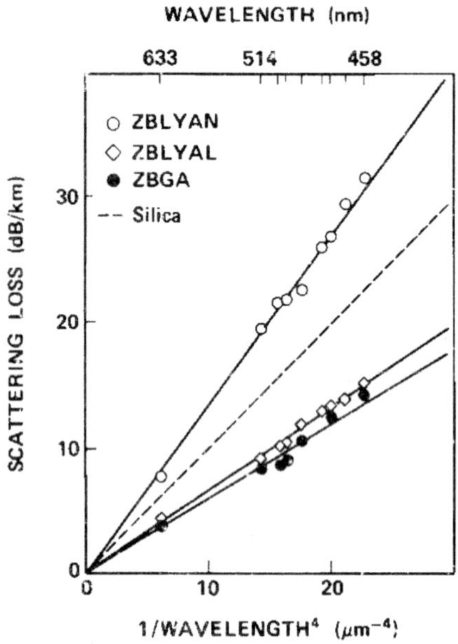

Figure 4.5 Rayleigh scattering in different fluoride glass compared to SiO_2 (dashed line). [Reprinted from Ref. 18, Copyright 1987, with permission from Elsevier.]

the fluoride glasses varies from 0.2 to 0.25 μm, compared to the cutoff wavelength for silica, which is about 0.16 μm. One conclusion from this is that fluoride glasses do not transmit well in the UV region of the spectrum. This could be inferred *a priori* by merely noting that, in general, the long wavelength transmission in the IR means less transmission in the UV region. Adding heavier elements to the glass composition serves to effectively shift the whole spectrum to longer wavelengths.

4.2.2 Extrinsic loss

As mentioned earlier, extrinsic losses are the dominant loss mechanisms in the HMFGs. There are two main types of extrinsic losses; those due to impurities and those arising from scattering centers. A wide variety of impurities and scattering defects have been identified in these glasses. The main ones identified as contributing most to the overall extrinsic loss are listed in Table 4.5.

One of the most common contaminants in fluoride glass, as well as in other IR fiber materials and even conventional silica fiber optics, is the ubiquitous hydroxyl ion (OH^-). This ion has its fundamental absorption peak near 2.9 μm. The OH^- impurities can lead to strong absorption in the HMFGs because the fundamental absorption peak at 2.9 μm contributes in the transparent regime of fluoride glasses, but only the weaker overtone at 1.4 μm contributes in silica fiber. An emphasis in fluoride glass chemistry is to reduce the amount of water present in the fluoride materials. One of the most successful methods of reduction is to use reactive atmosphere processing (RAP).[19] RAP chemistry methods involve passing reactive gases such as NF_3, HF, SF_6, CF_4, and CCl_4 through the melt to reduce the

Table 4.5 Some common impurities in floride glasses and the position of thier peak absorption.

Impurity	Peak, μm
OH^-	2.9
Cu^{2+}	1.0
Cr^{3+}	0.68
Ni^{2+}	0.88, 1.53
Fe^{2+}	1.14, 1.85
Co^{2+}	2.5
Nd^{3+}	1.59
NH_4	3.0
CO_3^{2-}	2.9

hydroxyl content in the glass melt as well as in the surrounding atmosphere. By bubbling the reactive gas through the melt, water impurities resulting from hydrolysis of the fluoride compounds can be largely eliminated. In particular, NF_3 gas has been shown to be especially effective in reducing not only impurity absorption, but also scattering losses, without contributing its own extrinsic absorption.

Table 4.5 also lists some of the most common transition metal and rare-earth impurities. Of these, the most important are the transition metal impurities such as Fe, Co, and Ni. Rare-earth ions do not contribute heavily to the loss because their absorption bands tend to be rather sharp and, therefore, they do not add significantly to the overall broad impurity band absorptions observed in these glasses.[8] In addition, rare-earth ion concentrations are usually less than those for divalent ions. It is difficult to reduce the concentration of the transition metals as they tend to be a common contaminant in the starting fluoride materials. RAP methods do not eliminate these ions, but they can help reduce absorption by creating valence changes. For example, under RAP conditions, Fe^{2+} can be changed to Fe^{3+}. This is good because the absorption of the Fe^{2+} ions is above 1 μm, whereas the absorption of the Fe^{3+} ions is less than 1 μm.

The scattering losses resulting from a wide variety of defects (see Table 4.5) can be significant. In the early days of HMFG fiber drawing, it was quite common to find crystallites of $BaZrF_6$, ZrF_4, and ZrO_2. One reason that crystallites form in the fiber drawing process is that the drawing temperatures are near the crystallization temperature, T_x. The formation of these small crystals can be controlled by a judicious choice of the core and clad compositions. It is also important to keep the preforms dry and free from moisture and oxygen. This can be done by preparing the glass melts and preforms in a dry box and then transferring the preform to a draw tower with a controlled atmosphere. Such a dry box assembly is shown in Fig. 4.6. Greater details on the exact nature of the various impurity and scattering defects present in fluoride glasses and fibers may be found in several books[4,7,10] and review papers[3,20,21] on HMFGs.

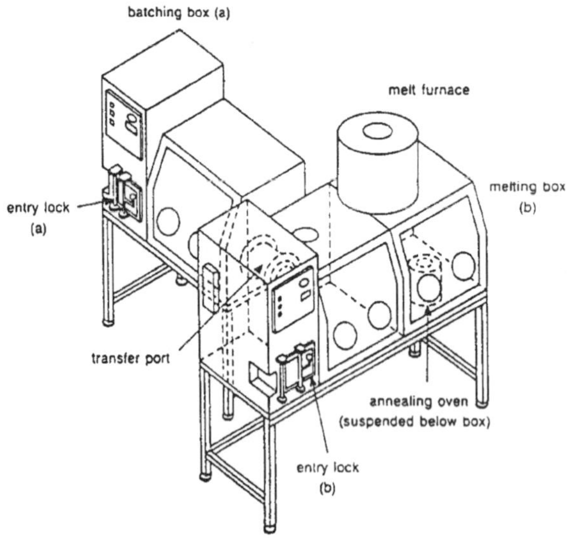

Figure 4.6 Dry glove box for batching and casting of fluoride glass preforms.

4.2.3 Refractive index and dispersion

The refractive indices of HMFGs are slightly higher than those for silica. The dispersive properties, or $dn/d\lambda$, are similar to those of silica. This may be seen from the Abbe numbers [see Eq. (2.17)]: V_D (ZBLAN) = 76 and V_D (silica) = 68. High Abbe numbers mean low dispersion, and Abbe numbers below ~45 indicate a more dispersive glass such as the flint oxide glasses. The dispersion for several fluorozirconates compared with that of silica are shown in the data of Gan[17] in Fig. 4.7. For the two glass compositions ZBLAN and ZBLA, in Fig. 4.7 it can be seen that ZBLA glass has higher refractive indices, as might be expected for a glass with heavier ions. Gan has also shown that the addition of Pb, Sn, Cd, and Ca also increase the refractive indices of the base glass compositions, whereas the addition of Li, Na, or Al lowers the refractive indices, as shown in Fig. 4.1. This is of obvious importance as one formulates core and clad glass compositions with suitable numerical apertures and dispersion properties for different applications.

The dispersive properties of optical materials are discussed in Sec. 2.4.1. In particular, the Sellmeier equation, Eq. (2.18), can be used to fit the dispersion data in Figs. 4.7(a) and (b). The Sellmeier constants, A_i and λ_{oi} in Eq. (2.18), have been obtained for many of the most common fluoride glasses used for fiber optics. Gan[17] gives tables for these constants so that one may use them to calculate $n = f(\lambda)$ to interpolate an accurate value of n over a wide wavelength range, and also to obtain the material dispersion coefficients, Eq. (2.21), needed for determining the pulse dispersion properties of the fibers.

Material dispersion leads to pulse spreading, as discussed in Sec. 2.4.1. The material dispersion coefficient, D_λ, for three different fluoride glasses is plotted as a function of wavelength in Fig. 4.8.[17] This data may be compared with that

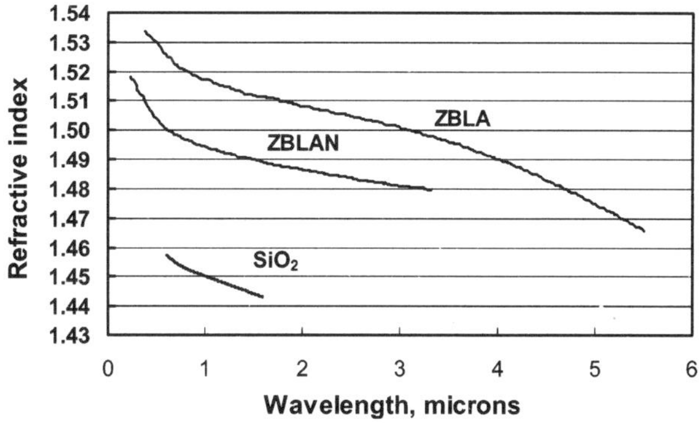

Figure 4.7 Dispersion for fluorozirconate glasses compared to that for SiO_2.[17]

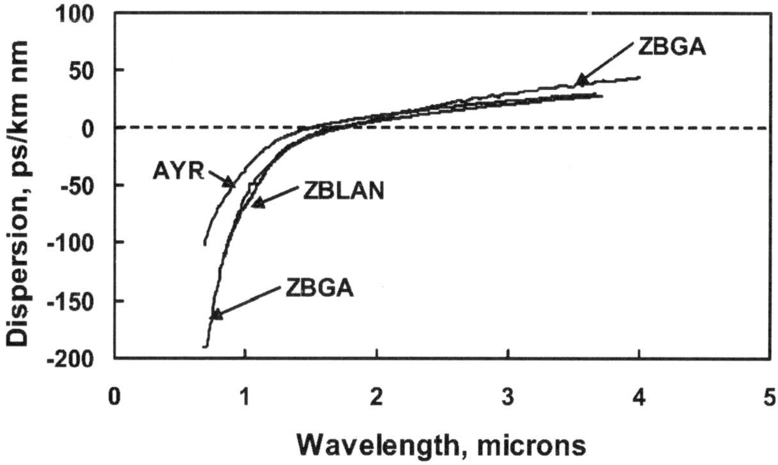

Figure 4.8 Dispersion coefficient for three fluoride glasses: ZBLAN; AYR (AlF_3-YF_3-MgF_3-CaF_2-SrF_2-BaF_2); and ZBGA (ZrF_3-BaF_2-GdF_3-AlF_3).[17]

given in Fig. 2.12 for silica and ZBLAN. One of the most important features of this data is the wavelength for which the material dispersion is zero, i.e., the ZMD wavelength. Since the HMFGs are heavier than the oxide glasses, there is a shift in the ZMD wavelength to longer wavelengths for the fluoride glasses. Specifically, the ZMD wavelength for silica, ZBLAN, ZBGA, and AYR are 1.312, 1.724, 1.71, and 1.494 μm, respectively. The addition of heavier cations shifts the ZMD to longer wavelengths, whereas the substitution of AlF_3 in the lighter AYR glass shifts the ZMD to shorter wavelengths. For example, CLAP (CdF_4-LiF-AlF_3-PbF_2) glass has a ZMD slightly more than 2 μm.

4.3 Fiber fabrication

The drawing of HMFGs into fiber is not generally as straightforward as it is for the oxide glasses. One of the main reasons for this is that it has been largely impossible to fabricate preforms using chemical vapor deposition (CVD), as is routinely done with oxide glasses. The typical CVD process would involve depositing a powder or soot of fluoride glass on the inside or outside of a tube. The soot is then condensed and sintered into a dense glassy layer and the tube collapsed into the final preform. This process does not work well with the HMFGs because the temperature at which the sintering is done is sufficiently high to cause crystallization. Therefore, while a variety of fluoride precursors have been tried, CVD for fabricating fluoride glass preforms is still in the future.

The primary method for making a preform is to begin with the core and clad glass compositions in either the melt or solid form. Normally, either the core or clad or both are in a viscous state. The final preform geometry is obtained using some general casting method. In the next section, a few common methods for preform fabrication are summarized. Since most fiber drawing is done using a preform rather than by a crucible technique, it is sufficient to describe only preform fiber drawing in Sec. 4.3.2.

4.3.1 Preform preparation

The most common method for preform fabrication is casting. The built-in casting method, shown schematically in Fig. 4.9, is a simple method in which the cladding glass is first poured into a metal (usually brass) mold and then the mold is upset to leave a cladding glass tube.[22] Next the core glass is poured into the mold to form the final preform. While this method is simple, there are inherent difficulties such as bubbles and core/clad irregularities. Also the preform is generally not too long as the cladding tube tends to be tapered rather than a straight cylinder. The problem of a tapered cladding geometry can be solved if a rotational casting method is used.[23] In this technique, shown in Fig. 4.10, the molten cladding glass is spun at a high

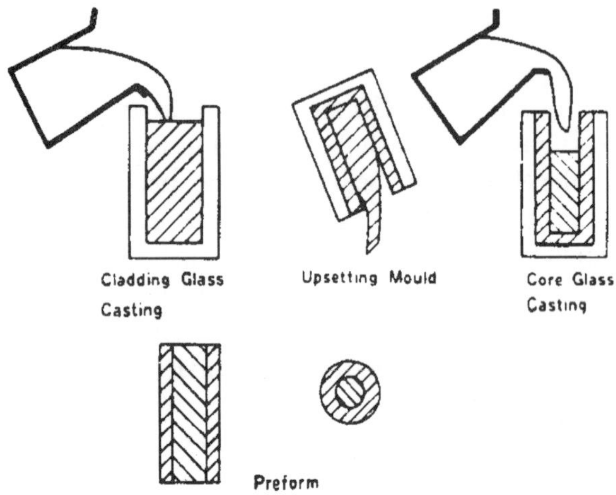

Figure 4.9 Casting a fluoride glass preform by upsetting the mold to first form the clad glass tube and then adding the core glass.[22] [Copyright 2003 IEEE. Reprinted with permission.]

speed to form a tube with a very uniform wall thickness. After the formation of the cladding tube, the core glass is either poured into the cladding tube or the tube is inserted into the molten core glass. Both of these methods are carried out in a dry box like the one shown in Fig. 4.6. Rotational casting can also have problems with bubbles and crystallites that form at the core/clad interface.

An improvement in both these casting techniques is possible if a vacuum is used to eliminate bubbles that form during casting. In the vacuum-casting process, the mold as well as the molten core and clad glasses are placed in a vacuum chamber, as shown in Fig. 4.11. A cladding tube is formed by rotational casting as in Fig. 4.10, but this time the process is carried out in vacuum. Then the core glass is added. This refinement in casting has greatly reduced bubble and other casting defects to the point that Carter et al.,[8] using this process, were able to prepare preforms that then have been drawn into some of the lowest-loss fluoride glass fiber made to date. Their fiber loss measurements are presented in Sec. 4.3.2.

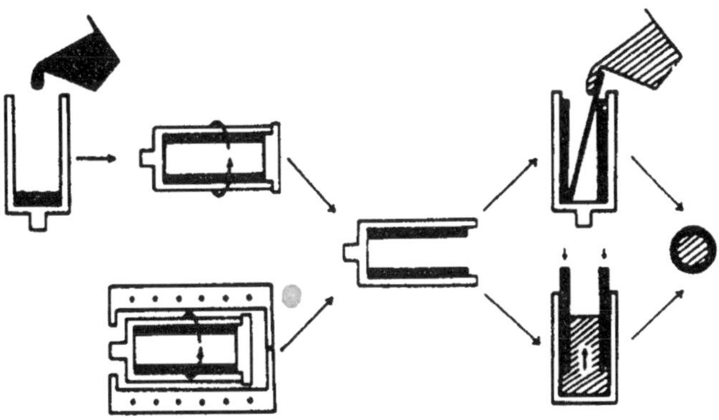

Figure 4.10 Rotational casting of fluoride glass preform by first spinning a clad glass tube and then adding the core glass.[23] [Reprinted with permission.]

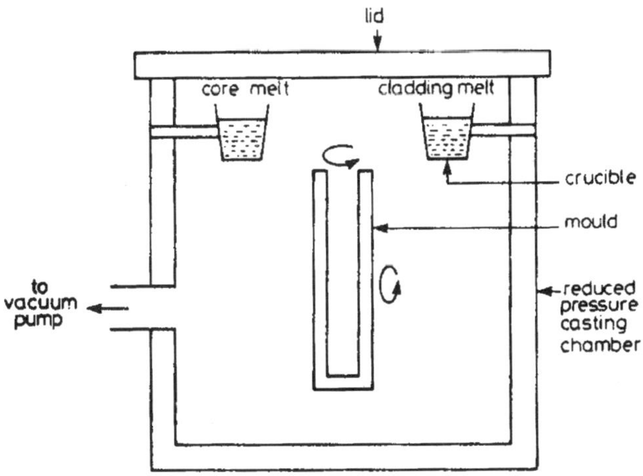

Figure 4.11 Reduced-pressure casting using an evacuated chamber and rotational casting methods to improve the core-clad interface.[8] [Reprinted with permission.]

There are several other techniques that have been used to make preforms. One that has proven particularly effective for fabricating SM fiber is a version of the rod-in-tube method. In this method, a polished core glass rod is inserted into a rotationally cast cladding tube. During fiber drawing, a vacuum surrounds the rod-in-tube preform to eliminate bubble formation. The rod-in-tube method has been applied in a stepwise fashion to fabricate SM preforms. The basic idea is shown schematically in Fig. 4.12.[7] The first step is to form a preform that has a large core typical of an MM fiber preform. This preform would have a diameter of about 10 mm. Next, this preform is drawn down to about a 1-mm diameter. Then, this small-diameter core/clad preform is inserted into a thick-wall cladding tube made by rotational casting, as shown in steps 2 and 3. The final step is drawing the small-core, rod-in-tube preform into SM fiber. This method is ideal for making SM preforms because the small core will be much more uniform than it would be if a one-step casting method were used.[22] That is, it is very difficult to rotationally cast a clad tube that has a wall thickness sufficiently large and uniform enough to be appropriate for an SM fiber. The two step, rod-in-tube process provides a much better means of obtaining the uniform SM core diameter in the final preform.

4.3.2 Fiber drawing

As mentioned, fluoride glass fibers may be drawn using either a preform or a crucible technique. The lowest-loss fibers have been fabricated using a preform, and today most fiber is drawn that way. The advantage of a single- or double-crucible method[24] over a preform is that much longer lengths of fiber may be drawn. Gen-

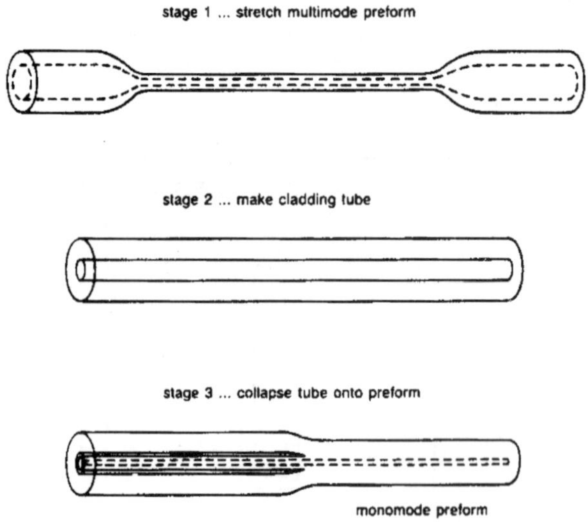

Figure 4.12 Method of successive drawdown to fabricate an SM fiber from HMFG. First, a preform is used to make a small-diameter core-clad preform to be inserted into the core of a larger cladding tube. [Reprinted with permission from Ref. 7. Copyright CRC Press, Boca Raton, Florida.]

erally the preforms are small, with diameters of approximately 1 cm and lengths of 10 to 15 cm. This is much smaller than silica fiber preforms. The disadvantage of crucible drawing is that the glasses must be drawn at a slightly higher temperature and lower viscosity than for a preform. This can lead to greater crystallite formation. Therefore, only preform drawing is discussed here and the reader is referred to other sources for details on crucible drawing.[4,24]

A schematic diagram of a fluoride glass-drawing tower is shown in Fig. 4.13. This is very similar to a standard oxide glass drawing tower, but there are several important differences. Since fluoride glasses are sensitive to moisture, it is essential that there be an environmental dry box surrounding the preform. This keeps the preform free from contamination by OH^-. Another important distinction is that the temperature dependence of the viscosity of the fluorides is much greater than for the oxides (see Fig. 4.2). Therefore, a sharp temperature gradient is required to draw this glass, and this is achieved by using a narrow, pancake-type heater. For ZBLAN fiber the drawing temperature is around 310°C, the viscosity is $10^{5.5}$ poise, and the drawing rate is 10 to 20 m/min. This is considerably slower than silica fiber drawing rates, which can be in excess of 20 m/sec.

The fibers are coated with a protective polymer coating as they are drawn. This is to protect against attack by moisture and to retain fiber strength. There are two ways to apply the buffer jacket to the fiber. One is to apply a UV acrylate coating just after the diameter control, as shown in Fig. 4.13. UV acrylates are the most common means of protecting silica fibers, and similar coatings and coating methods apply for the fluoride glass fibers. The other method is to apply a heat-shrink, Teflon (FEP) tube to the outside of the preform before drawing. The Teflon softens and it can be drawn down with the preform to form a tightly bonded polymer buffer coating. Teflon coatings can be applied to those HMFGs that can be drawn

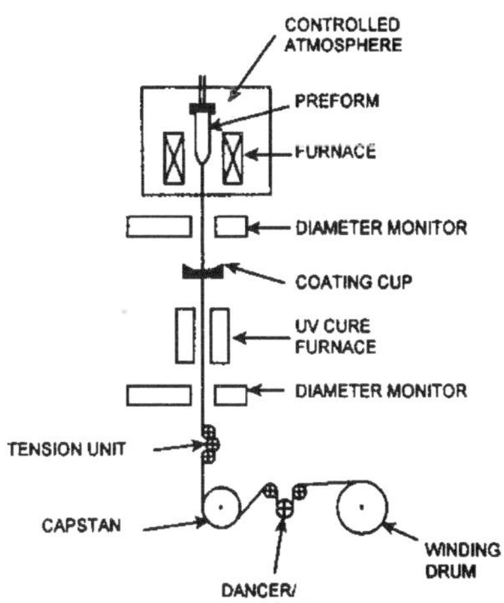

Figure 4.13 Typical fiber drawing tower for drawing HMFG fibers. Included in the setup is a laser micrometer for fiber diameter control and a coating cup for applying the UV curable acrylate buffer coating. [Reprinted with permission from Ref. 4. Copyright CRC Press, Boca Raton, Florida.]

at temperatures less than about 320°C. Higher-temperature fluoride glasses need to have the acrylate coating applied.

4.4 Optical properties of fiber

Measured losses for HMFG fibers are the lowest of any IR fiber made to date. While the losses are still well above the intrinsic loss, they are low enough for many short-haul applications. In Sec. 4.4.1, the lowest-loss fiber results will be discussed. It is important, however, to also describe the loss that one can expect to obtain from production fluoride glass fiber. Unlike silica fiber, which in production is essentially of equal optical quality to the best laboratory results, fluoride glass fibers that are commercially available can have a significantly higher loss than the best lab results.

4.4.1 Low-loss fiber

Very low loss fiber was prepared over a decade ago at British Telecom Research Laboratories (BTRL) by Carter et al.[8] The spectral results for a 70 μm/150 μm, MM ZBLAN fiber with a length of 110 m are shown in Fig. 4.14. Their data gives a minimum total fiber loss of 0.65 ± 0.25 dB/km at 2.59 μm. Following a very careful analysis, they were able to separate this total loss into a contribution due to a total extrinsic absorption of 0.33 dB/km and a contribution due to total scattering of 0.30 dB/km. As indicated in the Fig. 4.14, the extrinsic absorption losses were mainly due to absorbing ions identified by Carter et al. as Ho^{3+} (80 ppb), Nd^{3+} (15 ppb), Cu^{2+} (18 ppb), dissolved CO_2, and OH^- (4 ppb). Of these impurities, they determined that the absorption tails of Cu^{2+}, with a peak at 0.97 μm, and

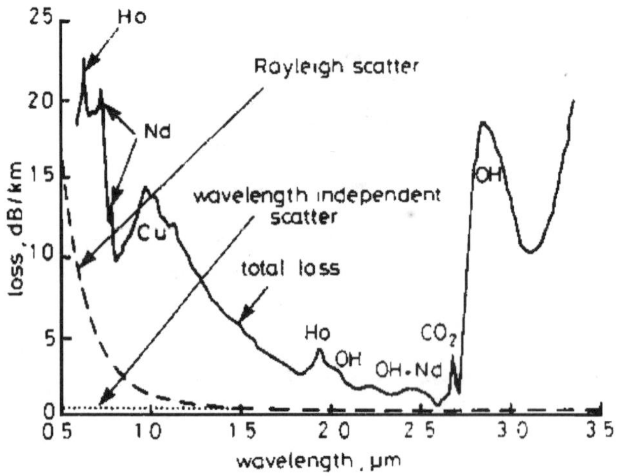

Figure 4.14 Lowest-loss ZBLAN fiber fabricated by British Telecom Laboratories in 1990. The minimum loss for this 110-m long fiber is 0.65 ± 0.25 dB/km at 2.59 μm.[8] [Reprinted with permission.]

of Nd^{3+}, with an absorption centered at 2.52 μm, were the main contributions to the extrinsic absorption at 2.59 μm. The scattering loss of 0.30 dB/km is largely independent of wavelength, as shown by the dotted line in Fig. 4.14. That is, the loss due to Rayleigh scattering at 2.59 μm is well below the measured scattering loss. Carter et al. hypothesize that the wavelength-independent scattering comes from Pt particles imbedded in the fiber. These submicron Pt particles probably originate from the crucibles used in glass melting.

Several years later, in 1993, the same group at BTRL fabricated a somewhat lower-loss fiber, which, to the author's knowledge, is the lowest-loss HMFG fiber fabricated to date. Szebesta et al.[9] at BTRL measured a total loss of 0.45 ± 0.15 dB/km at 2.35 μm for a 70 μm/150 μm MM ZBLAN fiber. This loss was obtained in a 60-m long fiber. This lower loss was achieved by reducing the OH^- content of the glass. The authors found that when the water peak was reduced, the minimum loss also shifted to shorter wavelengths. This is the reason for the shift in the minimum loss from 2.59 μm to 2.35 μm.

It is interesting to note that in the last 10 years since this work was reported by BTRL, there has been no reduction in loss, even though the measured loss is still 10 times higher than the intrinsic loss. On the surface it would seem that further work to purify these materials to achieve losses less than silica fiber would be worthwhile. Assuming that these fibers could be prepared with ultralow loss, there would still be many practical problems to overcome if the fluoride fibers were to be a replacement for silica fibers in telecommunication systems. Some of the main problems would be the stability of these glasses compared to silica, their lower strength, and their susceptibility to attack by moisture. For these reasons HMFG fibers are not likely soon to replace silica telecommunication fibers, but rather they have found other applications requiring shorter lengths. For these short-haul applications such as fiber amplifiers, sensors, and laser power delivery, losses even above 1 dB/km can be tolerable. Therefore, most research involving achieving a lower loss for the purpose of long-haul telecommunications has greatly diminished.

4.4.2 Production-grade fiber

Virtually all of the commercial-grade fluoride glass fiber is based on fluorozirconates, with ZBLAN being the most popular. It might seem surprising that this one composition has prevailed, given the literally hundreds of HMFG compositions studied over the past 25 years. In fact, ZBLAN was discovered early in the search for a fiberizable fluoride glass. The primary reason that ZBLAN has become one of the best fluoride fibers is that it is one of the most stable of all compositions studied. It has a high ΔT and low loss. Fibers made from this glass tend to remain strong and low loss if kept free of attack by moisture, i.e., they do not exhibit significantly aging. The other fluoride glass fiber that has been relatively successful is that made from the fluoroaluminates.[5] The advantage of these fibers is their higher operating temperature and concomitant higher laser-induced damage threshold.

4.4.2.1 Fluorozirconate fiber

The fluoride glass fibers that are commercially available do not have losses as low as those described in Sec. 4.4.1 above. In fact a typical low loss for production-grade ZBLAN fibers is approximately 0.05 dB/m, or 50 dB/km, at ~2.5 μm. Clearly this is well in excess of the minimum loss of 0.45 dB/km obtained by BTRL. That is, commercial sources of fluoride fiber have about 100 times greater loss than the best values obtained in a laboratory setting. Nevertheless, the fibers are quite useful for the many short-haul applications mentioned above. In particular, fiber amplifiers are only meters in length, so commercial fiber is quite acceptable.

The losses for two commercial sources of ZBLAN fiber are shown in Fig. 4.15 (note particularly that the ordinate scale is in dB/m and not dB/km as in Fig. 4.14). Absorption by OH^- and various metallic ions are identifiable in the spectra. The minimum loss for both sources is ~0.05 dB/m. This unit is appropriate here, as most applications require lengths measured in meters rather than kilometers. Recall that a loss of 1 dB/m is about 20% per meter and a loss of 0.5 dB/m is about 11% per meter. Also, remember that these losses are in addition to any Fresnel loss, which is ~0.18 dB/fiber end or 0.36 dB for two uncoated fluoride fiber ends.

The ZBLAN fibers shown in Fig. 4.15 are MM fibers available in sizes 70/140 μm, 150/200 μm, 200/250 μm, plus larger sizes useful mostly for laser power delivery. The NA of the fibers varies from 0.12 to about 0.2. This range of NAs is quite typical for oxide glass fibers as well. The fibers are coated with an acrylate coating. Most manufacturers of the fibers state a transmission range between 0.5 and 4.5 μm. Over this range the fibers will have losses less than 1 dB/m.

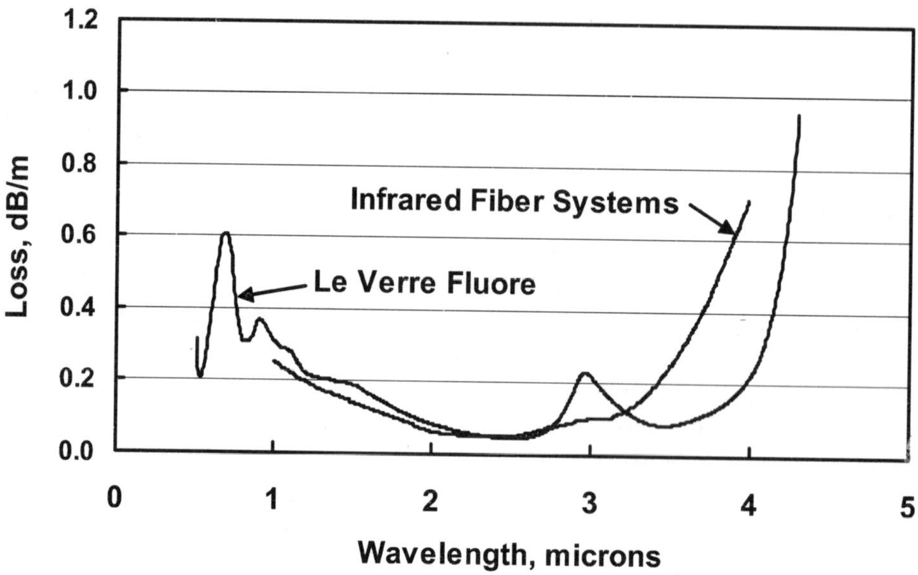

Figure 4.15 Commercial ZBLAN fiber losses. Note the higher loss in commercially available fiber compared to the lowest loss shown in Fig. 4.14.

Applications for these commercial fibers, as stated by the manufacturers, include the following:

- Temperature sensing
- IR imaging
- Remote spectroscopy
- Fiber amplifiers and lasers
- Laser power delivery.

These applications will be discussed in some detail in the later chapters on applications of IR fibers.

4.4.2.2 Fluoroaluminate fibers

The fluoroaluminate fibers are more durable than the ZBLAN. This means that they are chemically more inert, with a particularly strong resistance to attack by moisture. As a result, the AlF_3-based fibers are more stable, both thermally and mechanically. Much of the work leading to commercialization of these fibers is due to that of Itoh et al.[5] at Hoya Corporation in Japan. Their core glass composition was AlF_3-ZrF_4-YF_3-RF_2-NaCl, where $R =$ Mg, Ca, Sr, Ba. The cladding glass was made by increasing the AlF_3 content from about 25% to 30% and adding 3% NaF.

One of the most unusual features of this work was the method that they used to prepare the preform. They did not use the usual casting methods; instead they developed a clever extrusion technique in which clad and core glass disks were placed into an extrusion chamber and extruded into the requisite preform geometry. Their setup is shown in Fig. 4.16. Normally, hot extrusion techniques are not

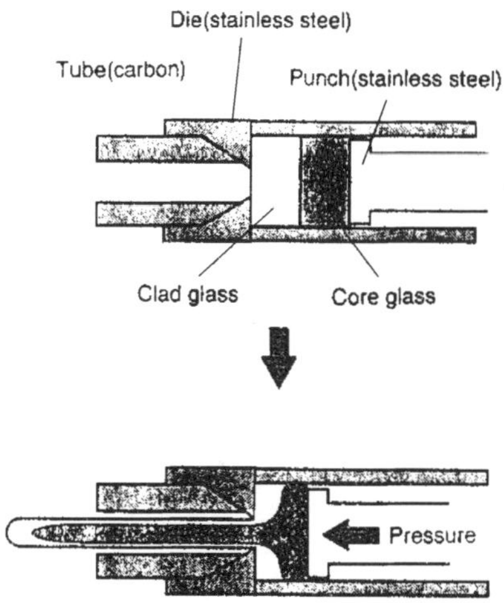

Figure 4.16 Extrusion of core/clad AlF_3-based fluoride glass fiber preform. This novel technique has been used to make a preform by pressing two or three plates into a cylindrically symmetric preform. [Reprinted from Ref. 5, Copyright 1994, with permission from Elsevier.]

used to fabricate glass fibers or performs; rather it is a method used extensively to fabricate crystalline materials into polycrystalline fibers (see Chapter 6). Nevertheless, Itoh et al. have found that if they first place a disk of cladding glass followed by a disk of core glass into a heated extrusion chamber, they can extrude a long core/clad preform. The core diameter in the first and last 20% of the preform can change dramatically, but over the remaining 60% of the preform they find that the maximum core diameter fluctuation is 3% per 10-mm length. Once the preform is made, they draw the fiber in the conventional fluoride glass drawing tower, as shown in Fig. 4.13.

The loss for their AlF_3-based fluoride fiber is shown in Fig. 4.17.[5] This fiber is a MM 440/500-μm fiber with a minimum-bending radius of 25 mm. It is interesting to note from the data in Fig. 4.17 that the minimum loss of 0.006 dB/m (6 dB/km) at 2.5 μm is quite low. The data represent, some of the lowest-loss fluoroaluminate fibers made to date.

The aluminum fluoride fibers studied by the group from Hoya have been commercialized and sold as part of Hoya's medical and dental Er:YAG laser system. The production fibers have a somewhat higher loss than that shown in Fig. 4.17. The minimum loss for the commercial fiber is 0.05 dB/m, with a loss of 0.1 dB/m at the Er:YAG laser wavelength of 2.94 μm. Their fiber, which is largely used for laser power delivery, is a MM 450/500/600 μm Teflon- or acrylate-coated fiber with a minimum-bending radius of 30 mm. The minimum loss of 0.05 dB/m is essentially identical to that for the commercial ZBLAN fiber.

4.5 Mechanical properties of fiber

The mechanical properties of fibers were discussed in general in Sec. 2.5. One of the most important mechanical properties of fluoride fibers is their strength, as this is related to the practical issues of reliability and bending. In fact, one of the

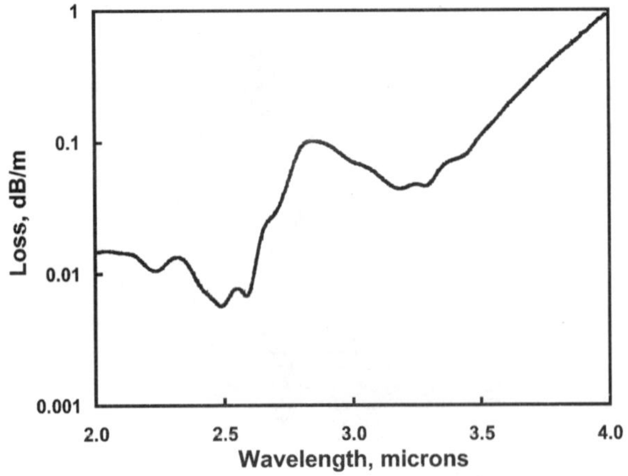

Figure 4.17 Loss for AlF_3-based fluoride glass fiber made using the extruded preform depicted in Fig. 4.16. [Reprinted from Ref. 5, Copyright 1994, with permission from Elsevier.]

questions most often asked is, how much can a fiber be bent before it fails? An excellent means of determining this is to make a two-point bending measurement using the apparatus shown in Fig. 2.13. Bending data taken on a group of fibers is then plotted on a Weibull plot. A typical Weibull strength distribution, shown in Fig. 2.14 for silica fibers, provides a convenient means of quickly determining the strain-to-failure and the safe bending limits for a fiber.

Some of the basic strength properties of fluoride glass compared to silica are listed in Table 4.4. Young's modulus, E, for ZBLAN as well as for the AlF_3-based fluoride glasses is less than that for silica. The values of E in Table 4.4, along with the simple expression for bending strength given by Eq. (2.29), can be used to estimate the minimum-bending diameter D. However, this diameter is only approximate, as all fiber failure is statistical in nature; in addition, it is very much dependent on environmental and fiber drawing conditions. In particular, fiber that is drawn in a RAP atmosphere free from moisture gives higher strength than one drawn in an inert atmosphere. Further improvement in the strength is obtained if the preforms are etched after mechanical polishing or casting. A chemical etching solution such as HCl plus small amounts of $BaZrF_6$ has been shown to remove polishing defects, small crystallites, and water on the surface of the preform. Strength data for ZBLAN fibers with various surface treatments and drawn in different environments is shown in Fig. 4.18.[10] Finally, it should be pointed out that the strength measured for fibers under tension is usually less than that measured in bending. This can be seen in Fig. 4.18. The two low-strength Weibull distributions (curves A and B) are for fibers under tension with a 20-cm gauge length, whereas the high-strength fibers (curves C and D) are for fibers under bending. The reason for this is that there are statistically more flaws in the long length of fiber used in the tensile tests compared to the short lengths encountered in bend testing.

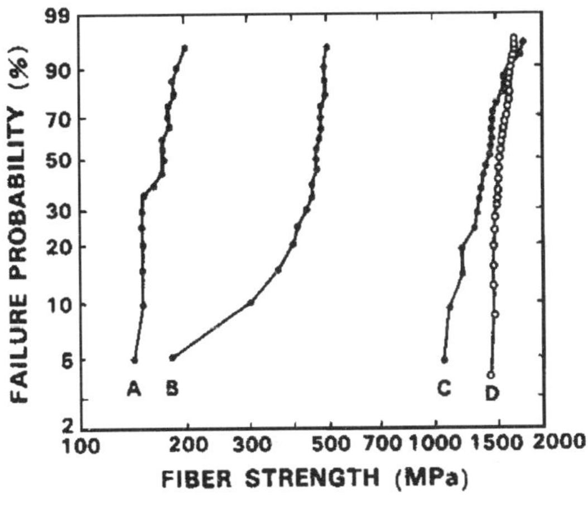

Figure 4.18 Weibull plots for tensile (A and B) and bending strengths (C and D) of (A) ZBLYALi fibers with FEP jacket, (B) ZBLYALi drawn in dry atmosphere, (C) ZBLA drawn in NF_3/N_2 atmosphere, (D) ZBLA fibers drawn in NF_3/N_2 atmosphere after etching in $ZrOCl_2$. [Reprinted from Ref. 10, with permission from Elsevier.]

4.6 Miscellaneous halide glasses

While ZBLAN and AlF_3-based fluoride glasses continue to be the most common HMFG fibers, there are other fluorides as well as mixed fluoride glasses that have been prepared and studied.[14,15,25] In general, the newer compositions with the most promise are those in which heavier elements like Ba, In, Ga, and Zn have been added in place of Zr and Al. One advantage of these glasses is that they transmit to slightly longer wavelengths than the ZBLAN glasses. In some cases they are also reported to be more stable than ZBLAN.

One example of such a glass is BIZYbT glass. This glass is BaF_2 (30%)-InF_3 (30%)-ZnF_2 (20%)-YbF_3 (10%)-ThF_4 (10%). The spectral losses for BIZYbT bulk glass and fiber are shown in Fig. 4.19.[26] From this data it can be seen that the transmission for this fiber is about 1 μm further into the IR compared to ZBLAN. This can be important for some applications requiring transmission into the 5-μm region. This material has a $\Delta T = 100°C$, and it is more resistant to corrosion than ZBLAN.

A family of glasses similar to BIZYbT is the BIG family of glasses.[27] Again, these glasses have heavy cations substituting for the lighter cations. A good example of a BIG glass is BaF_2 (30%)-InF_3 (18%)-GaF_2 (12%)-ZnF_2 (20%)-YbF_3 (10%)-ThF_4 (6%)-ZrF_4 (4%). Clearly the glass compositions are getting more complex as different fluorides are added to increase ΔT or to stabilize the glass. The BIG glasses have been drawn into fiber, but the losses are considerably higher than those for ZBLAN. The reported loss is about 5 dB/m at 4 μm.

There is another class of halide glasses in which the other halides (Cl, Br, and I) have been added in addition to or in place of fluorine.[15] This has been done largely in an effort to increase the IR transparency of the fluoride glasses. These are the so-called mixed halide glasses. For example, NaCl is a common additive to the

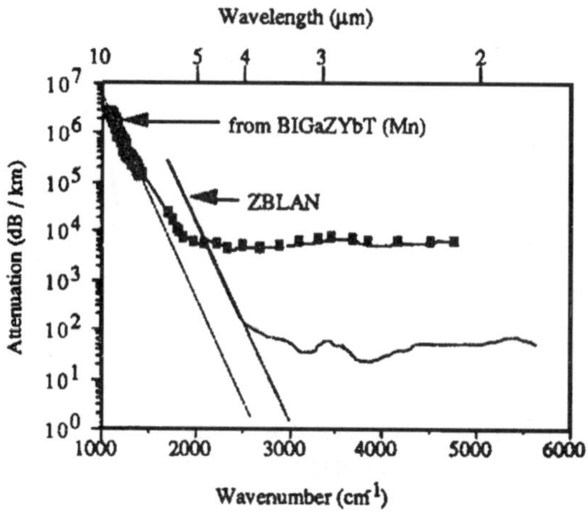

Figure 4.19 Special category of HMFGs in which heavier elements like gallium are added to increase the long wavelength transmission. The glass shown is a member of the so-called BIG glass family.[26]

fluoride host most often loosely based on the BIG glass family. Other halides added to the glasses are $CdCl_2$ and $PbCl_2$. Few of these glasses have been drawn into fiber even though they have an IR transmission that extends about 1 μm further than the BIG glasses. The reason for a declining interest in these glasses for fiber optics is that the chalcohalide glasses have become more popular. Chalcogenide halide glasses are discussed in Chap. 5. These glasses are a mixture of the chalcogen elements, most especially Te, and halides such as I^-. The IR transparency of the chalcohalide glasses is much greater than that of the mixed halide glasses, and they are suitable for CO_2 laser transmission at 10.6 μm.

References

1. M. Poulain, M. Poulain, J. Lucas, and P. Brun, "Verres fluores au tetrafluorure de zirconium proprietes optiques d'un verre dope au Nd^{3+}," *Mat. Res. Bull.*, Vol. 10, pp. 243–246 (1975).
2. M. Poulain, M. Chanthanasinh, and J. Lucas, "New fluoride glasses," *Mat. Res. Bull.*, Vol. 12, pp. 151–156 (1977).
3. D. Tran, G. H. Sigel, and B. Bendow, "Heavy metal fluoride glasses and fibers: a review," *J. Lightwave Tech.*, Vol. LT-2, pp. 566–586 (1984).
4. J. Sanghera and I. Aggarwal, *Infrared Fiber Optics*, CRC Press, Boca Raton, FL (1998).
5. K. Itoh, K. Miura, M. Masuda, M. Iwakura, and T. Yamashita, "Low-loss fluorozirco-aluminate glass fiber," *J. Non-Cryst. Solids*, Vol. 167, pp. 112–116 (1994).
6. P. W. France, S. F. Carter, M. W. Moore, and C. R. Day, "Progress in fluoride fibres for optical communications," *Br. Telecom. Tech. J.*, Vol. 5, pp. 28–44 (1987).
7. P. France, M. G. Drexhage, J. M. Parker, M. W. Moore, S. F. Carter, and J. V. Wright, *Fluoride Glass Optical Fibres*, Blackie and Son, London (1990).
8. S. F. Carter, M. W. Moore, D. Szebesta, D. Ransom, and P. W. France, "Low loss fluoride fibre by reduced pressure casting," *Elect. Lett.*, Vol. 26, pp. 2115–2117 (1990).
9. D. Szebesta, S. T. Davey, J. R. Williams, and M. W. Moore, "OH absorption in the low loss window of ZBLAN(P) glass fibre," *J. Non-Cryst. Solids*, Vol. 161, pp. 18–22 (1993).
10. I. Aggarwal and G. Lu, *Fluoride Glass Optical Fiber*, Academic Press, New York (1991).
11. D. Tran, U.S. Patent No. 5,274,728 (Issued 12-28-1993).
12. G. Cao, F. Lin, H. Hu, and F. Gan, "A new fluorogermanate glass," *J. Non-Cryst. Solids*, Vol. 326–327, pp. 170–176 (2003).
13. M. Poulain, "New trends in halide glass compositions," *J. Non-Cryst. Solids*, Vol. 184, pp. 103–108 (1995).
14. A. Akella, E. Downing, and L. Hesselink, "New fluoroindate glass compositions," *J. Non-Cryst. Solids*, Vol. 213/214, pp. 1–5 (1997).

15. J. L. Adam, C. Ricordel, and J. Lucas, "New compositions of low phonon energy fluoride and chloro-fluoride glasses," *J. Non-Cryst. Solids*, Vol. 213/214, pp. 30–35 (1997).
16. G. Zhang, J. Jiang, and M. Poulain, "Indium and gallium fluorochloride glasses," *J. Non-Cryst. Solids*, Vol. 221, pp. 78–83 (1997).
17. F. Gan, "Optical properties of fluoride glasses: a review," *J. Non-Cryst. Solids*, Vol. 184, pp. 9–20 (1995).
18. S. Takahashi, "Optical properties of fluoride glasses," *J. Non-Cryst. Solids*, Vol. 95/96, pp. 95–106 (1987).
19. M. Robinson, "Processing and purification techniques of heavy metal fluoride glass (HMFG)," *J. Cryst. Growth*, Vol. 75, pp. 184–194 (1986).
20. T. Miyashita and T. Manabe, "Infrared optical fibers," *IEEE J. Quantum Electron.*, Vol. QE-18, pp. 1432–1450 (1982).
21. S. Sakaguchi and S. Takahashi, "Low-loss fluoride optical fibers for midinfrared optical communication," *J. Lightwave Tech.*, Vol. LT-5, pp. 1219–1228 (1987).
22. Y. Ohishi, S. Mitachi, and S. Takahashi, "Fabrication of fluoride glass singlemode fibers," *J. Lightwave Tech.*, Vol. LT-2, pp. 593–596 (1984).
23. D. Tran, C. F. Fisher, and G. H. Sigel, "Fluoride glass preforms prepared by a rotational casting process," *Electron. Lett.*, Vol. 18, pp. 657–658 (1982).
24. H. Tokiwa, Y. Mimura, T. Nakai, and O. Shinbori, "Fabrication of long single mode and multimode fluoride glass fibres by the double-crucible technique," *Electron. Lett.*, Vol. 21, pp. 1131–1132 (1985).
25. G. Zhang, B. Friot, and M. Poulain, "New gallium and indium based fluoride glasses," *J. Non-Cryst. Solids*, Vol. 213–214, pp. 6–10 (1997).
26. J. Lucas, I. Chiaruttini, G. Fonteneau, P. Christensen, and S. Mitachi, "New multicomponent fluoride glasses with low critical cooling rates for optical fibers," in Proc. SPIE. Vol. 1228, pp. 56–62 (1990).
27. N. Rigout, J. L. Adam, and J. Lucas, "BIG fluoride glass optical fibers with improved NA," *J. Non-Cryst. Solids*, Vol. 161, pp. 161–164 (1993).

Chapter 5

Chalcogenide Glass Fiber Optics

5.0 Introduction

Chalcogenide glasses are composed of two or more chalcogen elements normally selected from the small group including As, Ge, Sb, P, Te, Se, and S. When these elemental materials are heated and mixed in an oxygen-free environment, some very stable and simple glasses can result. One of the oldest chalcogenide glasses studied is the binary glass arsenic trisulfide, As_2S_3. This glass is deep red in color, and it is very stable. In the mid 1960s, this glass was drawn into the first IR fiber by N. S. Kapany.[1] In his 1967 book on fiber optics,[2] Kapany includes a discussion of As_2S_3 fibers and some of their potential applications. It was not until some 10 to 15 years later, however, that these materials were studied seriously as viable IR fiber candidates. The reticence to pursue these materials in the early days came in part from the toxic nature of some of the elements used in the glasses. Today they are a popular IR fiber material as they are readily drawn into fiber with a broadband IR transmission. While they are much more delicate in nature than the oxide glass fibers, they are finding many applications in chemical and temperature sensor systems and as IR image bundles.[3]

Chalcogenide fibers fall into three categories: sulfide, selenide, and telluride. Within these categories, one usually finds that the binary and ternary glasses are excellent choices for fiberization. That is, unlike the fluoride glasses where it is commonplace to have five or more components, most chalcogenide glasses have only two or three elemental components. In general, these glasses have softening temperatures comparable to fluoride glass. They are very stable, durable, and largely insensitive to moisture. A distinctive difference between these glasses and the other IR fiber glasses is that they do not transmit well in the visible region and their refractive indices are quite high. Additionally, most of the chalcogenide glasses, except for As_2S_3, have a rather large value of dn/dT.[4] This fact limits the laser power handling capability of the fibers.

In brief summary, chalcogenide glass is made by combining highly purified (>6 nines purity) raw elements in a sealed ampoule that is heated and mixed in a rocking furnace. After melting and mixing, the glass is quenched and a glass preform fabricated using rod-in-tube or rotational casting methods. Fiber can be drawn using a preform or from a double crucible. As in the fluoride glass fibers, a buffer polymer coating is applied over the cladding using a UV acrylate or by applying a Teflon heat-shrink tube over the preform and then drawing.

The transmission range for chalcogenide fibers depends heavily on the mass of the constituent elements. The lighter-element glasses such as arsenic trisulfide have a transmission range from 0.7 to ~6 μm.[5] This glass and some phosphorous-containing and Ge-S based glasses are the only ones transmitting visible radiation.[4] Longer wavelength transmission is possible through the addition of heavier elements like Te and Se. When these elements are present, the glasses take on a silvery, metallic appearance, and they become essentially opaque in the visible region. A key feature of essentially all chalcogenide glasses is the strong extrinsic absorption resulting from contaminants such as hydrogen, H_2O, and OH^-. For example, there are invariably strong absorption peaks at 4.0 and 4.6 μm due to S H or Se H bonds, respectively, and at 2.78 μm and 6.3 μm due to OH^- (2.78 μm) and/or molecular water. As a result, typical chalcogenide loss spectra are normally replete with extrinsic absorption bands. This would seem at first glance to be sufficiently deleterious that the applications for these fibers would be limited. However, many applications for these fibers are possible simply by working outside these extrinsic bands. Another important feature of most of the chalcogenide fibers is that their losses are usually much higher than those of the fluoride glasses. In fact, at the important CO_2 laser wavelength of 10.6 μm, the lowest loss is still slightly above 1 dB/m for the Se-based fibers.[6]

5.1 Glass composition and key physical properties

5.1.1 Chalcogenide glass compositions

Some of the oldest and simplest chalcogenide glasses are the sulfide glasses. The most important sulfide glasses are the binary glasses based on As-S.[5,7–9] Arsenic trisulfide with the normal composition, $As_{40}S_{60}$ (or As_2S_3 for short), has been well studied. In general, the stoichiometry of the chalcogenide glasses can be changed rather dramatically within a broad glass-forming region. Thus, it is quite easy to increase the concentration of, say, As and decrease the concentration of S to slightly increase the refractive index of arsenic trisulfide. This is one reason why these glasses are particularly good choices for fiber optic applications. As-S glasses have low T_g and they are quite stable against devitrification. These glasses are reasonably easy to form by merely mixing As and S, but S has a high vapor pressure; therefore, it is necessary to contain the high temperature mixture in a vessel that will not explode during glass mixing. More details on glass formation may be found in Sec. 5.3.1.

The selenide glasses are also easy to form, and as in the case of the sulfide glasses, it is possible to vary the composition of the various components over a wide range.[10–12] Selenide glasses are usually made with three or more components as the binary glasses, such as As-Se, are of less interest than the binary sulfide glasses. In particular, the As-Se glasses with either Ge or Sb added to form the ternary glass are very popular. For example, the glass-forming region for Ge-As-Se glasses covers the compositional range: 0–20 at.% Ge, 0–45 at.% As, and 55–95 at.% Se. A typical composition for fiber drawing is $Ge_{10}As_{20}Se_{70}$.[5,12] With the addition of Se and Ge, these glasses become a deep red color in transmission, with a metallic, silvery color in reflection. They transmit at longer wavelengths than the sulfide glasses as a result of the substitution of Se for S.

The telluride glasses are those glasses for which Te is added, usually in conjunction with Se. A primary reason to add Te to the glass is to improve the long wavelength transmission of the glass.[13] In fact, the telluride glasses have the lowest loss at the CO_2 laser wavelength of 10.6 μm.[4,14] In general, these glasses are very stable against crystallization, but the T_g decreases as the Te content increases. They also have a somewhat lower vapor pressure than the sulfide and selenide glasses. The most popular telluride glasses studied for fiber fabrication are the Ge-Se-Te and Ge-As-Se-Te systems. An excellent fiber material is the glass, $Ge_{20}As_{30}Se_{30}Te_{20}$.

A good review of the composition and properties of chalcogenides can be found in two books devoted exclusively to these glasses.[15,16] There are also several good review articles on chalcogenide glass fiber optics.[4,5,9,12,17]

5.1.2 Key physical properties

Chalcogenide glasses are low-temperature glasses similar in this regard to the fluoride glasses. They can be quite stable against devitrification even though the glass transition temperature for some glasses can be only a little above 100°C. For example, an As-Se-Te glass made by Amorphous Materials, Inc., Garland, TX (glass C1) has a $T_g = 136°C$. In general, they do not transmit well in the visible region, and they give the outward appearance more of a semiconductor than a glass. In particular, the telluride and selenide glasses take on a metallic appearance when polished.

The largest ΔT and the greatest resistance to crystallization occur for As_2S_3. There is a greater tendency to crystallize in the As_2Se_3 and $GeSe_2$ glasses. As the mass of the chalcogen increases, the value of T_g decreases. The value of T_g is also related to other physical properties such as the thermal expansion coefficient and hardness. Glasses with low T_g have greater thermal expansion and smaller hardness. For example, $Ge_{28}Sb_{12}Se_{60}$ glass has a T_g 100°C greater than As_2S_3. The thermal expansion coefficient and hardness for this Ge-Sb-Se glass are $15 \times 10^{-6}/°C$ and 150 Knoop, respectively, compared to $25 \times 10^{-6}/°C$ and 109 Knoop, respectively, for As_2S_3. The data in Table 5.1 summarizes some of the key physical parameters for two chalcogenide glasses in comparison to silica.

The physical properties of the chalcogenide glasses, just as for the fluoride glasses, pale in comparison with silica. The thermal properties of the chalcogenides are poorer than those of silica. For example, the thermal expansion is more than 20 times higher for the chalcogenides than for silica. Also, the softening temperature of silica is many times that of a typical chalcogenide glass. The chalcogenide glasses, however, are more chemically durable, especially against attack by moisture, than the fluoride glasses.

Table 5.1 Key physical properties of some chalcogenide glasses.

Property	As_2S_3	$Ge_{33}As_{12}Se_{55}$ Amtir 1	Silica glass
T_g, °C	180	362	1175
Specific heat, cal/g °C	0.109	0.07	0.179
Thermal conductivity, W/m °C	0.37	0.25	1.38
Expansion coefficient	21.4×10^{-6}	12×10^{-6}	0.55×10^{-6}
Density, g/cm^3	3.20	4.40	2.20
Knoop hardness, kg/mm^2	109	170	600
Toughness (K_{1C}), MPa m$^{1/2}$	0.2	0.34	0.72
Young's modulus, GPa	15.9	22.1	72.2
Poisson's ratio	0.24	0.27	0.17
Refractive index	2.41 (5 μm)	2.49 (5 μm)	1.458 (0.59 μm)
dn/dT, °C^{-1}	9.3×10^{-6} at 5 μm	72×10^{-6} at 10.6 μm	11.9×10^{-6}

5.2 Optical properties of bulk glass

A key feature of the chalcogenide glasses is that they transmit much longer wavelengths than the fluoride glasses. Specifically, they transmit wavelengths beyond 3 to 4 μm, where most HMFGs cut off. This has great practical importance for fiber optic sensors operating in the sensitive fingerprint region (>4 μm) of the spectrum and for the delivery of IR laser power for lasers operating at wavelengths longer than the Er:YAG laser at 2.94 μm. The transparency of these glasses depends heavily on the composition of the glass. As the mass of the ions increases, so does the long wavelength transmission, as explained in Sec. 2.3.[13] A key optical feature of the chalcogenide glasses is their very strong extrinsic absorption due to impurities.[18] In fact, this one aspect alone stands out as a distinguishing feature of these fibers compared to other IR glasses and crystalline materials. The subject of intrinsic and extrinsic loss mechanisms is discussed in general in Secs. 2.3.1 and 2.3.2.

5.2.1 Intrinsic loss

The intrinsic losses at IR wavelengths for chalcogenide glass are determined by Rayleigh scattering at short wavelengths and by multiphonon absorption at long wavelengths. The multiphonon or IR edge for a sulfide and a selenide bulk glass, shown in Fig. 5.1,[18] illustrates not only the long wavelength behavior of these glasses, but also shows a multiphonon structure not usually seen in fluoride or oxide glasses. The first thing to note from the data in Fig. 5.1 is that the As_2S_3 glass (A) does not transmit as far into the IR as the $As_{38}Ge_5Se_{57}$ glass (B). This is a result of the heavier ion composition of glass B. It may also be seen in Fig. 5.1 that there are several broad absorption bands in both glasses. For example, the bands at 6.9, 7.6, 10.2, and 14.5 μm in As_2S_3 are due to multiphonon processes. That is, the IR or multiphonon edge in these glasses is not a featureless, exponentially increasing absorption with increasing wavelength as shown in Figs. 2.9 and 2.10. Instead there

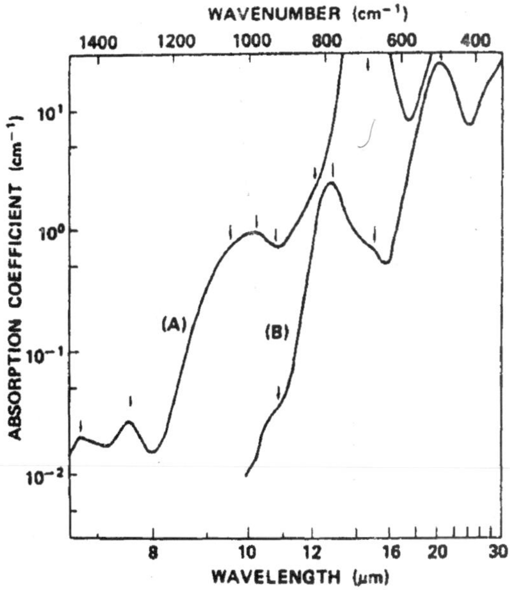

Figure 5.1 Multiphonon or IR edge of bulk (a) As_2S_3 and (b) $As_{38}Ge_5Se_{57}$ glasses showing the shift to longer wavelengths for the heavier Se-containing glasses compared to the S glasses. [Reprinted from Ref. 18, Copyright 1985, with permission from Elsevier.]

is an absorption structure superimposed on this edge. The multiphonon structure observed in chalcogenide glasses results from linear combinations of the fundamental phonon modes. At the longest wavelengths, two and three LO and/or TO phonons combine in an *n*th order process to the total intrinsic absorption. In fact, this occurs in all of the IR optical materials considered in this book, but normally these high-order phonon processes are very broad, and thus they overlap to give a smooth exponential multiphonon tail, as shown in Fig. 2.10. But the chalcogenide glasses are strongly covalently bonded, and this leads to less overlap in the phonon bands; therefore, a well-defined multiphonon structure results.

The total intrinsic losses for chalcogenide glass comprise the usual loss mechanisms: the electronic or Urbach tail, Rayleigh scattering, and multiphonon absorption. For these glasses, however, there is an additional absorption present due to a so-called weak absorption tail (WAT). An excellent example of how these losses contribute to the total and minimum loss may be seen in the data of Kanamori et al.[18] given in Figs. 5.2 and 5.3. The data in Fig. 5.2 for As_2S_3 and in Fig. 5.3 for $As_{38}Ge_5Se_{57}$ show the dominating influence of the WAT in the shorter wavelength region. This tail results from electronic transitions to lower-lying bandgap states. These bandgap states can be due to impurities or defects in the glass structure such as dangling bonds. If these glasses were of high purity, then only the defect states formed when the glass is cooled would contribute to the weak absorption tail. This tail is not seen in oxide glasses, which suggests that the formation energy for defect states is lower in chalcogenides than for oxide glasses. Given that the defect energy is low in chalcogenide glasses, Kanamori et al. speculate that even in the purest glass there will still be an intrinsic WAT that will limit the losses in these glasses.

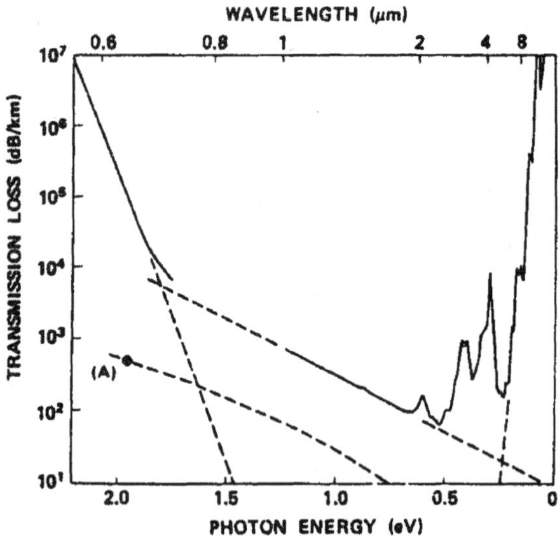

Figure 5.2 Losses for As_2S_3 glass. The intrinsic short wavelength losses (Urbach tail) are shown along with the extrinsic absorption peaks due to impurities and the IR edge. Curve (A) is an estimate of the intrinsic scattering losses. [Reprinted from Ref. 18, Copyright 1985, with permission from Elsevier.]

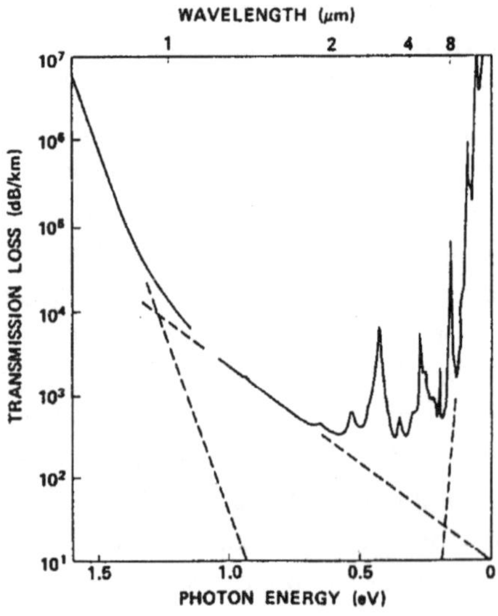

Figure 5.3 Losses for $As_{38}Ge_5Se_{57}$ glass similar to that shown in Fig. 5.2 for the S glass. This data shows the intrinsic and extrinsic losses for this long-wavelength transmitting glass. [Reprinted from Ref. 18, Copyright 1985, with permission from Elsevier.]

Several groups have studied the minimum loss in chalcogenide glass.[18,19] Unlike other transparent solids, the minimum loss is most accurately determined by extrapolating the WAT rather than the scattering loss. In Fig. 5.2, Kanamori et al.[18] plot the estimated Rayleigh scattering loss for As_2S_3. The scattering losses are significantly lower than the WAT losses. The minimum theoretical losses for these two glasses are given in Table 5.2. The most realistic values are those obtained by extrapolation of the WAT to the multiphonon curve. For As_2S_3 this gives a minimum loss of 23 dB/km at 4.6 μm, whereas if the scattering loss is extrapolated to the multiphonon curve, the loss turns out to be about 0.05 dB/km at 5 μm. Kanamori et al. indicate that the attenuation produced by the defect states even in the purest glasses will limit the minimum loss to about 10 dB/km for As_2S_3 glass in fiber form.

As indicated in Fig. 5.1, the IR edge is determined by the composition of the glass. An excellent example of how this edge is affected by compositional changes within the telluride glasses may be found in the data of Katsuyama and Matsumura.[13] They altered the Te concentration from 0% to 80% in a Ge-Se-Te glass. Their attenuation data is shown in Fig. 5.4. It can clearly be seen from Fig. 5.4 that the addition of the heavier Te in place of Se shifts the IR edge to

Table 5.2 Minimum loss for several chalcogenide glasses.

Glass	Theoretical loss, dB/km (wavelength, μm)	Comments
As_2S_3	23 (4.6)	Using WAT
As_2S_3	0.05 (5.0)	Using Rayleigh scattering
$As_{38}Ge_5Se_{57}$	30 (6.3)	Using WAT
GeS_2	169 (3.6)	Using WAT

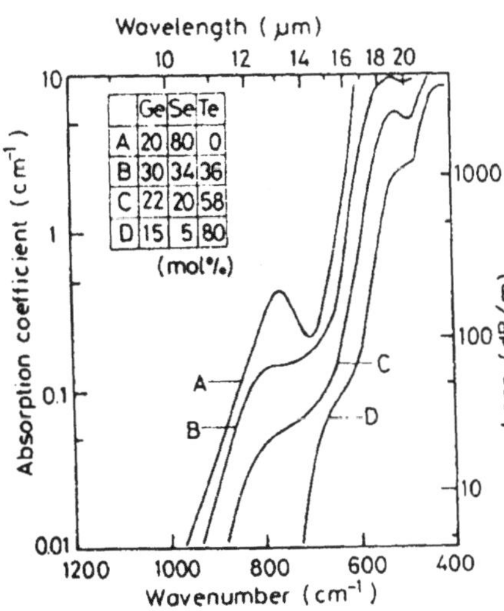

Figure 5.4 Multiphonon edge for Ge-Se-Te ternary glass with four different compositions. Note the shift to longer wavelengths with the addition of high concentrations of Te.[13] [Reprinted with permission.]

longer wavelengths. The absorption peaks in their data again reflect multiphonon absorption similar to that shown in Fig. 5.1. By adding increasing amounts of Te, it is possible to shift the IR edge sufficiently far that this glass would be a good candidate for 10.6-μm transmission. The highest Te-content glass, $Ge_{15}Se_5Te_{80}$, has the lowest projected loss at CO_2 laser wavelengths. However, the high Te-content glasses are much weaker than those containing no Te and, therefore, when high Te-content glass is drawn into fiber, the fiber is relatively fragile.

5.2.2 Extrinsic loss

As mentioned earlier, extrinsic losses are the dominant loss mechanisms in the chalcogenide glasses as they are for the fluoride glasses. There are two main types of extrinsic losses: those due to impurities and those arising from scattering centers. Unlike the fluoride glasses, impurity absorption is by far the most important extrinsic loss. Impurity absorption for As-S and As-Ge-Se glass is easily seen in the spectral data in Figs. 5.2 and 5.3, respectively. The impurity spectra are very rich and, in some cases, certain impurities lead to very strong attenuation. The major impurities for these two glasses are listed in Table 5.3.[18]

One of the strongest extrinsic absorptions results from hydrogen bonding to either S or Se. In the sulfide glasses the S—H fundamental stretching absorption occurs near 4.0 μm, whereas in the selenide glasses the Se—H absorption is shifted to 4.6 μm. It is extremely difficult to remove all of the hydrogen impurity during

Table 5.3 Absorption OH and H-bonded impurities in two important chalcogenide glasses. The strength* is a ratio of the intensity of the combination and overtone bands normalized to the strength of the fundamental stretching vibrations with strength 100.

Glass	Impurity	Wavelength, μm	Strength*
As_2S_3	OH	2.91	100
		2.29	1.5
		1.92	1.2
		1.44	0.5
	SH	4.03	100
		3.69	4.4
		3.11	4.8
		2.54	0.2
		2.05	0.5
$As_{38}Ge_5Se_{57}$	OH	2.92	100
		2.32	1.8
		1.92	1.4
		1.45	0.6
	SeH	4.57	100
		4.15	5.9
		3.55	6.6
		2.32	0.6

glass formation, and therefore it is commonplace to observe this absorption even in the most purified glass.[20]

Another common contaminant in these glasses is OH. The fundamental OH absorption is near 2.9 μm in both glasses. When the glasses are highly purified, it is possible to reduce the attenuation due to OH substantially. The data in Fig. 5.5 for an essentially As_2S_3 glass (actual glass $As_{40}S_{55}Se_5$) show how the losses due to impurities in general can be lowered substantially when purified chemicals, curve (B), are used as starting materials.[11] Devyatykh et al.[9] have more recently shown that the main OH absorption can be reduced to <50 dB/km when the As_2S_3 glass is purified using distillation methods.

5.2.3 Refractive index and dispersion

The refractive indices of the chalcogenide glasses are considerably higher than for fluoride or oxide glasses. The chalcogen elements are more polarizable than these glasses, and this leads to the higher refractive indices. Table 5.4 lists selected refractive index values for some of the key chalcogenide glasses used in fiber drawing. Also in Table 5.4 is the value of dn/dT for these glasses. The lowest refractive indices are for the sulfide glasses, with values of n approximately 2.4. As the Se and/or Te content is increased, the refractive index increases to values near 3.0.

The dispersion of bulk sulfide and selenide glasses is shown in Fig. 5.6. The data in Fig. 5.6 for four glasses clearly shows the trend to higher index values for increasing Se content. The lowest index material is $As_{25}S_{70}Ge_5$, and this glass has the highest sulfur content (70 wt%). When the sulfur content is reduced to zero and the Se content is increased, the index reaches its highest value. This can be seen for the $As_{40}Se_{60}$ glass in Fig. 5.6.

There is little on the material dispersion effects in these glasses. Some early work by Miyashi and Manabe[14] on As_2S_3 glass shows that the material dispersion coefficient, D_λ, from Eq. (2.21) is zero at 4.89 μm, and that the slope of the

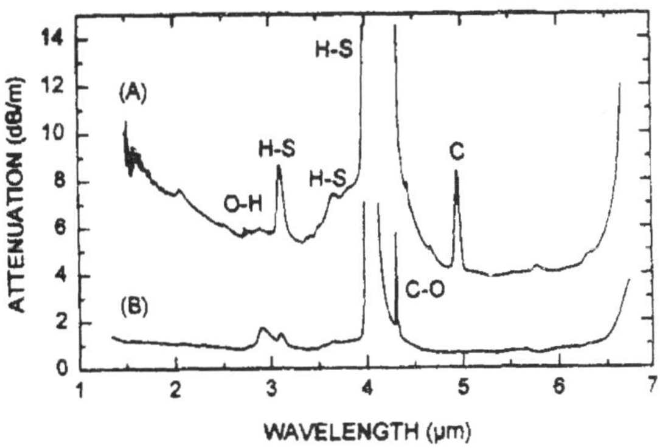

Figure 5.5 Reduction of extrinsic losses for As_2S_3 fibers resulting from the use of higher-purity starting materials: (a) as received chemicals and (b) using purified starting chemicals.[11] [Reprinted with permission.]

Table 5.4 Refractive indices at CO and CO_2 laser wavelengths for chalcogenide glass fibers. Note the low value of dn/dT for the As_2S_3 glass compared to the other glasses. This is a key reason why As_2S_3 glass has a high laser power damage threshold.

Glass	Wavelength, μm	n	dn/dT, °C^{-1}
As_2S_3	5.0	2.407	9.3×10^{-6}
$Ge_{33}As_{12}Se_{55}$	10.6	2.497	72×10^{-6}
$Ge_{28}Sb_{12}Se_{60}$	10.6	2.602	56×10^{-6}
$Ge_{27}Se_{18}Te_{55}$	10.6	2.90	80×10^{-6}

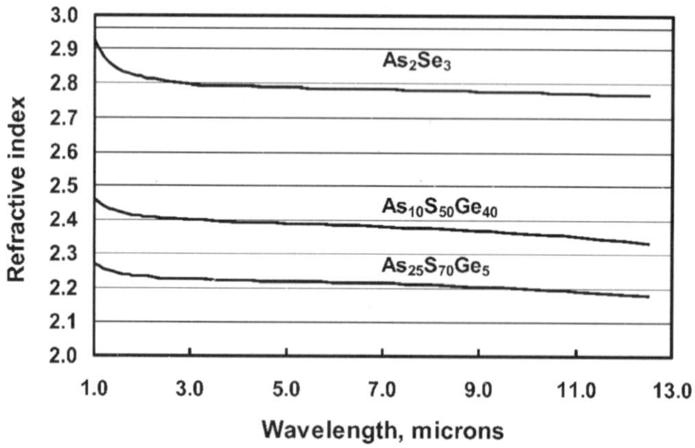

Figure 5.6 Dispersion for some representative chalcogenide glasses. The indexes are highest for the heavier Se glasses, with the lowest for the high S-content glasses. Data taken from literature supplied by Amorphous Materials, Inc., Garland, TX.

material dispersion curve is less than it is for fluoride and oxide glasses. The ZMD wavelength is close to the minimum loss for these fibers (see Table 5.2). One reason that material dispersion has not been so thoroughly studied in chalcogenide glasses is that these glasses have not really been strong candidates for telecommunications applications. This may change in the future, as there have been recent attempts to dope these fibers with rare earths for use as fiber amplifiers. The chalcogenides have excellent potential as fiber amplifiers because of their low phonon energies.

Many of the studies of dispersion in these materials have been on thin chalcogenide films. The films have largely been made using As_2S_3 and measurements of the dispersive and nonlinear properties of the films are made at wavelengths below approximately 1 μm.

5.3 Fiber fabrication

As for fluoride glass, the drawing of chalcogenide glass into fiber is not as straightforward as it is for silica glass. One of the main reasons for this is that it has been largely impossible to fabricate preforms using CVD as is routinely done with oxide glasses. Instead, chalcogenide glass preforms must be made by combining the raw materials to form the glass and then casting the glass into a preform. It is also possible to use a crucible technique to draw fiber. In the next section, a few common methods for preform fabrication are summarized. Since most fiber drawing is done using a preform rather than by a crucible technique, it is sufficient to describe only preform fiber drawing in Sec. 5.3.2.

5.3.1 Glass and preform preparation

It is a relatively simple procedure to make glasses from the chalcogen elements. High-purity starting materials are placed inside a sealed, evacuated silica ampoule. The ampoule can be a simple tube or a more sophisticated mixing chamber, as shown in Fig. 5.7. In the three-part ampoule shown in Fig. 5.7, the different elements can be placed in different chambers and sublimed/distilled into a common chamber for improved melting of the high-purity glass. The frit is used to filter out particulate matter. The glass mixture is heated between 700°C and 900°C for about 10 hours. During heating, the glass mixture is rocked back and forth in a rocking furnace. Very large plates of glass of up to 5 kg can be made using this method.

Once the glass is formed it can be cast into a preform using casting methods similar to those used for the fluoride glasses. A rotational casting method employed by Nishii et al.[4] is shown in Fig. 5.8. One difference between their method and that described in connection with Fig. 4.10 for the HMFGs is that the chalcogenide glasses have a higher vapor pressure than the fluorides. This necessitates the use of sealed ampoules to contain the glass at the high temperatures needed for casting. The core glass ampoule is spun in a vertical position during cooling, while the clad glass tube is formed by rotation in the horizontal direction. Once the clad tube and core glass are formed, they are formed into a rod-in-tube preform.

Another method for preparing preforms involves extrusion of core and clad glass disks into long core/clad preforms. This method has been pioneered by Seddon and it is similar to the work done by Hoya[21] extruding AlF_3-based, fluo-

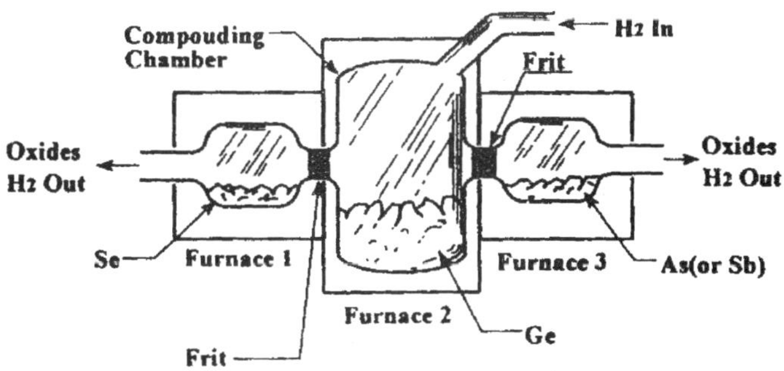

Figure 5.7 Mixing chambers for the synthesis of chalcogenide glasses similar to that used by Amorphous Materials, Inc. This setup illustrates the preparation of the ternary Ge-Sb-Se glass, but the arrangement is similar for the other glasses. The chamber is heated and rocked so that the compounded glass resides in the middle chamber. [Reprinted from A. Hilton, et al., *J. Non-Cryst. Solids*, **17**, 319 (1975), with permission from Elsevier.]

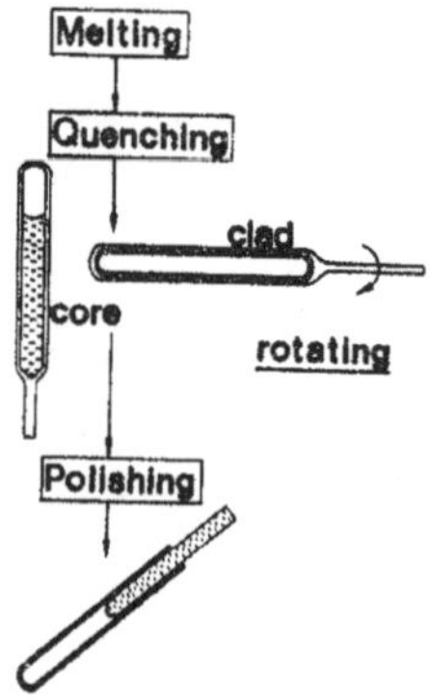

Figure 5.8 Rotational casting method for preparation of clad/core preforms. The core rod is polished before insertion into the cladding tube. [Reprinted from Ref. 4, Copyright 1992, with permission from Elsevier.]

ride glass disks into preforms. The properties of these preforms are discussed by Seddon,[22] and Furniss and Seddon[23] further explore the possibility of an SM preform using extrusion techniques.

5.3.2 Fiber drawing

Chalcogenide glass, like the fluoride glass fibers, may be drawn using either a preform or a crucible technique. Both methods yield good quality fiber, but since most fiber made today is drawn from a preform this method will be emphasized. The tower for drawing preforms used by Nishii et al.[24] is illustrated in Fig. 5.9. The preform was protected during the draw by Ar gas. The drawing speed was between 0.5 and 3 m/min. Two methods were used to apply a polymer buffer coating to the fiber. One method is the standard UV acrylate coating applied on line immediately after drawing. This is shown in Fig. 5.9. It is also possible to draw a preform that has been coated with Teflon FEP. This is a simple means of drawing the coating along with the fiber. This was a common technique in the early work, as it was less complicated than using a UV coater. In fact, fiber was often drawn as core-only fiber with a Teflon jacket. Unfortunately, the Teflon coating on bare core fibers leads to significantly higher loss as a result of absorption by the polymer.

5.4 Optical properties of fiber

Over the years there have been many improvements in the losses of chalcogenide fibers. Most of the reduction in loss has come as a result of careful purification of the starting materials rather than as improvements in drawing techniques. In general, the losses for these fibers are well above those for the HMFG fibers. In fact, most commercial fiber has a loss between 0.5 and 1.0 dB/m. Even the lowest-loss fiber is far from the intrinsic limit. Yet the losses are low enough for many short-haul applications. In Sec. 5.4.1, the lowest-loss fiber results will be discussed.

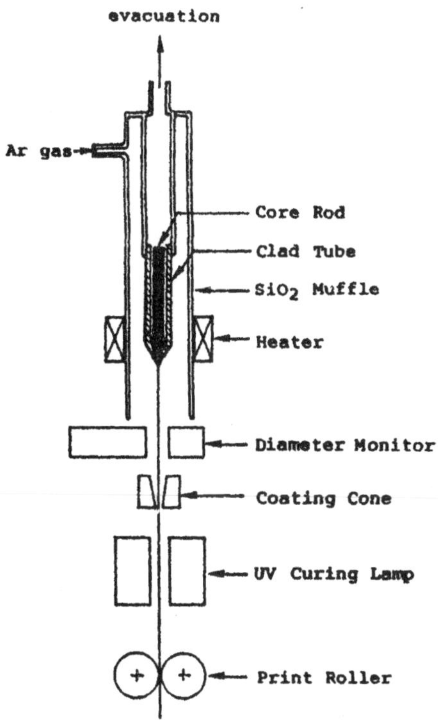

Figure 5.9 Preform drawing tower for fiber drawing of the chalcogenide preform fabricated as shown in Fig. 5.8.[24] [Reprinted with permission.]

It is also important, however, to describe the loss that one can expect to obtain from production-grade chalcogenide glass fiber. This is discussed in Sec. 5.4.2.

5.4.1 Low-loss fiber

The lowest-loss chalcogenide fiber has been obtained for As_2S_3 glass. Somewhat surprisingly, these losses were measured on unclad As_2S_3 fibers or on fibers with only a Teflon coating over the core. Additionally, some of the lowest losses were obtained on fibers drawn over two decades ago. The losses for As_2S_3 fibers are shown in Fig. 5.10. The 1984 data by Kanamori et al.[5] in Fig. 5.10 for unclad fiber indicates a minimum loss of 78 dB/km at 2.40 μm. In some further work on the same fiber in 1985, they reported a slightly lower minimum loss of 35 dB/km at 2.44 μm.[18] These values can be compared with the minimum projected losses given in Table 5.2. From the data in Table 5.2 it is seen that the projected loss based on an extrapolation of the WAT is 23 dB/km at 4.6 μm, or 0.05 dB/km if the Rayleigh scattering loss curve is extrapolated instead of the WAT. The measured loss of 35 dB/km is close to the projected loss of 23 dB/km, but it occurs at shorter wavelengths. It would seem, therefore, that for even higher purity glass with less structural defects, a lower loss could be obtained in the 4.6 μm region. But this theoretical loss wavelength is seen from Fig. 5.10 to be very near the broad

S—H impurity absorption that is present in every As_2S_3 glass. The suggestion by Kanamori et al.[5] in 1984 that 10 dB/km is the practical lowest limit is probably very reasonable. More recently, Sanghera et al. have demonstrated a minimum loss of 33 dB/km in core/clad As-S fibers.[25]

The data in Fig. 5.10 also shows the losses for Teflon-coated, core-only fiber and a fiber with a glass cladding made by Nishii et al.[4] Their glass-clad fiber was made using purified sulfur. This resulted in greatly improved transmission compared to core/clad fiber drawn from material made from unpurified sulfur. In general, the losses were reduced threefold or more for the higher purity fiber. It is interesting to note that their Teflon-coated fiber has a fairly low loss. The extrinsic absorption from OH at 2.91 μm and S—H at 4.0 μm is also quite evident in Fig. 5.10. In some more recent work by Devyatykh et al.[9] they were able to reduce the peak OH absorption to less than 50 dB/km. For Devyatykh et al., fiber losses were similar to those shown in Fig. 5.10, varying from 23 to 45 dB/km within 2.2 to 2.7 μm, and from 50 to 70 dB/km within 3.2 to 3.6 μm. However, their S—H peak was more than 1000 dB/km at 4.0 μm.

Selenide glass fiber exhibits higher loss than sulfide glass fiber. The losses for two different As-Se-Ge fibers are given in Fig. 5.11 (note that the loss is plotted in dB/m in Fig. 5.11 and not dB/km as in Fig. 5.10). The lowest-loss data (dashed curve) from Nishii et al.[4] was taken on a core/clad fiber for which the Se was purified in a manner similar to that used for purifying sulfur in the As_2S_3 glass. The data (solid curve) by Kanamori et al.[18] is for unclad $As_{38}Se_{57}Ge_5$ fiber. The lowest loss is for Nishii's clad fiber with a minimum loss of 0.2 dB/m at 6 μm.

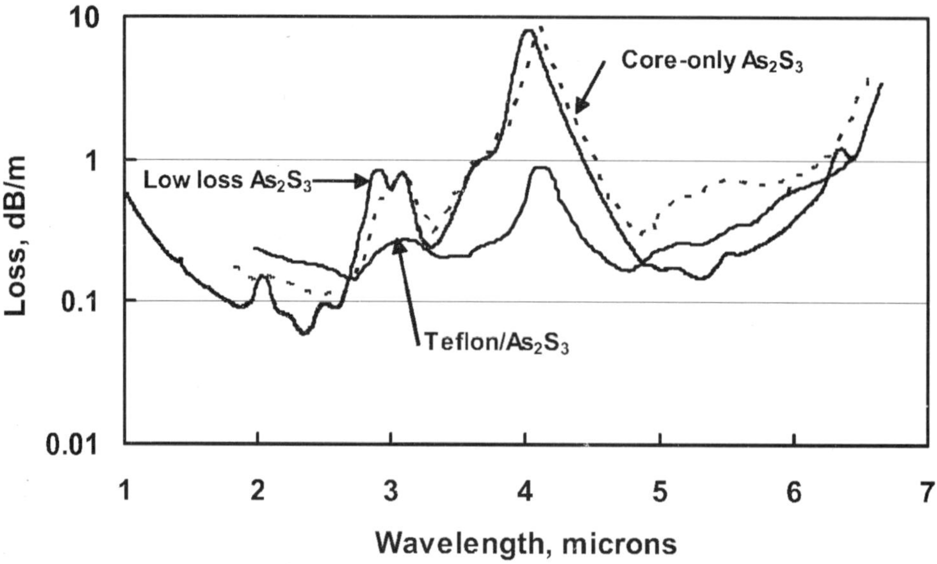

Figure 5.10 Loss for As_2S_3 fibers from several sources. The peak absorption near 4 μm is due to the strong S—H bonded impurities.[4,8]

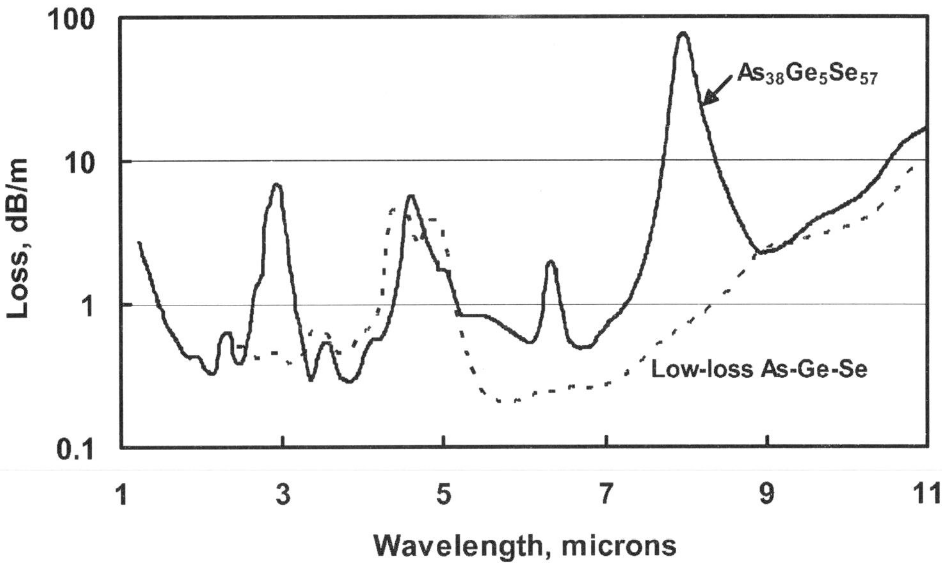

Figure 5.11 Se-glass fiber loss for unclad (solid) and clad (dashed) fibers.[4,8]

As in the sulfide glasses, the data in Fig. 5.11 for the selenide glases shows a rich extrinsic loss spectrum, which dominates the losses for both fibers. Many of the impurity absorption peaks are identified in Table 5.3. Particularly strong are the usual losses due to OH and Se—H, but there is also a strong absorption from H_2O at 6.3 μm and a very strong oxide peak at 7.9 μm.

The telluride glass fibers have been studied primarily for their potential application in the delivery of CO_2 laser power at 10.6 μm. As with the selenide fibers, telluride fibers have high loss compared to the sulfide fibers. Representative low-loss telluride fiber data[24] is shown in Fig. 5.12. The fiber is core/clad $Ge_{25}Se_{23}Te_{52}$. The lowest loss obtained for these fibers at 10.6 μm was 1.7 dB/m. The lowest losses varied from 0.2 to 0.7 dB/m at about 8 μm. An interesting aspect of this data is that there is not an obvious extrinsic absorption spectrum. The 4.6-μm peak due to Se—H is seen in some of these glasses but, in general, the WAT is sufficiently strong in these glasses to mask details of individual impurity absorption. Nishii et al. have also reported a very strong oxygen related peak at about 8 μm. They attribute this absorption to a Ge—O bond.

5.4.2 Production-grade fiber

There are few commercial sources of chalcogenide glass fiber. One source, Amorphous Materials, Inc. (AMI), has manufactured most of this fiber for the past 10 years. There are two fiber materials from AMI. One is an As_2S_3 fiber, type C2, for short wavelength transmission and the other an As-Se-Te fiber, type C1, for long wavelength transmission. The loss data for both fiber types is shown in Fig. 5.13. The data in Fig. 5.13 is for clad fiber protected with a polymer buffer coating.

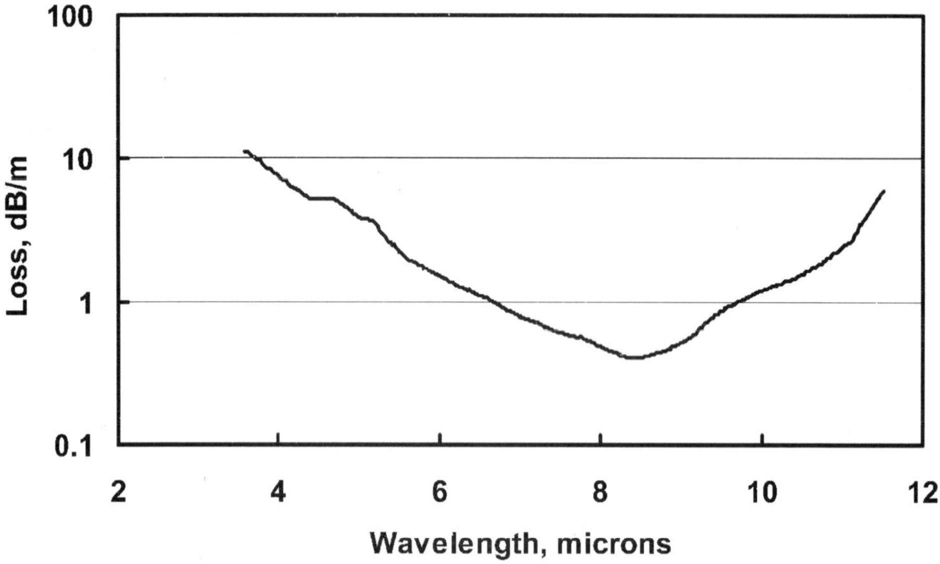

Figure 5.12 Lowest-loss Ge-Se-Te fiber at 10.6 μm; GeSeTe core and GeAsSeTe clad.[24] [Reprinted with permission.]

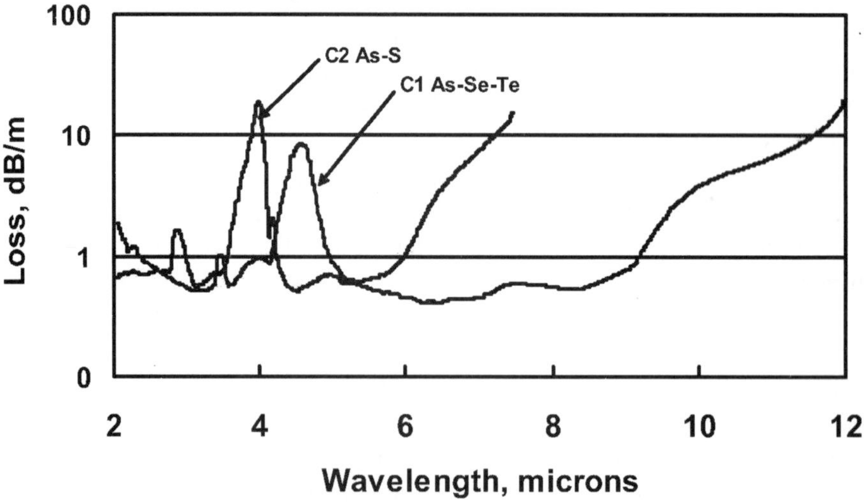

Figure 5.13 Losses for commercial fibers manufactured by Amorphous Materials, Inc. Fibers have a soft chalcogenide glass cladding that has been extruded over the core fiber during drawing.

The losses for both fibers are above the best losses for similar fiber discussed in the last section. Nevertheless, except for the intense S—H and Se—H bands at 4.0 and 4.6 μm, respectively, the losses remain below 1 dB/m for the As_2S_3 fiber from 1.5 to 6.5 μm and for the As-Se-Te fiber from 2 to 10 μm. The lowest loss was measured for As_2S_3 fibers. These losses ranged from 0.1 to 0.2 dB/m at 5.25 μm. The losses for the C1 telluride fiber are higher at 10.6 μm, 4 to 5 dB/m, than similar fibers made by Nishii et al. This rather high loss in the 10 μm region makes CO_2 laser power delivery problematic. In fact, these fibers have a CO_2 laser damage threshold of only several watts. Both fiber types are stable against crystallization, resistant to attack by moisture and weak bases and acids, and unaffected by exposure to UV radiation. Generally they need to be protected by an outer cable for most applications. Sometimes they are placed in protective stainless-steel tubing or conventional strain-relieved fiber jackets.

There are a variety of applications for these commercial fibers. Some of these are

- Temperature sensing
- Remote spectroscopy
- IR image guides
- Laser power delivery at 5 μm using As_2S_3 (C2) fibers.

These applications will be discussed in some detail in the later chapters on applications of IR fibers.

5.5 Mechanical properties of fiber

The mechanical properties of fibers were discussed in general in Sec. 2.5. An important mechanical property is the strength of the chalcogenide fiber, as this is related to the practical issues of reliability and bending. It is immediately evident to anyone handling these fibers that their strength, in particular their flexure and shear strength, is much less than that for silica. Therefore, it is important for many applications to know the bending strength to determine how much the fiber can be bent. Bending strengths are most easily determined by means of a two-point bending measurement, using the apparatus shown in Fig. 2.13. Bending data taken on a group of fibers is then plotted on a Weibull plot like that shown in Fig. 2.14.

A Weibull strength distribution for an As_2S_3 fiber is given in Fig. 5.14. This recent data by Devyatykh et al.[9] was taken on a 400-μm diameter, clad fiber with a fluoropolymer buffer coating. About 30 samples were measured at room temperature. Using a strength value for a particular probability of failure taken from Fig. 5.14, it is possible to obtain a minimum-bending diameter using Eq. (2.29) and Young's modulus, E. Young's modulus averages about 18 GPa for either the As_2S_3 fiber (C2) or for the telluride fiber (C1). When using Eq. (2.29), the value used for the theoretical strength is often taken as $E/5$ rather than E. This provides an additional safety factor. Be sure to remember that the values for minimum bending calculated using Eq. (2.29) are only approximate, as all fiber failure is statistical in nature; in addition, it is very much dependent on environmental and fiber drawing conditions. Nishii et al.[4] showed that a heat treatment at 80°C in 85% relative humidity for 24 hours weakened As_2S_3 fibers. The strength of the fiber was particularly degraded for Teflon-coated core-only fiber. This is due to the penetration of moisture through the Teflon. The average strength decreased after heat treatment

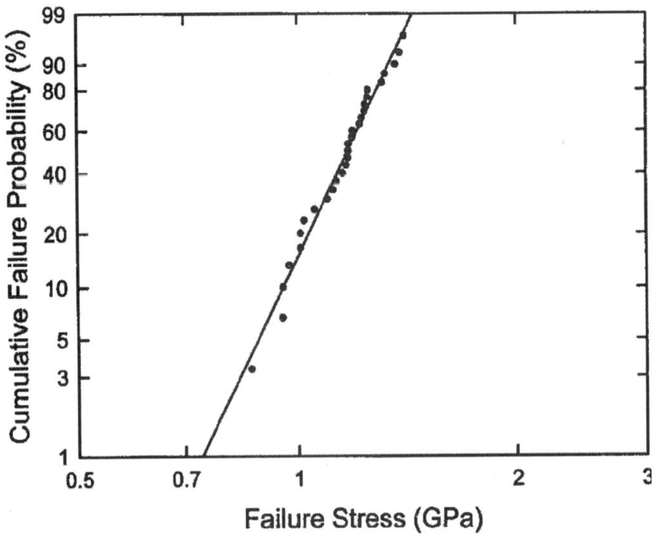

Figure 5.14 Weibull plot for As_2S_3 glass fibers. [Reprinted from Ref. 9, Copyright 1999, with permission from Elsevier.]

by as much as 20%. The situation improved somewhat for Teflon-coated glass-clad fiber.

Recently, Quinn et al.[26] measured the strength of As-S-Se and the weaker Ge-As-Se-Te glass fibers. They obtained a mean strength of 545 MPa and a Weibull modulus of 8.5 for the sulfide glass fibers, and a mean strength of 427 MPa and a Weibull modulus of 11 for the telluride glass fibers. Their fractographic analysis of the fibers indicated that the sources of failure in these fibers were inclusions, microbubbles, and microcracks. Further fracture mirror measurements revealed that the lower strength of the telluride fibers was due to more severe flaw populations.

5.6 Other chalcogenide glasses

Over the many years of chalcogenide glass development, there have been a variety of other glass compositions studied. One of the most important of these is a combination of two or more chalcogen elements, such as S, Se, and Te mixed with one or more halides. These glasses form the chalcohalide glass family. They are becoming increasingly popular largely owing to their long wavelength transmission and the ease of fabrication. In effect this technology is loosely based on combining the halogen from the heavy-metal halide glasses and the chalcogen from the chalcogenide glasses to form new glasses that have some properties of both classes of glass. That is, adding a heavy halogen like Br^- or I^- shifts the IR edge to much longer wavelengths, as would be expected based on the long wavelength transmission of ionic alkali halide materials such as KBr. Adding the chalcogen elements tends to make the glass less sensitive to moisture compared to an alkali halide material. A motivation for forming the chalcohalide glasses is to get long wavelength performance, but in a glass that is considerably more durable than a halide crystal. Some of the best examples of these glasses are the so-called TeX glasses such as the Te-Se-I and Te-Se-Br systems. The "X" represents the halide component of the glass.

Another glass of interest is the sulfide glasses not containing arsenic. The most common of these are the germanium sulfide glasses[27–29] and they have been studied for many years. The problem has been to make high-quality fibers from these glasses.

5.6.1 Chalcohalide glasses

While many different chalcohalide glass compositions have been studied, the most popular today are the quaternary systems based on tellurium. These are the TeX glass systems, and one of the most popular is Te-Se-As-I. For this TeX glass, the halide $X = I^-$. In thin-window-type samples, TeX glasses transmit from 1 to 20 μm. Thus, their transparency range extends further than the standard telluride glasses fabricated without any halogen. Most of these glasses have rather low values of T_g—usually about 150°C or less.

The preparation of the TeX glasses is very similar to that for the chalcogenides. The starting materials are first purified and then a rod-in-tube preform is made by a rotational casting method identical to that shown in Fig. 5.9. Lucas' group at the University of Rennes has done much of the recent work on these glasses and on fiber drawing.[30] For the core glass composition they have chosen $Te_2Se_{3.9}As_{3.1}I$ and for the clad glass $Te_2Se_4As_3I$. The refractive indices for the two glasses are 2.8271 and 2.8205 at 10.6 μm for the core and clad, respectively. This gives an NA of about 0.2. Fiber drawing from the preform was done at about 200°C with a drawing speed of 0.5 to 3.5 m/min. An online UV acrylate coating was applied to the fiber for protection against moisture and, of course, to improve its strength.

The losses for a core/clad TeX fiber made by Lucas' group at the University of Rennes are presented in Fig. 5.15(a).[30] These losses are very similar to core-only TeX fiber (not shown) fabricated by the same group. From the data in Fig. 5.15(a), it can be seen that the minimum attenuation occurs between 7 and 9 μm. The minimum loss for the core/clad fiber is about 1 dB/m at 9 μm. The core-only fiber has a slightly lower loss of 0.5 dB/m at 9 μm, but the acrylate coating on this fiber exhibits absorption, especially in the 8- to 10-μm region. The core/clad TeX fiber shows some strong extrinsic absorption bands. Referring to Fig. 5.15(a), there is strong absorption due to OH^- (3 μm), Se−H (4.6 μm), and H_2O (6.3 μm). The data also shows that the loss at 10.6 μm is rather large, but it is significantly better at the 9.3 μm. This is one of the strong lasing wavelengths of the CO_2 laser. Using a 9.3 μm CO_2 laser, they were able to transmit 2.6 W of laser power though a 600-μm diameter, 1-m-long fiber.

Another TeX glass that has been well studied is the ternary system Te-Se-I. Inagawa et al.[31] investigated these ternary glasses in the range $Te_{10\text{-}40}Se_{20\text{-}70}I_{20\text{-}45}$. As the Te concentration increased from 10 at.% to 40 at.%., the T_g decreased from

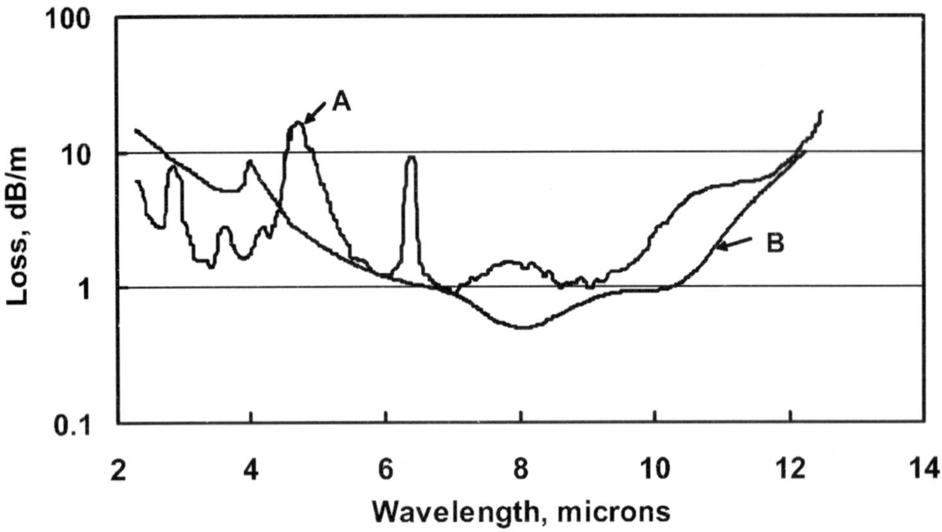

Figure 5.15 Chalcohalide or TeX glass fibers: (a) is a core/clad Te-Se-As-I fiber and (b) is an unclad Te-Se-I fiber. [Reprinted from Ref. 30, Copyright 1995, with permission from Elsevier.]

53 to 43°C and the refractive index of the glass increased from 2.75 to 2.92 at 10.6 μm. This is a very low T_g that can lead to poor CO_2 laser power delivery as well as to some long-term instability in the fibers. The losses for a $Te_{30}Se_{25}I_{45}$ unclad fiber are shown in Fig. 5.15(b). This fiber composition has the lowest loss of 0.9 dB/m at 10.6 μm of any chalcohalide fiber. The lowest loss of 0.5 dB/m at 8.8 μm is comparable to the loss for the unclad Te-Se-As-I fiber made by Blanchetiere et al.[30] Inagawa et al. were able to transmit only about 1 W of CO_2 laser power through their fiber. This is commensurate with the lower T_g of their fiber compared to the 2.6 W of CO_2 laser power and a $T_g = 150°C$ for the Te-Se-As-I fiber.

5.6.2 Ge-S based glasses

Germanium sulfide glasses are based on the glass-former GeS_2, and the compositions of interest here contain no arsenic. One of the earliest ternary compositions studied and drawn into fiber was the Ge-P-S glass by Shibata et al.[27] In general, most Ge-S glasses, including the systems GeS_2-BaS-ZnS and GeS_2-Ga_2S_3-BaS, have considerably higher values of T_g compared to the TeX glasses. Specifically, the glass transition temperatures are higher than 400°C. Most of the glasses are stable, but they are not always easy to fiberize. One glass that has been fiberized by Shibata et al. is $Ge_3S_{7.5}P$. The lowest loss for an unclad fiber of this glass was 0.4 dB/m at 2.5 μm. Again, these fibers exhibit strong extrinsic absorption peaks due to water and S—H impurities. One advantage to these glasses is that the solubility of rare-earth ions is greater in these glass systems because they do not containing arsenic.

References

1. N. S. Kapany and R. J. Simms, "Recent developments of infrared fiber optics," *Infrared Phys.*, Vol. 5, pp. 69–75 (1965).
2. N. S. Kapany, *Fiber Optics, Principles and Applications*, Academic Press, New York (1967).
3. J. Nishii, T. Yamashita, T. Tamagishi, C. Tanaka, and H. Sone, "As_2S_3 fibre for infrared image bundle," *Int. J. Optoelectronics*, Vol. 7, pp. 209–216 (1992).
4. J. Nishii, S. Morimoto, I. Inagawa, R. Iizuka, T. Yamashita, and T. Yamashita, "Recent advances and trends in chalcogenide glass fiber technology: a review," *J. Non-Cryst. Solids*, Vol. 140, pp. 199–208 (1992).
5. Y. Kanamori, Y. Terunuma, and T. Miyashita, "Preparation of chalcogenide optical fiber," *Rev. Electrical Comm. Lab*, Vol. 32, pp. 469–477 (1984).
6. J. Nishii, S. Morimoto, R. Yokota, and T. Yamagishi, "Transmission loss of Ge-Se-Te and Ge-Se-Te-Tl glass fibers," *J. Non-Cryst. Solids*, Vol. 95-96, pp. 641–646 (1987).
7. T. Miyashita and Y. Terunuma, "Optical transmission loss of As-S glass fiber in the 1.0 to 5.5 micron wavelength region," *Jap. J. Appl. Phys.*, Vol. 21, pp. L75–L76 (1982).
8. M. Saito and M. Takizawa, "Teflon-clad As-S glass infrared fiber with low-absorption loss," *J. Appl. Phys.*, Vol. 59, pp. 1450–1452 (1986).
9. G. G. Devyatykh, M. F. Churbanov, I. V. Scripachev, G. E. Snopatin, E. M. Dianov, and V. G. Plotnichenko, "Recent developments in As-S glass fibers," *J. Non-Cryst. Solids*, Vol. 256/257, pp. 318–322 (1999).
10. J. Sanghera and I. Aggarwal, "Active and passive chalcogenide glass optical fibers for IR applications: a review," *J. Non-Cryst. Solids*, Vol. 256 & 257, pp. 6–16 (1999).
11. J. Sanghera, V. Nguyen, P. Pureza, R. Miklos, F. Kung, and I. Aggarwal, "Fabrication of long lengths of low-loss IR transmitting $As_{40}S_{(60-x)}Se_x$ glass fibers," *J. Lightwave Tech.*, Vol. 14, pp. 743–748 (1996).
12. J. Sanghera, L. B. Shaw, L. Busse, V. Nguyen, P. Pureza, B. Cole, B. B. Harbison, and I. Aggarwal, "Development and infrared applications of chalcogenide glass optical fibers," *Fiber and Integrated Optics*, Vol. 19, pp. 251–274 (2000).
13. T. Katsuyama and H. Matsumura, "Low-loss Te-based chalcogenide glass optical fibers," *Appl. Phys. Lett.*, Vol. 49, pp. 22–23 (1986).
14. T. Katsuyama and H. Matsumura, *Infrared Optical Fibers*, Adam Hilger, Bristol, UK (1989).
15. M. Popescu, *Non-Crystalline Chalcogenicides*, Kluwer Academic Publishers, Boston, MA (2000).
16. V. F. Kokorina, *Glasses for Infrared Optics*, CRC Press, Boca Raton, FL (1996).
17. J. Sanghera, L. B. Shaw, and I. Aggarwal, "Applications of chalcogenide glass optical fibers," *C. R. Chimie*, Vol. 5, pp. 873–883 (2002).

18. Y. Kanamori, Y. Terunuma, S. Takahashi, and T. Miyashita, "Transmission loss characteristics of $As_{40}S_{60}$ and $As_{38}Ge_5Se_{57}$ glass unclad fibers," *J. Non-Cryst. Solids*, Vol. 69, pp. 231–242 (1985).
19. E. M. Dianov, M. Petrov, V. G. Plotnichenko, and V. K. Sysoev, "Estimate of the minimum optical losses in chalcogenide glasses," *Sov. J. Quantum Electron.*, Vol. 12, pp. 498–499 (1982).
20. M. F. Churbanov, "High-purity chalcogenide glasses as materials for fiber optics," *J. Non-Cryst. Solids*, Vol. 184, pp. 25–29 (1995).
21. K. Itoh, K. Miura, I. Masuda, M. Iwakura, and T. Yamashita, "Low-loss fluorozirco-aluminate glass fiber," *J. Non-Cryst. Solids*, Vol. 167, pp. 112–116 (1994).
22. A. B. Seddon, "Chalcogenide glasses: a review of their preparation, properties, and applications," *J. Non-Cryst. Solids*, Vol. 184, pp. 44–50 (1995).
23. D. Furniss and A. B. Seddon, "Towards monomode proportioned fibreoptic preforms by extrusion," *J. Non-Cryst. Solids*, Vol. 256, pp. 232–236 (1999).
24. J. Nishii, T. Yamashita, and T. Yamagishi, "Chalcogenide glass fiber with a core/cladding structure," *Appl. Opt.*, Vol. 28, pp. 5122–5127 (1989).
25. J. Sanghera, I. Aggarwal, L. B. Shaw, V. Nguyen, P. Pureza, L. Busse, P. Thielen, F. Kung, and S. Bayya, "Fabrication, properties, and applications of chalcogenide glass optical fibers," *J. Non-Cryst. Solids* (2004) in press.
26. J. B. Quinn, V. Q. Nguyen, J. S. Sanghera, I. K. Lloyd, P. C. Pureza, R. E. Miklos, and I. D. Aggarwal, "Strength and fractographic analysis of chalcogenide As-S-Se and Ge-As-Se-Te glass fibers," *J. Non-Cryst. Solids*, Vol. 325, pp. 150–157 (2003).
27. S. Shibata, Y. Terunuma, and T. Manabe, "Ge-P-S chalcogenide glass fibers," *Jap. J. Appl. Phys.*, Vol. 19, pp. L603–L605 (1980).
28. Y. Meresse, G. Fonteneau, and J. Lucas, "New germaium sulfide-based glasses," *J. Non-Cryst. Solids*, Vol. 184, pp. 254–256 (1995).
29. B. B. Harbison, C. I. Merzbacher, and I. Aggarwal, "Preparation and properties of BaS-Ga_2S_3-GeS_2 glasses," *J. Non-Cryst. Solids*, Vol. 213/214, pp. 16–21 (1997).
30. C. Blanchetiere, K. LeFoulgoc, H. L. Ma, X. H. Zhang, and J. Lucas, "Tellurium halide glass fibers: preparation and applications," *J. Non-Cryst. Solids*, Vol. 184, pp. 200–203 (1995).
31. I. Inagawa, T. Yamagishi, and T. Yamashita, "Transmission-loss spectra of chalcohalide Se-Te-I glass fibers and its delivery of CO_2 laser power," *Jap. J. Appl. Phys.*, Vol. 30, pp. 2846–2849 (1991).

Chapter 6

Crystalline Fiber Optics

6.0 Introduction

Crystalline IR fibers are an attractive alternative to glass IR fibers, because most nonoxide crystalline materials can transmit longer wavelength radiation than IR glasses; in the case of sapphire, they exhibit some superior physical properties as well.[1,2] One disadvantage over glass fibers is that crystalline fibers are somewhat difficult to fabricate. This is due in part to an important distinction between crystals and glasses: glasses are mostly covalently bonded, amorphous materials lacking long-range order, whereas crystals are ionically bonded materials possessing long-range order. Furthermore, glasses have a large glassy region for which the glass is in a softened or supercooled state. This temperature region is between the glass transition temperature, T_g, and T_x, the crystallization temperature. It is within this temperature region that fiber drawing occurs. Crystals, however, possess no softening region or T_g under normal conditions. Therefore, crystals cannot be drawn like fibers because they do not have a temperature that would give an appropriate viscosity for fiber drawing. In short, crystalline fibers must be fabricated either using modified crystal-growth techniques in which a fiber is pulled from the melt, or by heating the crystal to temperatures below the melting point and then applying significant pressure to force the material through a die.

There are two types of crystalline fiber: polycrystalline (PC) and single-crystal (SC) fiber.[3,4] Historically, the first crystalline fiber made was the PC fiber, KRS-5 or TlBrI. This fiber was fabricated by a hot-extrusion technique at HRL in 1976.[3] KRS-5 was chosen because it is very ductile and because it can transmit beyond the 20-μm range required for the intended military surveillance satellite application. In fact crystalline fibers such as KRS-5 and other halide crystals were initially thought to hold great potential as the next generation of ultralow-loss fibers because their intrinsic loss was predicted to be as low as 10^{-3} dB/m.[3] Unfortunately, this loss was not only never achieved, but not even approached experimentally. Today, the best PC fibers, made from silver halide crystals, have losses in the 0.2-dB/m range or well above the intrinsic limit. Nevertheless, the Ag-halide PC fibers continue to be popular today for short-length applications in sensor systems and for limited use in low-power laser delivery.

There has been comparatively less work on SC fiber optics. One reason for this is that they are much harder to fabricate than PC fibers. Only a few crystalline host materials have been studied, with the most important being the refractory oxides. Of these, sapphire fiber (Al_2O_3) is the most-studied SC fiber, and it has the lowest loss.[1,2,5,6]

Crystalline fibers have some unusual properties not shared by their glass counterparts. One of these is the difficulty in properly cladding the fibers. Simply stated, it is not easy to grade the index of a crystal as it is for a glass, or to apply the cladding. That is, most crystals want to retain their stoichiometry, and therefore it is difficult to change the composition sufficiently to alter the refractive index. In addition, even when a proper crystalline cladding material exists, it is challenging to apply the clad layer to the fiber either in the form of a core/clad preform or in a post-cladding operation on the core-only fiber. Somewhat surprisingly, core-only fiber can work quite well in many applications. This is in part due to the high refractive index of many crystalline fiber materials. A high refractive index results in better confinement of light energy to the core, so that with proper handling, core-only fibers can transmit well with little leakage of energy through the sides of the fibers. Another aspect of crystalline fibers is that they are more fragile than glass fibers. The Ag-halide fibers, for example, are very ductile and easily plastically deformed if improperly handled. SC sapphire fibers have a modulus six to seven times that of silica, which makes them more brittle and less flexible than glass fibers. Because of these limitations, crystalline fibers have been relegated to MM fibers usually with core diameters in excess of 300 to 500 μm.

This chapter is divided into two sections; the first is on PC fibers and the second is on SC fibers. More emphasis is placed on PC fiber optics, as these fibers have been studied longer and more extensively than the SC fibers. There is little similarity between the two categories of fibers other than both are made from crystalline materials. The Ag-halide PC fibers are very soft and ductile, low temperature, and with transmission beyond 18 μm; the sapphire SC fibers are extremely hard, high temperature, and with transmission limited to less than 3.5 μm.

6.1 Polycrystalline fiber optics

6.1.1 Composition and key physical properties

There are many metal-halide crystals that have excellent IR transmission, but only a few have been fabricated into PC fiber optics. The primary reason for this is that these fibers are fabricated by a hot extrusion technique. As a result, only the silver, thallium, and some alkali halides have the requisite physical properties such as ductility, low melting point, and independent slip systems to be successfully extruded into fiber. In the hot extrusion process, a single-crystal preform or billet is placed in a heated chamber and the fiber extruded to net shape through a diamond or tungsten carbide die at a temperature equal to about one-half the melting point. If the crystalline material cannot be easily plastically deformed, then this method fails to produce good quality fiber.

There are three classes of halide crystals that have been extruded into PC fiber. Historically, the first halides to be extruded were the thallium halides, notably KRS-5 and TlBr. KRS-5 is a mixed thallium halide salt with the composition 54.3 mol% TlBr and 45.7 mol% TlI. This eutectic composition has a cubic structure with a melting point of 414°C. The most successful halide fibers to date have been the silver halides. In some sense they are quite similar to the thallium halides in that the simple salts such as AgCl and AgBr have been extruded along with the preferred mixture of approximately 50% AgCl and 50% AgBr. The alloys of silver salts, as well as those of thallium, are preferred because they have better mechanical properties than the monohalide crystals. Finally, there have also been reasonably successful attempts to extrude the simple alkali halide salts such as KBr, KCl, and NaCl.

The purification, growth, and properties of all of these halide crystals are very well established. In fact, the infrared properties of many simple salt crystals were studied back in the early 1900s. Some relevant physical data for the PC fiber materials is given in Table 6.1. The only relatively "recent" crystalline material in Table 6.1 is KRS-5. This material was discovered, i.e., synthesized, by Smakula during WWII for use as a long-wavelength (>40 μm) transmissive optical element.

The thermal properties of all these salt crystals, as given in Table 6.1, indicate that the materials generally have low melting points and high coefficients of thermal expansion. The Ag and Tl salts are quite ductile even at relatively low temperatures. For example, it is quite easy to physically deform a small-diameter Ag-halide rod by bending it between one's fingers. It is this property of plastic or anharmonic deformation that makes these materials suitable for extrusion into fiber. Normally, hot extrusion is carried out at a temperature equal to about half the melting point, but it can be done even at room temperature if sufficient pressure is used. The high thermal expansion of these ductile crystals generally does not present a problem during extrusion or in applications of the fibers.

Table 6.1 Properties of crystalline compounds used in the extrusion of PC fiber optics.

Property	AgCl	AgBr	$AgBr_{0.5}Cl_{0.5}$	KRS-5
Bulk crystal transmission range, μm	0.4–28	0.45–35	0.4–30	0.5–40
Refractive index, (wavelength, μm)	1.98 (10.6)	2.25 (10.6)	—	2.371 (10.6)
Melting point, °C	457	419	418	414
Density, g/cm^3	5.59	6.44	—	7.37
Thermal conductivity, W/m °C	1.17	1.21	—	0.54
Thermal expansion, °C^{-1}	30×10^{-6}	35×10^{-6}	—	60×10^{-6}
Hardness, kg/mm^2	9.5	7	13	40
Solubility, g/100g H_2O	8.9×10^{-5}	1.2×10^{-5}	7×10^{-5}	0.02

All of these crystalline materials are cubic. Normally, a single-crystal preform is used as the starting material. Once this preform is extruded into fiber, the fiber will have a small-grain, polycrystalline structure. Grain sizes vary from a few microns to hundreds of microns in size. The strength of PC fiber is greater than the rupture strength listed in Table 6.1 for the single crystal material. This is discussed further in Sec. 6.1.5. Additionally, crystalline materials are quite soft, as may be seen from the hardness values listed in the table. This will be appreciated by anyone who has handled these crystals, as they are easily scratched and deformed.

The optical properties of halide crystals are quite attractive from the viewpoint of fabricating a long-wavelength transmissive fiber. In fact, it was this property that motivated the initial research in KRS-5 fiber by HRL in the mid 1970s. They wanted a fiber that could transmit beyond 20 μm for a military sensor application. This requirement alone essentially limits one only to a few select crystalline materials and hot extrusion methods for fiber fabrications. The data in Table 6.1 indicates a very large transmission window for these crystals. This is a particularly appealing feature of ionically bonded materials. The alkali halides are water-clear, with refractive indices close to those of oxide and fluoride glasses. The silver salts are yellow, whereas KRS-5 is red, with refractive indices varying from 2.0 to 2.4. The greater covalent bonding in these hosts compared to the alkali halides and their higher density lead to the higher values of the refractive indices.

In addition to being more difficult to form into fiber, these crystalline materials have some other less-than-ideal properties. The alkali halides are quite soluble, and great care must be taken to prevent moisture from attacking the fibers. KRS-5 is also sensitive to water, but less so than the alkali halides. Fortunately, the silver halides are insoluble in water. Thallium salts are also quite toxic, so it is important that they are not vaporized or broken off in tissue. This property was recognized early in the KRS-5 fiber development, and it surely hindered their acceptance as an IR fiber-delivery system for CO_2 laser surgery. The silver and alkali halides are not toxic. On the other hand, the silver salts are well known to be photosensitive. This deleterious property leads to solarization or darkening of the Ag-halide fibers if they are exposed to visible, or especially UV, radiation. This darkening effect is often visible as a distinct graying of the fiber that results from the formation of colloidal silver in fibers exposed to light for long periods of time. Solarization leads to increased losses not only in the visible region as expected, but also in the IR region. Thallium salts are also light sensitive, but to a lesser extent than the silver halides. The alkali halides are essentially insensitive to normal visible radiation. A simple solution is to keep the fibers in the dark by either protecting them with a dark buffer jacket or packaging them in opaque tubing. Finally, the silver halides have a rather annoying problem in that they can be rather corrosive when they contact many metals. Specifically, they will corrode stainless steel fiber connectors, brass, copper, and other similar metals. They do not attack gold or titanium, so connectors made from these materials or from ceramic materials circumvent corrosion problems.

6.1.2 Optical properties of crystals

Halide crystals feature a broad IR transmission, which makes them almost ideal candidates for IR fiber optics. Furthermore, these crystals have a simple structure, and they have relatively high purity. Since they have been studied and grown for many years, they are readily available commercially. The challenge is to find a means to convert the usual high-purity starting single crystals into PC fiber without significantly altering the optical properties of the starting material.

The three most important halide crystals used to fabricate IR fibers all have a cutoff or IR edge that extends well into the infrared. In Fig. 6.1, the transmission spectra of millimeter-thick window samples of KBr, KRS-5, and AgCl are shown. In general, these crystals transmit to wavelengths well beyond 20 μm and, therefore, much further than the IR glasses. This feature of crystalline materials is important more from a fundamental rather than a practical viewpoint, as there have been few applications for IR fibers operating beyond about 12 μm. Nevertheless, if there were a need for very long wavelength transmission, then these crystalline fibers as well as many hollow waveguides (see Chapter 7) would be a good choice.

The intrinsic losses for halide crystals can be extraordinarily low.[7] As discussed in Sec. 2.3.1, the theoretical scattering loss for a perfect crystal is governed by Brillouin (inelastic) scattering as opposed to Rayleigh (elastic) scattering in glassy hosts.[8,9] Both scattering mechanisms have a $1/\lambda^4$ wavelength dependence, but Brillouin scattering is much less intense. The contribution to the attenuation due to Brillouin scattering can be calculated using Eq. (2.15). The intrinsic, or V-curves, are plotted in Fig. 6.2 for three halide salts.[7] Note from the data that the minimum loss for the thallium and silver halides is in the range of 10^{-3} dB/km, whereas it is even lower for KCl or KBr (not shown). The reason that the minimum in the V-curve is about 10 times less for the alkali halides is that the Brillouin scat-

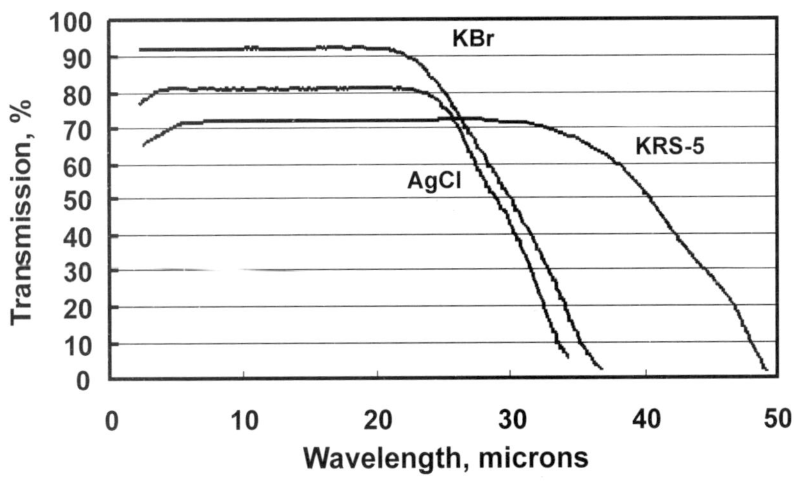

Figure 6.1 Infrared transmission of three of the most common crystalline materials used to make PC fibers.

tering losses are less. This is a result in part from the lower refractive index for KCl compared to the Ag and Tl halides. From Eq. (2.15) it can be seen that scattering losses depend on the eighth power of the index, so low-index materials can have a much smaller loss.

The minimum loss depends on the composition of the halide crystal. For the mixed silver halides, for example, Saar and Katzir[10] showed for the $AgCl_xBr_{1-x}$ mixture that α_{min} varied from 2.3×10^{-3} dB/km to 6.0×10^{-3} dB/km as x varied from 0 (pure AgBr) to 0.5. Over this same compositional range, λ_{min} varied from 10.2 to 9.0 μm. Similar considerations would hold for KRS-5 as well. In practice these differences are relatively insignificant, as the measured losses for crystals or fibers are considerably higher than these intrinsic losses. The difference between bulk crystal intrinsic loss and measured fiber loss is many orders of magnitude, as may be seen from the $AgCl_{50}Br_{50}$ fiber data also plotted in Fig. 6.2.

Extrinsic losses clearly dominate the losses in both the crystalline hosts and their fibers. While this is abundantly evident from the data (see Fig. 6.2), what is not so clear is the source of this high extrinsic absorption. In Fig. 6.3, the losses for the $AgCl_{50}Br_{50}$ fiber extruded by Moser et al.[11] (also shown in Fig. 6.2) clearly show some extrinsic absorption bands. There are obvious impurity bands due to OH^- at 2.9 μm, H_2O at 6.28 μm, and CO^{3-} at 7.15 μm. In addition, there is a strong extrinsic scattering loss indicated by the increasing loss as the wavelength decreases from 10 to 2 μm. This extrinsic scattering loss varies as $1/\lambda^2$ in these fibers. In fact, this unusual $1/\lambda^2$ wavelength dependence is seen in all PC fibers.[12,13] This

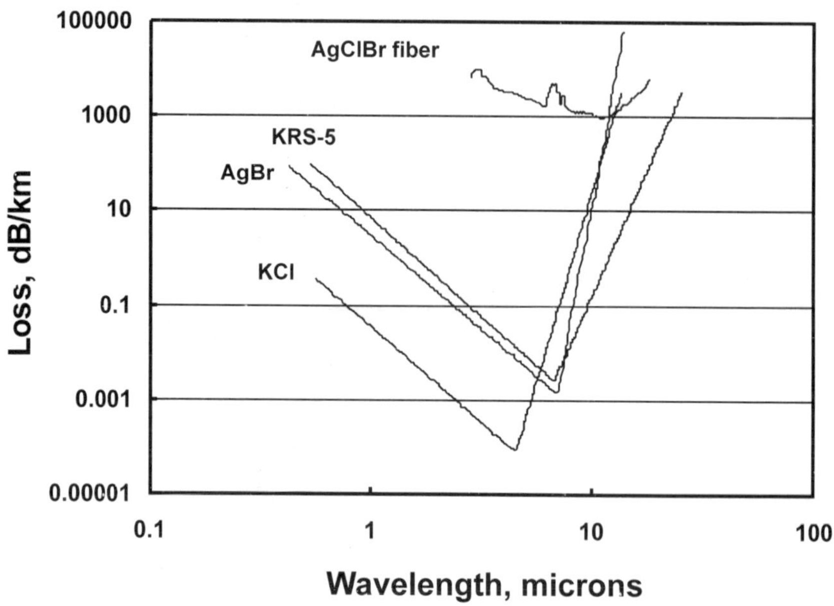

Figure 6.2 Intrinsic loss for same three crystalline materials shown in Fig. 6.1, with the addition of the measured loss for the Ag-halide PC fiber. Note the very low minimum for each V-curve, especially compared to the actual loss for the fiber.[11] [Copyright 2003 IEEE. Reprinted with permission.]

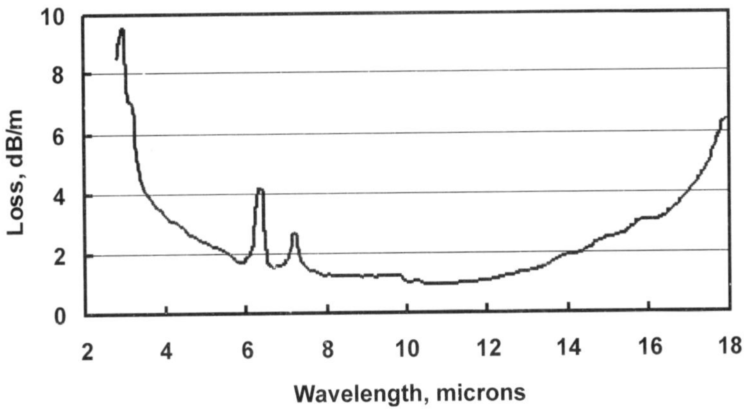

Figure 6.3 Extrinsic losses for AgBrCl PC fiber. This is the same fiber data shown in Fig. 6.2. Several impurity absorption bands due to OH^- are present, along with one at 7.15 μm resulting from carbonate absorption.[11] [Copyright 2003 IEEE. Reprinted with permission.]

unusual scattering effect will be discussed further in Sec. 6.1.4. In general, the extrinsic losses due to impurities result from low levels of both anionic and cationic impurities that reside in the crystal even after purification. The most effective purification methods, applied with great success to the alkali halides, are the RAP techniques developed by Pastor and Pastor[14,15] at HRL many years ago. Using RAP techniques to purify KCl significantly reduced the losses in bulk crystals.[15] Allen and Harrington[16] found that a RAP-treated single-crystal bar of KCl had a measured loss of 8×10^{-5} cm^{-1} or 34.7 dB/km at 10.6 μm. This is one of the lowest losses measured in a bulk crystal at CO_2 laser wavelengths to the author's knowledge. Furthermore, this loss in KCl at 10.6 μm is intrinsic. Belousov et al.[17] report that single crystals of KRS-5 have been measured with losses in the range of 0.8 to 1.4×10^{-5} cm^{-1}, or 3.5 to 6 dB/km, at 5.54 μm.

It is normally quite difficult to attain an impurity level in crystalline materials as low as it is in oxide glasses. This is due to the strong ionic bonding of impurity ions to the host lattice. Such bonding makes it difficult to use standard chemical methods to reduce the impurity level to parts per billion. Therefore, it would seem that low-level impurities would be an important limiting loss mechanism for PC fibers. In practice this is not the case, as there are more serious losses ultimately affecting fiber loss. The major problem seems to be the extrusion process itself. Bunimovich et al.[18,19] show that the loss increases significantly when a reasonably low loss crystal is deformed during extrusion. Their absorption data[18] for an AgBr crystal and the corresponding fiber is shown in Fig. 6.4. The losses in the crystal are quite low but not intrinsic. Their data reveals that in addition to a more than tenfold increase in loss for the fiber over the starting preform, there is a new impurity band appearing at 9.8 μm. They attribute the absorption that results as the crystal is deformed to cation vacancies localized on charge dislocations. That is, the plastic deformation of crystals leads to a significant increase in the concentration of vacancies and dislocations. This in turn contributes to the IR absorption.[19]

The refractive indices for the crystalline hosts are listed in Table 6.1. As expected, the refractive index for the mixed crystals differs from that for the monohalide host crystal. A good example of this may be seen in data for $AgCl_xBr_{1-x}$. In Fig. 6.5, the refractive index at 10.6 μm is shown for the complete range $0 \leq x \leq 1$.[20] For AgBr the refractive index is 2.25, decreasing to 1.98 for AgCl. Since it is possible to grow crystals of any mixture or value of x, this means that a suitable clad and core composed of Ag-halide materials is a logical choice for a

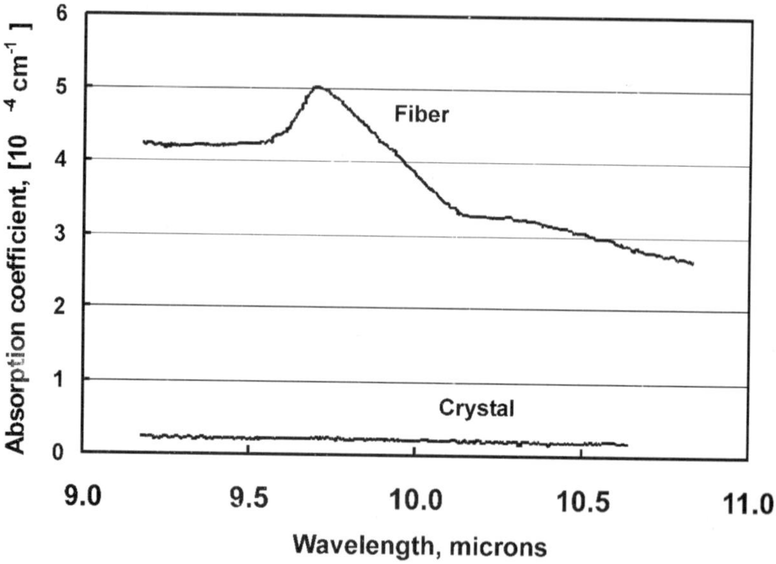

Figure 6.4 Absorption of AgBrCl crystal before and after deformation into fiber.[18] [Reprinted with permission.]

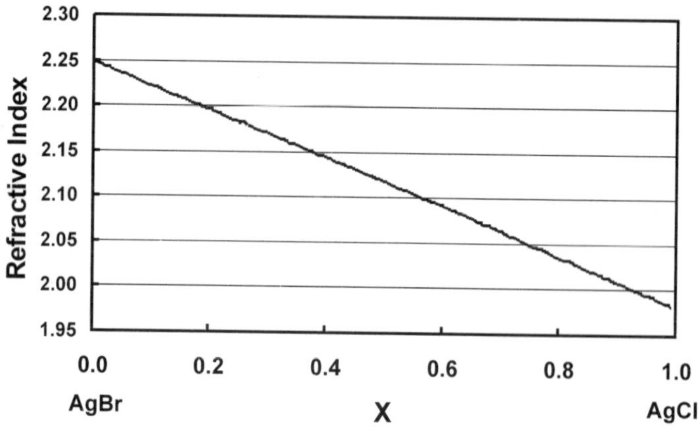

Figure 6.5 Refractive index variation for mixed silver halide crystal, $AgCl_xBr_{(1-x)}$.[20] [Reprinted with permission.]

rod-in-tube preform. In fact this is what is done, as will be discussed in Sec. 6.1.4 on fiber loss.

6.1.3 PC fiber fabrication and structure

The starting preform or billet for extrusion is normally a single crystal. While the preform could be formed by compacting granular crystalline material, it is found that this approach introduces too many voids and other impurities into the starting billet. The best and most common approach is to use single-crystal billets, as they are grown from the highest-purity starting materials, and they are easy to obtain. Purification methods such as RAP chemistry are normally used to purify the salts just prior to crystal growth. For example, in the case of KCl, the molten salt is purged with a carrier gas containing CCl_4.[15] The hydrocarbon undergoes pyrolosis and effectively removes OH-containing impurities. After treatment the crystal is grown by slowly lowering the crucible containing the melt in an inert atmosphere into a cooler temperature zone where the crystal freezes. This is the commonly used method of Bridgman growth. It has been used to produce very large ingots (>30 cm in diameter) of alkali halides. Another method of crystal growth is the Czochralski, or crystal pulling, method. This method generally produces smaller-diameter boules. Czochralski growth is the basis for SC fiber fabrication discussed in Sec. 6.2.

The starting core-only or core/clad billet is placed in a heated die for extrusion into fiber. A typical extrusion press is shown schematically in Fig. 6.6.[21] Many variations of extrusion presses have been used. Most use hydraulic pressure to drive the piston and force the crystalline material through the extrusion die. Another

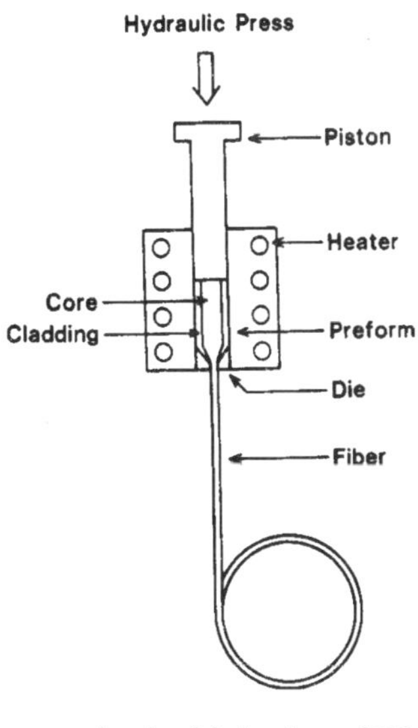

Figure 6.6 Hot extrusion press for the fabrication of PC fibers starting with single-crystal billets or preforms. The extrusion pressure may come from either a hydraulic or a screw-driven press.[21]

approach is to use a screw-driven piston. This is the type of extruder used by the author and also at HRL where the first PC fibers were extruded. Generally though, the type of press used is not as important as the die chamber and the extrusion dies used.

As a specific example of an extrusion press and the conditions used to extrude fiber, the extrusion apparatus used by the author will be described.[22] The extruder had a screw-driven tungsten carbide piston and a tungsten carbide die chamber. The piston and die chamber were approximately 1 cm in diameter, and both the die chamber and piston were honed and lapped to a tolerance of 10 millionths of an inch. This is important because a tight-fitting die and piston assembly ensures that there will be little or no flow-by of material up the piston during extrusion. Flow-by can lead to the piston jamming in the die cylinder after extrusion, and sometimes this results in a cracked die. The fiber dies were diamond, wire-drawing dies with diameters ranging from 250 to 1000 μm. The chamber was heated with rod heaters embedded in the die holder. While it is possible to extrude halides at or near room temperature, it is more typical to use an extrusion temperature from one-half to three-quarters of the melting point of the crystal. For KRS-5, a normal range of extrusion temperature was from 200°C to 300°C. Similar temperatures were used for the silver and alkali halides. In general, the lower the extrusion temperature, the smaller the grain size of the resultant PC fiber. Therefore, it is often desirable to extrude at lower temperatures because a smaller grain size yields stronger fiber. Further discussion of the effect of grain size on fiber strength is included in Sec. 6.1.5. The rate of extrusion is considerably slower than the rate for glass fiber drawing. Typical extrusion rates are 5 to 20 cm/min.

The PC fiber that results from extrusion is quite different in many respects from glass fiber. The most important difference is the polycrystalline structure itself. This structure is shown for KRS-5 fiber in Fig. 6.7.[23] The KRS-5 fiber has a diameter of 500 μm. The broken end of the fiber in Fig. 6.7 and the microphotograph of the end surface clearly show that the fiber is composed of small crystallites with grain sizes ranging from 10 to 30 μm. The grain size that results is a function of the extrusion temperature (higher temperatures produce larger grain sizes), the ex-

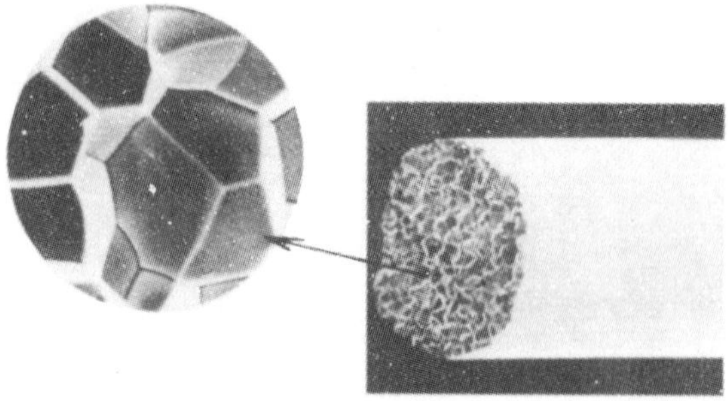

Figure 6.7 Broken end of a KRS-5 fiber showing the polycrystalline grain structure of these extruded fibers. The end face is magnified 1000× in the photo on the right.[23] [Reprinted with permission.]

trusion rate, and the extrusion ratio (ratio of billet to fiber area). The smallest grain size for PC fibers is about 1 μm. The important point to remember is that crystalline materials extrude into polycrystals and not amorphous or glassy fibers. PC fibers do not cleave like glass; rather they fracture along grain boundaries, as indicated in Fig. 6.7. When properly extruded, PC fibers are fully dense, with minimal separation and voids at the grain boundaries.

The microstructure of PC fibers presents some interesting problems. Of particular concern is the possibility of recrystallization and grain growth. Grain growth can lead to weaker fiber as well as increased scattering and absorptive losses. After extrusion, PC fibers have a small grain size, but it is possible for the grain size to increase as the fiber ages under normal conditions. But if the fiber is heated and/or left at an elevated temperature, then recrystallization and concomitant grain growth will occur. Bunimovich et al.[24] studied the effect of thermal treatment on AgClBr fibers with an initial grain size of 1 μm. They found that for temperatures above 170°C there was substantial grain growth and an increased loss. Their data in Fig. 6.8 shows that the grain size can grow from a few microns in the starting fiber to well over 400 μm when the fiber is heated for 1 h at temperatures approaching the melting point of the silver halide. Bunimovich et al. also found that the absorptive losses increased exponentially with temperature, while the scattering losses increased linearly with increasing temperature. They explain the increased losses for temperatures above 170°C as arising from cation vacancies localized at charged dislocations or divalent impurities. Curiously, they found that below 170°C there was little grain growth and the overall absorption decreased. One explanation for the decrease in loss is annealing of strain energy induced in the extrusion process. Induced strain birefringence is responsible for much of the scattering loss in these fibers, so a reduction in the birefringence can lead to lower loss. Similar effects have also been observed in KRS-5 fibers by Kimura et al.[21] They observed little grain growth below about 150°C and a threefold decrease in fiber loss for fiber annealed at 150°C for 30 min.

It is important to stress that a major difference between glass fiber drawing and PC fiber extrusion is that PC fiber is configured by a die rather than being freely

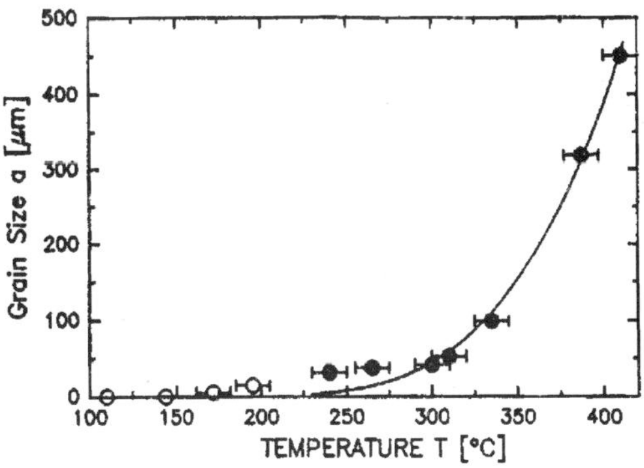

Figure 6.8 Effect of thermal treatment on grain size.[24] [Reprinted with permission.]

formed in a glass draw down. The fact that a die, albeit one with a smooth surface, must be used to shape the fiber inherently means that the surface of the PC fiber will not be as smooth as a glassy fiber surface. In practice, it is often observed that the fiber surface has elongated lines along the fiber axis. Sometimes these imperfections are attributed to die marks even though the die has no visual marks or imperfections that can lead to these fiber defects. In addition, the stress placed on the fibers during extrusion, along with the friction of the die, lead to induced strain and to surface imperfections. As a result, the best PC fibers are those made from the most ductile crystalline materials, and these seem to be the silver halides. The surface of these fibers is relatively smooth compared to KRS-5 and alkali halide fibers.

Most PC fiber fabricated to date is core-only. The extrusion of core/clad fiber is particularly challenging because the forces needed to extrude fiber can easily distort the core/clad interface and lead to high absorption. Nevertheless, a proper core/clad structure has been made from both silver and thallium halide fibers. The clad fiber is extruded from a core/clad preform made using a rod-in-tube technique. Unlike in glass preform fabrication, the cladding tube cannot be formed by rotational casting; rather it is made by the brute force technique of drilling a hole in the lower-index cladding billet and then inserting a higher-index rod into the tube. The composite preform is then extruded into fiber.[11,21,25,26] The choice of materials for a core/clad silver halide preform may be selected based on the variation in refractive index, with composition as given in Fig. 6.5 for the Ag halides. In general, the core material will be bromine rich and the clad will be chlorine rich. A typical cross section of a core/clad PC fiber is shown in Fig. 6.9. From the picture the integrity of the core/clad interface appears reasonable, but some irregularities, such

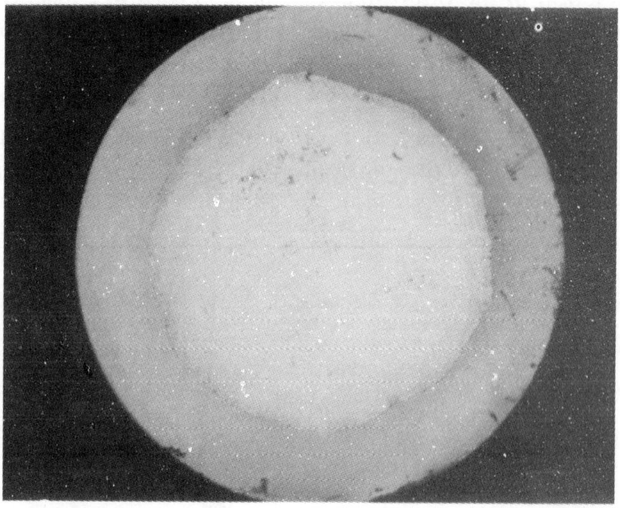

Figure 6.9 Core/clad Ag-halide fiber made by coextrusion of core/clad preform. Note the irregularities in the core/clad interface. Irregularities such as these can lead to clad fibers with losses higher than their core-only counterparts.[11] [Copyright 2003 IEEE. Reprinted with permission.]

as undulations in the core/clad interface, are visible. Other problems are voids at the interface, which result from the fact that the clad layer is only physically and not chemically bonded to the core preform. To overcome the core/clad bonding problem Saito et al.[27] used a post-cladding technique to clad KRS-5 fiber. They applied a coating of KRS-6 (TlBrCl) to the as-extruded fiber using physical vapor deposition. While this method did produce a lower-index cladding, the cladding layer was quite thin for most applications. The best core/clad PC fiber today has been extruded by Artjushenko and his colleagues at the General Physics Institute (GPI) in Moscow.[25,28] They have achieved clad Ag-halide fibers with losses nearly as low as the core-only Ag-halide fiber.

6.1.4 Optical properties of PC fibers

The optical losses in PC fibers are well above the intrinsic loss of the bulk material. In general, the best PC fibers made have losses between 0.3 and 0.5 dB/m around 10 μm. This is typically the lowest-loss region, which is fortuitous since many applications for these fibers involve the transmission of CO_2 laser radiation. The loss for both core-only and core/clad silver halide fibers are shown in Fig. 6.10.

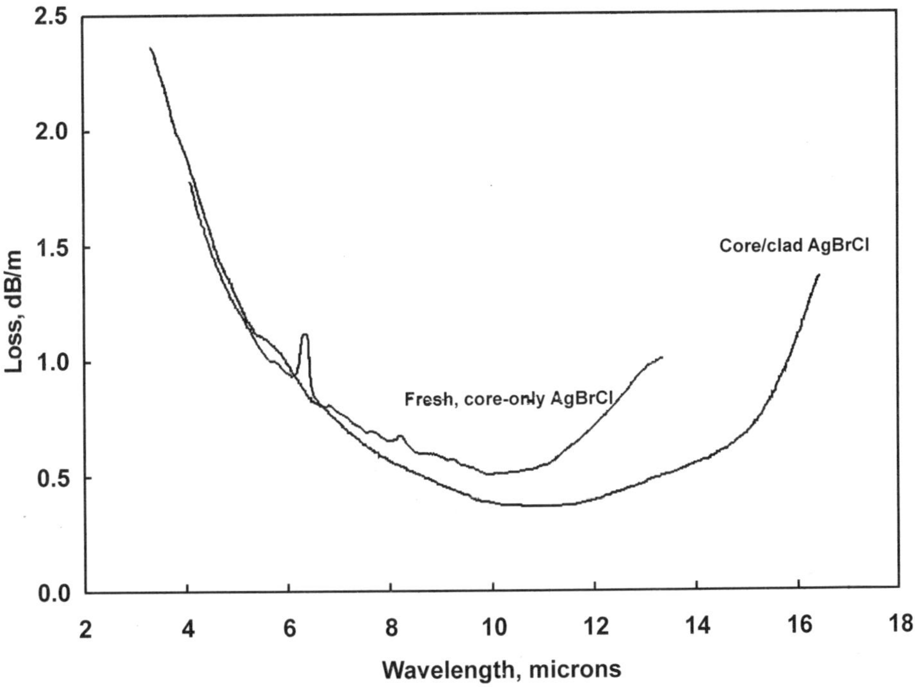

Figure 6.10 Losses for two of the best Ag-halide fibers made to date: (a) AgBrCl fiber extruded by Katzir at Tel Aviv University and (b) core/clad AgBrCl fiber fabricated by Artjushenko and coworkers at the General Physics Institute. The core/clad fiber loss is very low compared to that usually obtained in a coextrusion of a clad preform.[11] [Copyright 2003 IEEE. Reprinted with permission.]

The core-only fiber (a) is the AgBrCl fiber extruded by Katzir and his group at Tel Aviv University,[11] while fiber (b) is a core/clad AgBrCl fiber fabricated by Artjushenko and coworkers at the GPI in Moscow and ART Photonics in Berlin.[29] Both loss curves represent the current technology of silver halide fibers, and, in the case of fiber (b), the quality of commercially available fiber. The core/clad fiber is available with an NA $= 0.15$ or 0.3. The core diameters of the fibers range from 500 to 900 μm. The lengths of the fibers typically do not exceed 3 or 4 m, but is possible to extrude fibers as long as 20 m.

As pointed out in Sec. 6.1.2, there are several impurity absorption bands present in the silver halide fiber loss spectra. These bands are seen in the core-only fiber but are less evident in the core/clad fiber. Presumably this is due in part to the presence of the clad layer, which protects the core from contamination by water and other ions during extrusion. Another reason is that purer starting materials are reportedly used to produce the core/clad fiber.

The losses for the thallium and alkali halide PC fibers are also well documented in the literature. These fiber systems have fallen out of favor compared to the silver halide fibers, and consequently there is much less emphasis on them today. The thallium halides, especially KRS-5, were heavily studied in the 1980s, but interest in this host lessened considerably due to the toxic nature of Tl salts.[21,30–32] In fact, applications involving CO_2 laser delivery systems using KRS-5 fiber vanished quickly once toxicology studies were done that showed the necrosis of ocular tissue imbedded with KRS-5 fiber. The alkali halide fibers developed by Harrington[22] initially showed promise but their KBr/KCl core/clad structure was inherently weak and hygroscopic. In summary, the losses for KRS-5 fiber are remarkably like those for the silver halides given in Fig. 6.10. The lowest loss for the KBr/KCl fibers at 10.6 μm was also in the 0.3- to 0.5-dB/m range.

There is an unusual scattering loss mechanism observed in PC fibers.[12,13] Instead of the normal Rayleigh-type λ^{-4} dependent scattering loss, the measured loss is found to vary as λ^{-2}. Rayleigh scattering results when the size of the scatterers is small compared to the wavelength. This is what is observed in glasses, as indicated schematically in Fig. 2.9. For example, for the best fluoride glass fibers, the measured scattering losses exactly fit a λ^{-4} dependence. From Fig. 6.2 it may be seen that that the PC fiber loss is well above the intrinsic scattering limit, and, therefore, intrinsic Brillouin scattering (λ^{-4} dependence) is negligible. What is observed from Figs. 6.3 and 6.10 is that there is increasing loss as the wavelength decreases from about 11 to 2 μm. In Fig. 6.11, the total loss for a KRS-5 fiber is shown. A fit to this short-wavelength data indicates a $\lambda^{-2.2}$ dependence.[13] Harrington and Sparks[13] have shown that a λ^{-2} wavelength dependence may be associated with Rayleigh–Gans scattering. Rayleigh–Gans scattering occurs when the size of the scattering center is comparable to the wavelength. In the case of PC fibers, it is proposed that the scattering centers are strain fields and surface irregularities. These types of centers are physically large (length l) in the transverse direction but optically thin. This means that the imperfections are much larger than the wavelength, or $kl \gg 1$, and that they have a refractive index close to that of the host material or

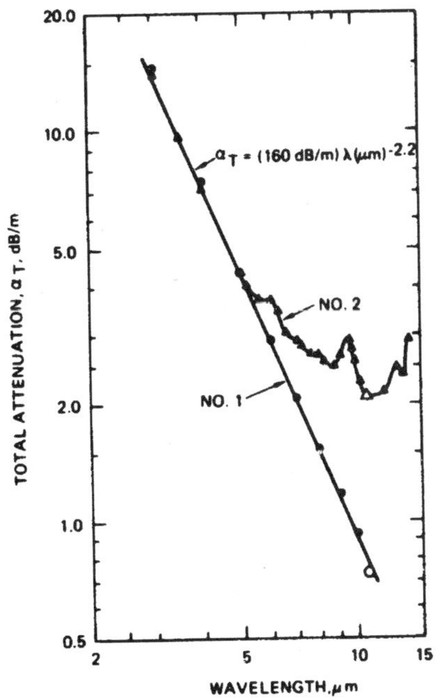

Figure 6.11 Total attenuation in core-only KRS-5 fiber, showing the approximate λ^{-2} dependence of the attenuation that results from Rayleigh–Gans scattering.[13] [Reprinted with permission.]

$kl|n_{scat} - n_{host}| \ll 1$. The strain fields are a result of the extrusion process, and their presence is easily detected as strain birefringence when the fiber is viewed under a polarizing microscope. The surface of the fibers is also rough compared to that of a pristine glass-drawn surface. These imperfections have the requisite geometry for Rayleigh–Gans scattering. Sa'ar and Katzir[12] have made a similar analysis for the silver halide fibers, and they also found a λ^{-2} wavelength dependence.

There is a deleterious aging effect that frequently occurs in PC fibers. Aging is a condition in which the attenuation of the fiber increases in time. This effect has been observed in some of the early KRS-5 fibers[33] as well as in the current silver halide fibers.[28,34,35] What is observed is that the fiber loss will increase over a period of months while the fiber is stored. The actual amount of aging depends on the fiber type and on the storage conditions. As an example, the attenuation at 10.6 μm for KRS-5 fiber was found to increase by about a factor of two over a period of six months. Wysocki et al. showed that the least amount of aging occurred when the KRS-5 fibers were stored in vacuum in the dark.[33] They also found that after a period of about one year that the aging process accelerated and the losses increased dramatically for fibers stored in any condition. These researchers attributed the aging process to the diffusion of water into the fiber. They reasoned that the OH impurities would most likely enter the fiber's surface and travel along grain boundaries. They used a variety of analytic methods to confirm that there was an uptake of water as the fiber was stored.

The aging in silver halide fibers appears similar to that seen in KRS-5 fiber except that the aging process appears to slow considerably after 9 to 12 months. The spectral data of Grigorjeva et al.[35] showing the increase in attenuation in $AgCl_{0.25}Br_{0.75}$ fiber with time is shown in Fig. 6.12. From this data it can be seen that there is an overall increase in loss from 3 to 13 μm from the time the fiber is first extruded (curve 1) up to and beyond 12 months (curve 4). For example, the loss at 4.0 μm was found to increase from 1.7 to 6.0 dB/m after 12 months of storage. The overall increase in attenuation indicates that increased scattering loss is responsible for much of the aging effect. The authors attribute increased scattering to microvoid buildup at grain boundaries, increasing grain size, and the formation of large impurity centers like AgBr. Just as in KRS-5 fiber, absorption by H_2O impurities is observed to be a strong contributor to the aging process. This may be seen in Fig. 6.12 as a disproportionate increase in the water band at 6.25 μm. Additionally, the impurity peaks at 7.3 μm (NO^{3-}) and 8.5 to 8.9 μm (SO^{2-}) also increase in time. Grigorjeva et al. also noted a linear increase in grain size even after 20 months of aging. Finally, Kupper et al.[34] also observed aging in their Ag-halide PC fibers. They saw aging for fibers stored in black plastic tubing made of PEEK or Teflon.

In summary, it seems that the most likely mechanism for the aging process is the increase in scattering resulting from the voids and other impurities decorating grain boundaries. As the fibers age there is a relaxation of the strain energy built up in the fibers during extrusion. The release of strain energy leads to grain growth and the buildup of voids along grain boundaries. Voids are strong scatterers because scattering depends on the index difference between the void and host material, and this is quite large. In some of the early work at HRL on KRS-5 fiber, the author

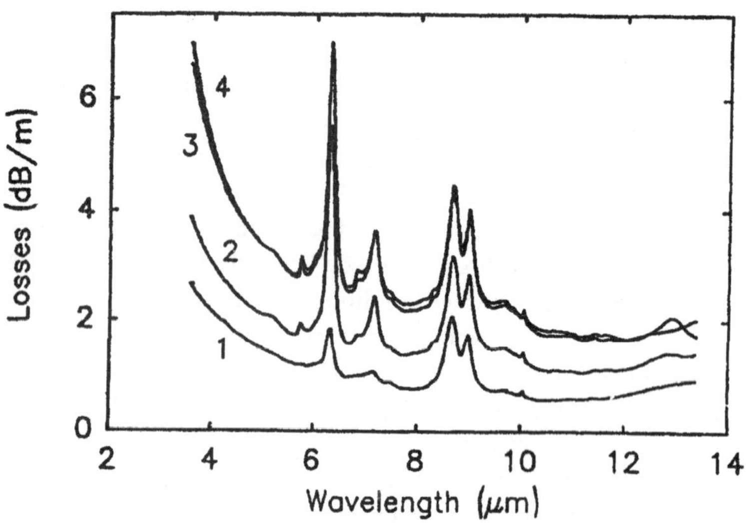

Figure 6.12 Aging effects in AgBrCl fiber, showing the increase in loss with time. (1) loss immediately after extrusion; (2) 5 months; (3) 9 months; and (4) 12 months after extrusion.[35] [Reprinted with permission from IOPP.]

noticed that there was a small but measurable change in length of a 1-m-long fiber over time. This observation led to a closer investigation of grain growth and void formation as a cause of the length change. The situation seems to be better with the silver halide fibers, as Grigorjeva et al. report that changing the composition of the silver halide to $AgCl_{0.5}Br_{0.5}$ or $AgCl_{0.75}Br_{0.25}$ greatly reduces aging.

There are several other miscellaneous difficulties with the silver halide fibers that can directly affect the optical properties of the fibers. A serious problem is the photosensitivity of silver salts. When these fiber materials are exposed to visible or UV radiation, colloidal Ag is formed. In many cases there is a visible darkening of the fiber, and this leads to increased losses in the visible as well as in the IR. Also, AgBrCl is corrosive to many metals. To overcome these problems, the fibers should be packaged in dark jackets and connectorized with materials such as Ti, Au, or ceramic materials. Silver compounds do not attack these materials.

6.1.5 Mechanical properties of PC fibers

The mechanical properties of PC fibers are quite different from those of glass fibers. The obvious difference is that these fibers plastically deform, whereas glass fibers fail well below their ultimate tensile strength. That is, glass fractures while the fibers are still in the elastic or Hooke's Law limit, but when PC fibers are deformed they can elongate well beyond the elastic limit. Furthermore, PC fibers are generally much weaker than glass fibers, with ultimate tensile strengths of about 90 MPa for an $AgCl_{0.5}Br_{0.5}$ fiber.[36] This is still a reasonably high failure stress. In practice, the stresses placed on these fibers should be limited to about half this value; otherwise the fibers will be permanently plastically deformed. Plastic deformation leads to increased loss as a result of increased scattering from separated grain boundaries.

A typical stress-strain curve for silver halide fibers is taken from the work of Barkay et al. and given in Fig. 6.13(a).[36] The initial straight portion of the curve represents the elastic region. If fiber stress or strain does not exceed this limit, the mechanical and optical properties of the fiber are unchanged in normal bending. Once this limit is exceeded, the fibers elongate plastically, as indicated by the increase in length with little increase in stress. Basically the fiber is merely stretching under load like a rubber band. It can also be seen from Fig. 6.13(a) and (b) that the strength depends on composition. Pure AgCl and AgBr have ultimate tensile strengths that are only about one-half as great as the strongest 50/50 mixture of AgCl/AgBr. The elastic strain limits of the fibers are summarized in Fig. 6.13(b).[37] The limits given in Fig. 6.13(b) should be regarded as the maximum strain that can be placed on the fiber without degradation of the optical properties. Also in this figure is the minimum-bending radius before plastic deformation. The strength, σ, also depends on the grain size (g.s.), through the well-known Petch relation, which says that $\sigma \sim (\text{g.s.})^{-1/2}$. Therefore, the smaller the grain size the stronger the fiber.

The mechanical properties of PC fibers can strongly affect the optical properties of the fibers. In contrast to glass fibers that generally retain their optical characteristics even when bent or continually flexed, PC fibers can be permanently damaged by bending and cyclic flexure. Several investigators have studied the effects of cyclic fatigue and the adverse impact on transmission. Barkay and Katzir[38] found that repeated mechanical bending of silver halide fibers in the plastic regime led to a increase in attenuation. Their data, shown in Fig. 6.14(a) for a 900-μm diame-

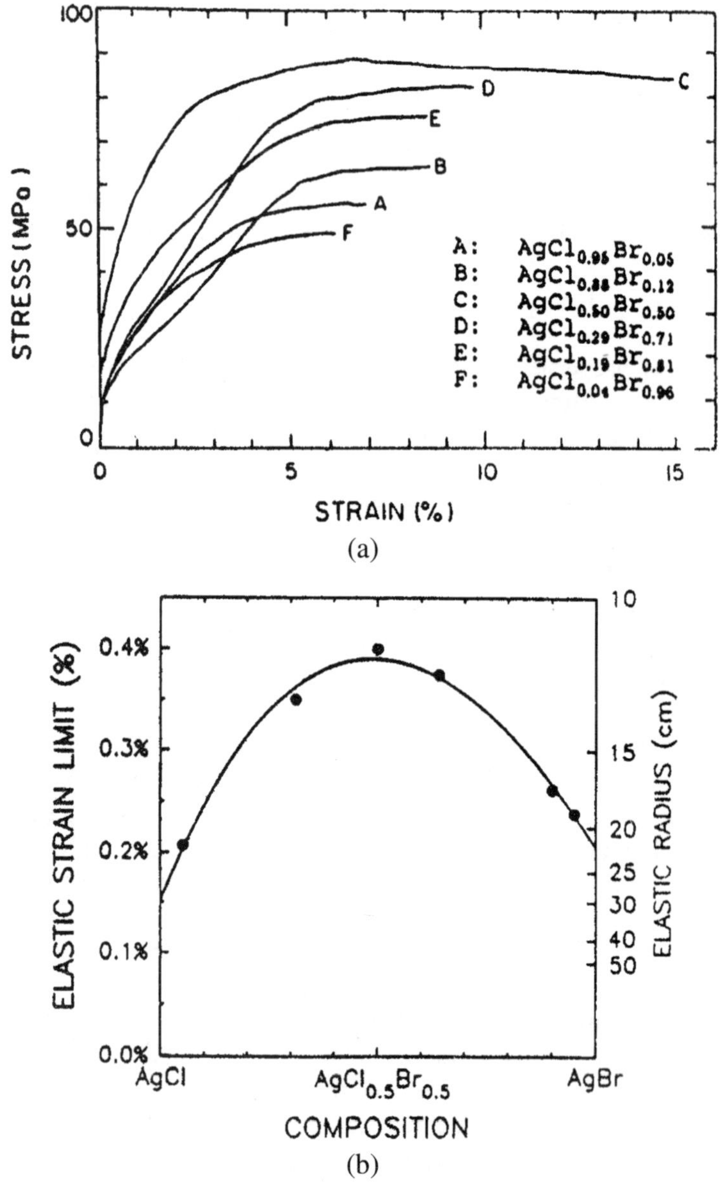

Figure 6.13 The strength of Ag-halide fibers for different compositions. The strain to failure in (a) is at the end point. The strongest fibers are those with approximately a 50/50 mix of AgBr and AgCl as shown in (b). The elastic limit for the different fibers is shown on the right ordinate scale in (b) for a 900-μm-diameter fiber.[36,37] [Reprinted with permission.]

ter, $AgCl_{0.5}Br_{0.5}$ fiber, indicates that the transmission falls dramatically for fibers cycled to a radius of 38 mm. The fibers are measured at 10.6 μm in a straight and 90° bend configuration. After about 900 cycles the fibers break. Before mechanical failure, however, the fibers fail optically. A microscopic photograph of the surface of the fatigued fiber is shown in Fig. 6.14(b).[39] The grain boundary separation is clearly evident on the surface of the fiber. The separation seen in Fig. 6.14(b) occurs in the bent portion of the fiber, and it leads to increased scattering losses. In related experiments on KRS-5 fiber, Harrington and Standlee[31] showed that when stress was applied to the fiber the loss increased. In their experiments a load was applied to the fibers under tension while the attenuation was measured at 10.6 μm. They found that $\alpha_t \sim \sigma^2$, which they explain results from an increase in the strain birefringence and, therefore, the scattering.

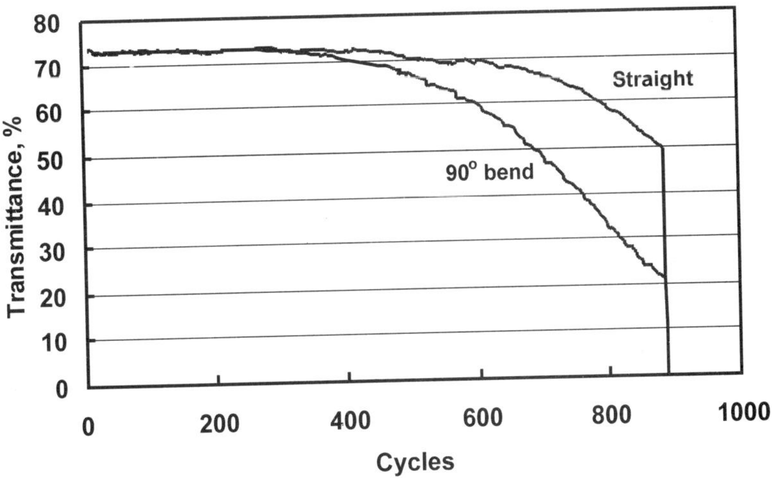

(a)

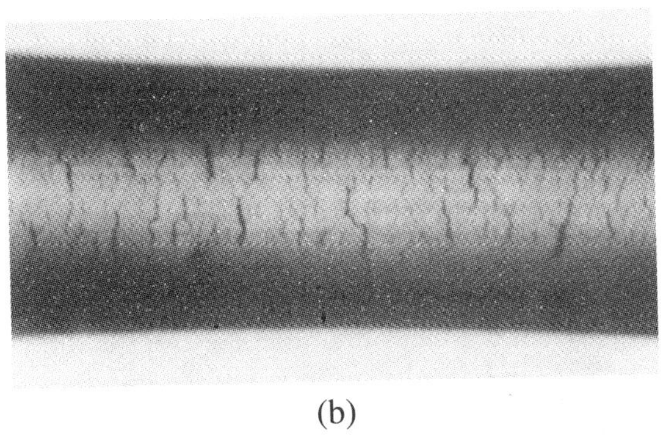

(b)

Figure 6.14 Results of cyclic-flexure testing on straight and bent Ag-halide, 900-μm-diameter fiber. The fiber was repeatedly bent to ±90° to a radius of 38 mm. The percentage loss (a) for the straight fiber and for fiber bent at 90° increases rapidly after about 500 cycles.[38] The separation of grain boundaries is shown for a fatigued fiber in (b). [Copyright 2003 IEEE. Reprinted with permission from Ref. 39.]

6.2 Single-crystal fiber optics

6.2.1 Single-crystal fiber materials and key physical properties

There are 80 IR transmissive crystalline materials, but only a small number have been grown into meter-long lengths of single-crystal (SC) fibers. One of the motivations for growing SC fibers is to make a crystalline fiber without grain boundaries. As shown in Sec. 6.1, grain boundaries can present problems that lead to increased attenuation in PC fibers. One way to circumvent this situation is to fabricate a SC fiber from some of the same materials used in the extrusion of PC fibers. Unfortunately, the extrudable metal halides turn out to be poor candidates for SC fibers. Losses of the metal halide SC fibers are high, and these materials are too ductile and delicate for most applications. Furthermore, the surface quality of the SC halide fibers is generally not as good as even the PC fibers. As a result, most of the crystalline materials chosen today for SC fiber fabrication are oxides. Oxide materials like Al_2O_3 (sapphire) have the advantage over halides of high melting points and chemical inertness, and they may be conveniently melted and grown in air. Currently, sapphire is the most popular SC fiber.[40–44]

In the late 1970s and early 1980s, SC fiber work concentrated on the fabrication of fibers from metal halides. Specifically, AgBr,[45] CsI,[46] CsBr,[47] $BaTiO_3$,[44,48] and KRS-5[49] fibers were pulled from the melt. Initially the fibers looked very good in that they were optically clear and could be grown in lengths of several meters. In most cases, the fibers were grown using a modified crystal-pulling technique. In this method a seed crystal is pulled at rates of several centimeters per minute from the melt through some sort of shaping element. The shaping element is a hollow die placed in the melt and used to help establish the diameter of the fiber. All of these fibers were core-only, with diameters ranging from 600 to 1000 μm. The losses for the fibers were usually quite high. The attenuation measured at 10.6 μm for AgBr[45] and CsI[46] SC fibers were 8.7 dB/m and 13 dB/m, respectively. The CsBr SC fibers were somewhat better, with losses reported by Mimura and Ota[47] of 0.3 to 0.4 dB/m at 10.6 μm. In all cases the fibers were very difficult to handle, as sometimes even the least amount of bending would produce obvious slip planes or other surface irregularities that would degrade the fiber's transmission. Finally, there was an attempt by Harrington et al. to recrystallize PC fibers into SC fibers using a floating-zone technique.[50] These authors converted PC KRS-5 fiber into the same size SC fiber by passing the fiber through a small heating element. The molten zone was contained by surface tension. Unfortunately, the losses for the SC fiber produced this way were greater than the starting PC fiber loss. For example, recrystallized SC AgBr fiber had a loss of 6.6 dB/m compared to an initial PC fiber loss of 2.6 dB/m.[40,51]

The most common and viable SC fiber developed to date is sapphire.[2] Sapphire is an insoluble, uniaxial crystal (trigonal system, hexagonal class) with a melting point of 2053°C. It is an extremely hard and robust material with a usable fiber transmission from about 0.5 to 3.2 μm. The key physical properties of sapphire are given in Table 6.2. Comparing the properties of the metal halides in Table 6.1 to

Table 6.2 Some key properties of sapphire used in pulling SC fibers.

Property	Sapphire–Al_2O_3
Bulk crystal transmission range, μm	0.24–4
Refractive index, (wavelength, μm)	1.71 (3)
dn/dT, °C^{-1}	1.71×10^{-6} at 3 μm
Melting point, °C	2040
Density, g/cm^3	3.98
Thermal conductivity, W/cm °C	36
Thermal expansion, /°C	5.8×10^{-6}
Young's modulus, GPa	430
Hardness, kg/mm^2 parallel to c-axis	1900
Solubility, g/100g H_2O	Insoluble

those of sapphire, it is easy to see that sapphire has some outstanding physical characteristics that rival even those of silica. For instance, Young's modulus is approximately six times greater than silica, the melting point is above 2000°C, and it is extremely hard. Also, it is very inert and it can be grown and stored in an ambient atmosphere. These properties make sapphire an almost ideal IR fiber candidate for applications less than about 3.2 μm.

6.2.2 Sapphire fiber fabrication

Sapphire fibers are fabricated by pulling a fiber from a molten reservoir of Al_2O_3. The extremely high melting point of sapphire requires high-temperature crucibles made from tungsten or molybdenum or conversely by supporting the molten material without a crucible. A crucible technique, initially proposed by Gasson and Cockayne[52] and also by Haggerty et al.,[53] was the first method developed for growing sapphire filaments. The modern version of this method is due to LaBelle et al.[54] working at Tyco and then later at Saphikon, Inc. This method is called the edge-defined, film-fed growth (EFG) technique. A crucibleless method, called laser-heated pedestal growth (LHPG),[54] was developed at Stanford University by researchers working for both Byer[40] and Feigelson.[2,44]

In the EFG method, a capillary tube is inserted into a crucible containing molten sapphire.[55] Capillary action propels the liquid Al_2O_3 up the tube, where it forms either a positive or negative meniscus at the tip. A seed fiber, which is normally oriented with the c-axis along the fiber axis, contacts the molten sapphire at the tip and a fiber is pulled vertically upward. The essential elements of the EFG process used originally by Saphikon, Inc., and now by Photran, Inc., are shown in Fig. 6.15. It is also possible to use a platinum wire to pull

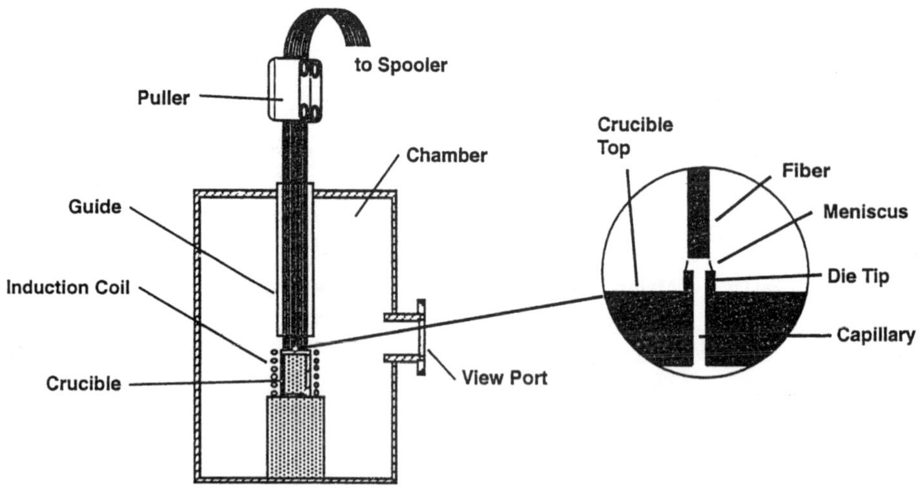

Figure 6.15 EFG apparatus for single-crystal growth of sapphire fibers. The key features for pulling a single EFG fiber are shown in the callout. The complete apparatus on the left, however, is shown with multiple die tips for pulling many fibers from one crucible into a ribbon of fibers. This schematic of the EFG apparatus is courtesy of Photran, Inc.

the fiber if a seed fiber is not available. Typical crystal-pulling rates are about 1 to 2 cm/min, which is much slower than glass fiber drawing. Slow growth rates are necessary to minimize the formation of voids, which can occur at the solidus-liquidus interface if the growth rate is too fast. There are several advantages to this method. First, it is inherently a continuous process, as the crucible can be recharged during growth to keep a constant supply of molten material. This approach is used to draw very long fibers. Second, the EFG method has the unique advantage of being capable of multiple fiber growth. This can be done by using a die with multiple holes placed in a single crucible. Saphikon, Inc., has used a multielement die to draw more than 25 fibers at one time. In addition, it is possible to use dies of different shapes to grow ribbons, squares, and other unusual cross-section filaments and plates. For example, Saphikon, Inc., draws continuous sapphire plates several millimeters thick by 20 cm in width.

The LHPG method is a crucibleless technique that closely resembles the float-zone method of crystal growth. In the float-zone method, the molten zone is freely supported between the two ends of the crystal rod. LHPG is inherently the best technique for growing high-optical-quality sapphire fibers because the molten zone is held in place by surface tension, eliminating the need for a crucible, which could be a possible source of contamination. Furthermore, a CO_2 laser beam, which provides a uniform, ultraclean heat source, is used to melt the starting rod. Unlike the EFG technique, this growth method allows only one fiber to be grown at a time, so commercialization is difficult. LHPG, however, is the method used to produce the lowest-loss sapphire fibers, with losses approaching theoretical values at 3 μm.

In LHPG SC sapphire fiber growth, a CO_2 laser beam is focused onto the tip of a source rod creating a small molten bead of sapphire.[40,56] A seed fiber is dipped into the molten region, shown schematically in Fig. 6.16, and slowly pulled upward, forming the single-crystal fiber. The source rod, which may be single crystal, polycrystalline, sintered, or a pressed powder, is simultaneously fed upward to replenish the supply of molten sapphire. Mass conservation dictates that the diameter reduction be equal to the square root of the fiber to source rod velocity ratio, that is,

$$\frac{d_s}{d_f} = \sqrt{\frac{v_f}{v_s}}, \tag{6.1}$$

where d_f and d_s are the diameters, and v_f and v_s are the velocities of the fiber and source rod, respectively.

The shape of the molten zone is a function of the laser power, the diameter reduction, and the material being grown. Fejer's[57] thermodynamic models of the molten-zone shape during LHPG growth of sapphire indicate a range of molten-zone lengths that will produce stable growth. His calculations, as well as those of Kim et al.,[58] suggest a maximum stable zone length of approximately three times the fiber diameter.

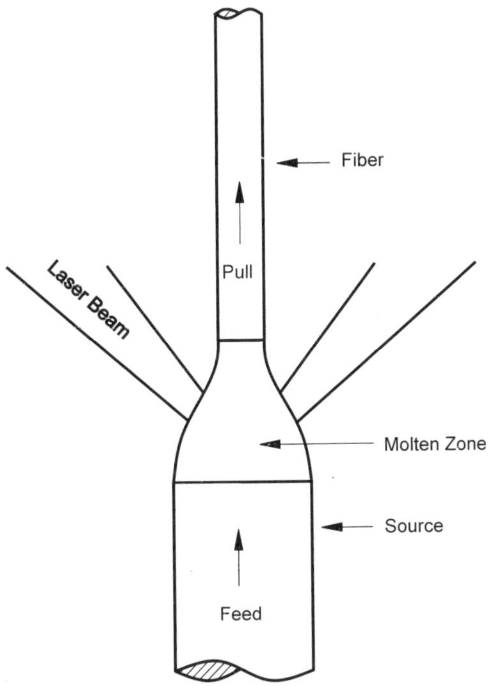

Figure 6.16 Schematic diagram of laser-heated pedestal growth illustrating the small molten zone produced in an approximate 3:1 source-rod-to-fiber diameter reduction.[56] [Reprinted with permission.]

It is much more difficult to produce SC fibers with smooth surfaces than for glass fibers with smoth surfaces. This is because the viscosity of a glass is very high during drawing, whereas the viscosity of crystalline material at the molten zone is very low; it is thus sensitive to any minor perturbations in the system. Since the molten region is held in place simply by surface tension, any air currents, vibrations, laser-power fluctuations, etc., will have enormous effects on the stability of growth. For this reason it is necessary to use a very stable laser source and to keep the molten zone small. The optimal molten-zone length is approximately equal to the diameter of the source rod, and the usual source-to-fiber reduction ratio is 3:1.

There have been several LHPG growth facilities established for sapphire fiber fabrication. Jundt et al.[40] and Fejer et al.[57] at Stanford report the details of their system used to produce 110-μm diameter fiber. Chang et al.[51] at the University of South Florida also grew fibers with similar diameters. All of these groups used essentially the same key components in their LHPG apparatus. To illustrate the details of LHPG, the system described by Nubling and Harrington,[56] shown in Fig. 6.17(a), will be described. This apparatus was used to grow fibers 1 m or greater in length with diameters of about 300 μm. Referring to Fig. 6.17(a), two drive motors smoothly translate the source rod and seed fibers along a low-friction V-groove in a hardened metal block. Using a sapphire guide tube on the end of the fiber translator minimized side-to-side motion of fiber. A laser micrometer was used to measure and control the fiber diameter during growth. A 50-W CO_2 laser beam was expanded and configured into a focused ring of light by the reflaxicon optical arrangement shown schematically in Fig. 6.17(b). Using feedback from the laser micrometer, it was possible to stabilize the laser power to $\pm 0.5\%$. Typically, 10 W was needed to melt a 1-mm-diameter source rod. Source rods were obtained from Saphikon, Inc., in lengths up to 2 m. Using this length of source rod and a 3:1 diameter reduction would, in principle, yield fibers as long as 15 m. Typical growth rates were 2 mm/min, so that a 1-m-long fiber took nearly 8.5 h to grow. The longest fiber grown by Nubling and Harrington was 5 m. Most fibers were 300 μm in diameter with peak-to-peak diameter fluctuations less than 2 μm. Chang et al.[51] and Phomsakha et al.[59] were able to increase the growth rate by a factor of 10 by growing the fibers in a helium atmosphere instead of air. They also studied the use of Ar and N_2 gas atmospheres, but they found that a helium pressure of 20 torr provided the optimal growth condition. Unlike the other atmospheres, helium gas was not entrapped in the fiber at high growth rates and the excellent exchange gas properties of helium also provided better thermal conditions for growth.

The sapphire fibers grown by either technique are generally pure Al_2O_3, with the *c*-axis oriented along the fiber axis. That is, there is no cladding layer. Recently there have been some limited attempts to produce a cladding. One method is to overcoat the fiber with Teflon AF, an amorphous Teflon that transmits over most of sapphire fiber's transmission range. A more attractive method is to overcoat the fiber with a sol-gel such as aluminum oxide. Aluminum oxide gels have lower refractive indices than sapphire due to their porosity. Finally, Al_2O_3 doped with

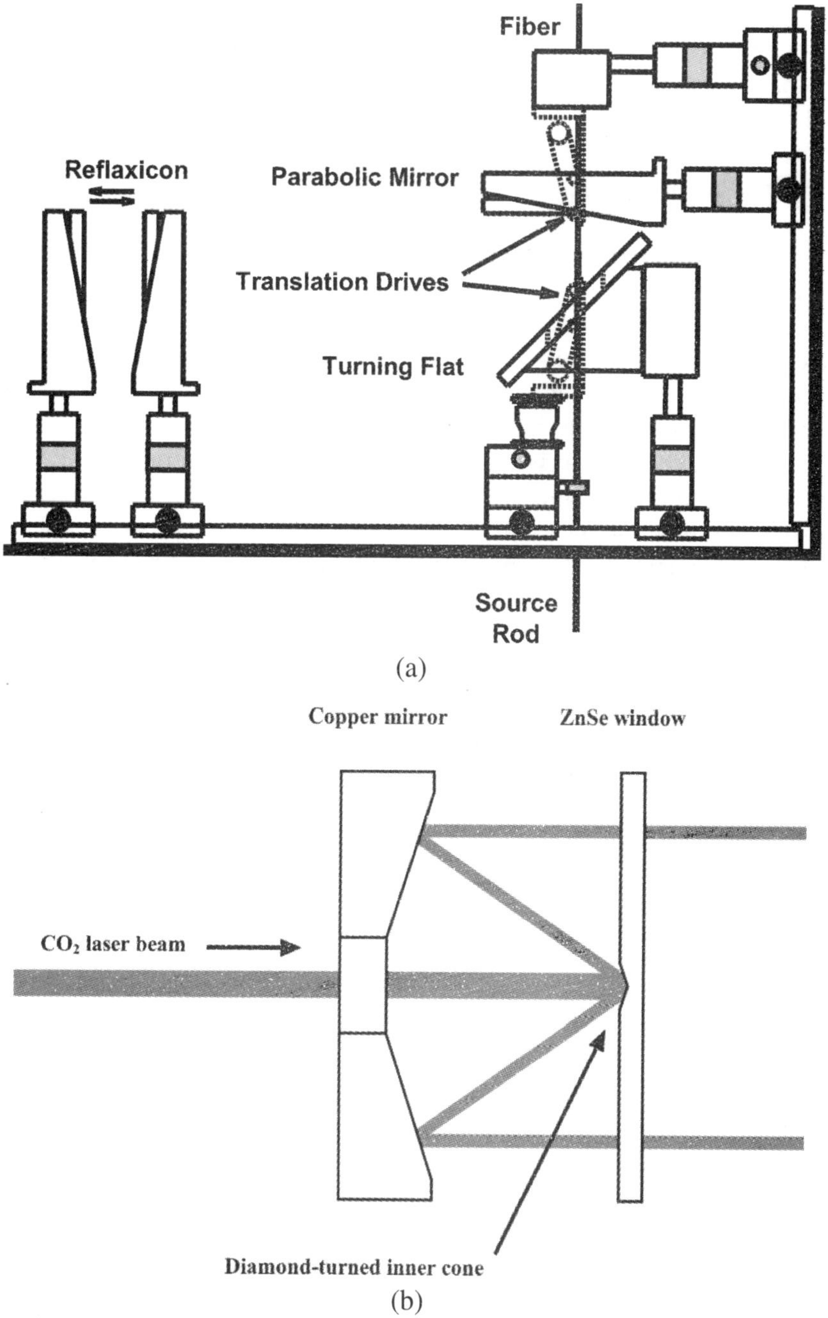

Figure 6.17 Schematics of (a) continuous-feed LHPG apparatus and (b) reflaxicon optical arrangement. The source rod (up to 2 m long) is fed through a hole drilled in the table, and the entire system is sealed within an air-tight Plexiglas enclosure.[56] [Reprinted with permission.]

Ti^{3+} has been grown into fiber by Sharp et al.[60] with the hope that the growth process itself will produce a gradient in the Ti^{3+} and, thus, a graded index fiber. They found that they did indeed produce a fiber with an index 12% higher in the center than the edge. The index profile of the fiber did prove to be parabolically graded.

6.2.3 Optical properties of sapphire fiber

The intrinsic losses for bulk sapphire are given in Fig. 2.11. From the data in Fig. 2.11, the minimum attenuation coefficient (minimum in the V-curve) for sapphire is found to be approximately 8×10^{-9} cm^{-1} or 3.4×10^{-3} dB/km at 1.78 μm. The measured loss for the bulk crystal as well as for the sapphire fibers is, however, many orders of magnitude higher than this minimum theoretical value. This may be seen in the measured loss for high-purity bulk Al_2O_3, also shown in Fig. 2.11. The solid curve through the experimental data for wavelengths shorter than 1.5 μm is a λ^{-4}-power-law fit. This suggests that Rayleigh scattering from inhomogeneities in the crystal contributes to the extrinsic loss. From Fig. 2.11 it is also possible to estimate the intrinsic loss at 2.94 μm. By extrapolation of the multiphonon absorption, this loss is determined to be ~0.13 dB/m at 2.94 μm.

Loss measurements for LHPG fibers are shown in Fig. 6.18. Two spectra are given in the figure. One spectrum is for a 100-μm diameter, 0.8-m-long fiber grown by Chang et al.[51] at Univerity of Southern Florida and the other a 300-μm diameter, 1-m-long fiber grown by Nubling and Harrington[56] at Rutgers. The USF fiber was grown in He at a rate of 20 mm/min, and the fiber was post-annealed at 1250°C

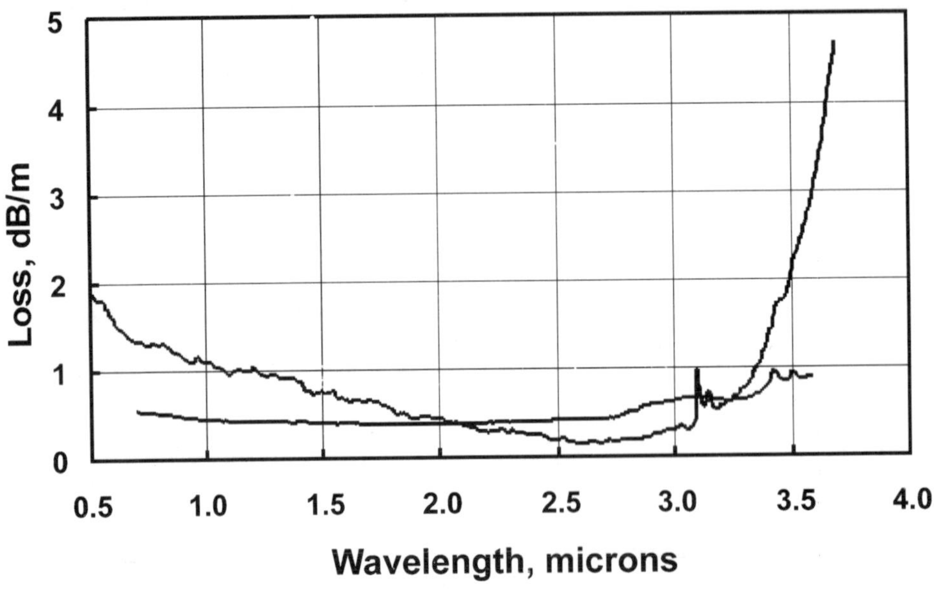

Figure 6.18 Losses for two of the best SC sapphire fibers grown to date. Both SC fibers were grown by the LHPG method.[56] [Reprinted with permission.]

for 24 h. The Rutgers fiber was grown at 2 mm/min in air, and it was post-annealed in air at 1000°C for 12 h. The Rutgers fiber had diameter fluctuations of less than ±0.5%. Figure 6.19 gives the spectral details in the 3-μm region of the Rutgers' fiber. The measured loss at 2.94 μm is ~0.3 dB/m, which is approaching the theoretical limit of ~0.13 dB/m. This is one of the lowest losses obtained at 2.94 μm in a SC sapphire fiber. Nubling and Harrington attribute this low loss to the elimination of crystal imperfections caused by instabilities in the growth process and good diameter control. Jundt et al.[40] measured a loss of 0.88 dB/m at 2.94 μm in a 110-μm-diameter sapphire fiber.

Nubling and Harrington have fit the Rutgers' fiber data, shown in Fig. 6.18, to a λ^{-2} dependence below 2.5 μm. As is the case for the PC fibers, the λ^{-2} dependence suggests that the scattering is dominated by Rayleigh–Gans scattering. In support of this, these authors measured a strong degree of forward scattered light in their fiber. Forward scattering is characteristic of Rayleigh–Gans scattering.

The attenuation in sapphire fibers is affected by defects formed during fiber growth. In particular, color centers, which have their peak absorption in the visible region, and voids, which strongly scatter light, are frequent problems. An absorption peak at 410 nm due to a V-center dominates the loss in the visible region. This V-center absorption peak has been seen in bulk sapphire as well as sapphire fibers, and it is associated with a trapped hole pair on an O^{2-} adjacent to an Al^{3+} vacancy. Chang et al.[51] have studied the effect of annealing on the visible loss and determined that the V-center defect was largely eliminated by annealing at a few hundred degress Celcius. By annealing in air or oxygen, they were able to reduce the losses in the 300- to 550-nm region. For instance, they reduced the 410-nm

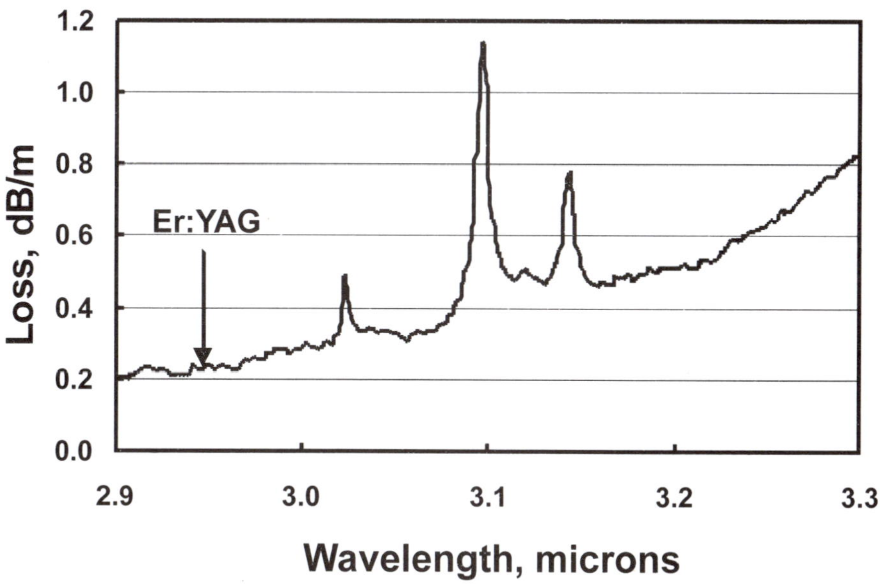

Figure 6.19 The lowest-loss SC sapphire fiber at 2.94 μm. Three distinct extrinsic absorptions are believed to be due to Ti or Fe.[56] [Reprinted with permission.]

loss of the as-grown fiber by a factor of 5 by a 24-h anneal at 1250°C. The Rutgers group also saw similar results at short wavelengths after an anneal in air at 1000°C. Somewhat surprisingly, they also saw a decrease in loss at 2.94 μm after the annealing process.

The data in Figs. 6.18 and 6.19 reveal extrinsic absorption peaks near 3 μm. These absorptions are believed to be due to transition metal impurities like Ti or Fe in the fibers. The intensity of these impurity bands is also drastically reduced through annealing. Eigenmann and Gunthard[61] observed the same peaks in Fe-doped sapphire and also demonstrated a reduction in peak intensity when the sample was annealed in O_2. There is also evidence in some of the fibers of OH^- near 3 μm. This impurity is not present in all fibers. Those that have been annealed show less absorption by water-related impurities.

Photran, formerly Saphikon, Inc., produces commercially available sapphire fibers grown by the EFG method. These fibers have losses that tend to be somewhat higher than fibers grown by the LHPG method. The reasons for this include contamination of the fiber by the tungsten crucibles used to contain the molten Al_2O_3 and a greater concentration of voids. The advantage of the EFG method is that multiple fibers can be grown at one time, in contrast with the LHPG method where only one fiber can be grown at a time. Since the growth rate for either technique is rather slow, the ability to grow multiple fibers is clearly the best commercial alternative. The loss spectrum for the EFG fibers is given in Fig. 6.20.[43] The commercial fibers are available in nominal sizes of 325 μm and 425 μm and in lengths

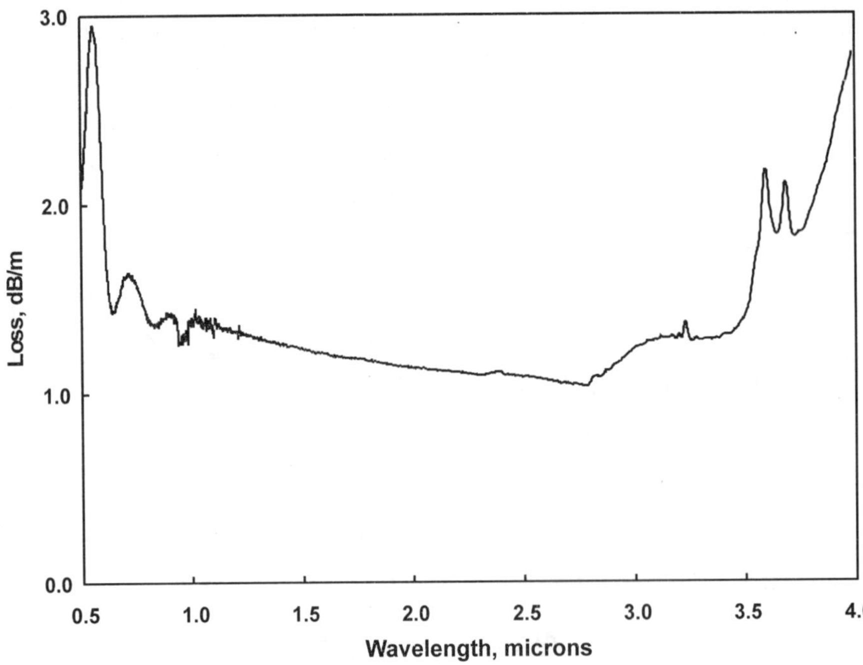

Figure 6.20 Commercial SC sapphire fiber fabricated by Photran, Inc., by the EFG process. The fiber diameter was 325 μm.

up to 2 m. The fibers are clad only in a loose-fitting PEEK plastic tube. Losses at 2.94 μm are between 1 and 2 dB/m, which is well above those grown using LHPG methods. Nevertheless, Photran's fibers are proving to be quite effective in Er:YAG laser power delivery and for use in very rugged fiber sensor systems. In the latter case, they may be used over a broad temperature range, from cryogenic to almost 2000°C.

The calculated NA of the core-only sapphire fiber with refractive index equal to 1.75 is 1. This would suggest that sapphire fibers would transmit all incident rays. In practice, however, this is not the case, since bulk and surface defects scatter light into large-angle lossy modes. As discussed in Sec. 2.1.2, it is more practical to use an effective numerical aperture, NA_{eff}, to describe the true launch constraints. NA_{eff} is defined as the sine of the launch angle at which the transmission drops to 50% of the peak transmission. The NA_{eff} at 2.94 μm was determined by measuring the transmission as a function of launch angle with a collimated input beam from an Er:YAG laser. The data for an unclad, 300-μm-diameter LHPG sapphire fiber by Nubling and Harrington[56] is shown in Fig. 6.21. From this data an $NA_{eff} \cong 0.35$ is obtained.

6.2.4 Mechanical properties of sapphire fiber

As may be seen from the data in Table 6.2, the Young's modulus for sapphire is very high. In fact, it can be seen that the modulus for sapphire is about six times greater than that for silica. In practice this means that SC sapphire fibers are rather stiff, a feature readily observed when one bends equal-diameter sapphire and silica fibers. There has been only limited strength data taken on the optical SC sapphire fibers. Wu et al.[62] have measured 110-μm-diameter LHPG fiber under tension. They measurements yielded failure strains between 1.20% and 1.85%. These are

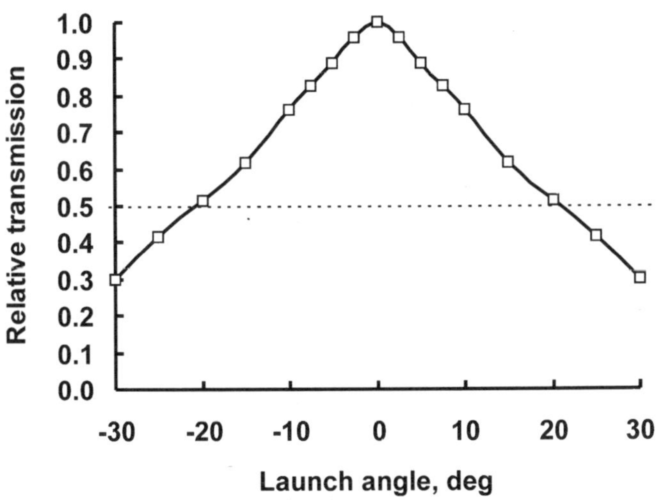

Figure 6.21 Effective NA of 300-μm-diameter sapphire fiber measured at 2.94 μm.

measurements based on only two samples, and thus their statistics are quite limited. Photran, Inc., claims that their 325-μm-diameter fiber can be bent into a 60-mm loop. Jundt et al.[40] indicate that their 150-μm-diameter LHPG fiber can be bent to a 4-mm radius.

References

1. J. A. Harrington, *Selected Papers on Infrared Fiber Optics*, SPIE Press, Bellingham, WA (1990).
2. J. Sanghera and I. Aggarwal, *Infrared Fiber Optics*, CRC Press, Boca Raton, FL (1998).
3. D. A. Pinnow, A. L. Gentile, A. G. Standlee, A. J. Timper, and L. M. Hobrock, "Polycrystalline fiber optical waveguides for infrared transmission," *Appl. Phys. Lett.*, Vol. 33, pp. 28–29 (1978).
4. J. A. Harrington, "Infrared fiber optics," in *Handbook of Optics; Fiber and Integrated Optics*, M. Bass, J. Enoch, E. Van Stryland, and W. Wolfe, Eds., McGraw-Hill, New York (2001).
5. R. Nubling and J. A. Harrington, "Single-crystal LHPG sapphire fibers for Er:YAG laser power delivery," *Appl. Opt.*, Vol. 37, pp. 4777–4781 (1998).
6. R. S. F. Chang, S. Sengupta, L. B. Shaw, and N. Djeu, "Fabrication of laser materials by laser-heated pedestal growth," Proc. SPIE, Vol. 1410, pp. 125–132 (1991).
7. J. A. Harrington, "Crystalline infrared fibers," Proc. SPIE, Vol. 266, pp. 10–15 (1981).
8. M. Lines, "Scattering losses in optic fiber materials. II. Numerical estimates," *J. Appl. Phys.*, Vol. 55, pp. 4058–4063 (1984).
9. D. A. Pinnow, T. Rich, F. W. Ostermayer, and M. DiDomenico, "Fundamental optical attenuation limits in the liquid and glassy state with application to fiber optical waveguide materials," *Appl. Phys. Lett.*, Vol. 22, pp. 527–529 (1973).
10. A. Sa'ar and A. Katzir, "Intrinsic losses in mixed silver halide fibers," Proc. SPIE, Vol. 1048, pp. 24–31 (1989).
11. F. Moser, D. Bunimovich, A. DeRowe, O. Eyal, A. German, Y. Gotshal, A. Levite, L. Nagli, A. Ravid, V. Scharf, S. Shalem, D. Shemesh, R. Simchi, I. Vasserman, and A. Katzir, "Medical applications of infrared transmitting silver halide fibers," *IEEE J. Selected Topics Q. Electron.*, Vol. 2, pp. 872–879 (1996).
12. A. Sa'ar and A. Katzir, "Scattering effects in crystalline infrared fibers," *J. Opt. Soc. Am. A*, Vol. 5, pp. 832–833 (1988).
13. J. A. Harrington and M. Sparks, "Inverse-square wavelength dependence of attenuation in infrared polycrystalline fibers," *Opt. Lett.*, Vol. 8, pp. 223–225 (1983).
14. R. C. Pastor and A. C. Pastor, "Solid solutions of metal halides under a reactive atmosphere," *Mat. Res. Bull.*, Vol. 11, pp. 1043–1050 (1974).

15. R. C. Pastor and A. C. Pastor, "Crystal growth of KC1 in a reactive atmosphere. II. The crystal growth technique," *Mat. Res. Bull.*, Vol. 19, pp. 1195–1199 (1984).
16. S. D. Allen and J. A. Harrington, "Optical absorption in KCl and NaCl at infrared laser wavelengths," *Appl. Opt.*, Vol. 17, pp. 1679–1680 (1978).
17. A. P. Belousov, E. M. Dianov, I. S. Litsitskii, T. M. Nesterova, V. G. Plotnichenko, and V. K. Sysoev, "Single crystals of thallium halides with optical losses below 10 dB/km," *Sov. J. Quantum Electron.*, Vol. 12, pp. 496–497 (1982).
18. D. Bunimovich, L. Nagli, S. Shalem, and A. Katzir, "Absorption spectrum of silver bromide crystals and fibers in the 9–11 micron wavelength range," *J. Appl. Phys.*, Vol. 81, pp. 1612–1613 (1997).
19. D. Bunimovich, I. Nagli, and A. Katzir, "Absorption measurements of mixed silver halide crystals and fibers by laser calorimetry," *Appl. Opt.*, Vol. 33, pp. 117–119 (1994).
20. D. Bunimovich and A. Katzir, "Dielectric properties of silver halide and potassium halide crystals," *Appl. Opt.*, Vol. 32, pp. 2045–2048 (1993).
21. M. Kimura, S. Kachi, and K. Shiroyama, "Characteristics of KRS-5 fiber with crystalline cladding," Proc. SPIE, Vol. 618, pp. 85–88 (1986).
22. J. A. Harrington, "Infrared alkali halide fibers," *Appl. Opt.*, Vol. 27, pp. 3097–3101 (1988).
23. M. Ikedo, M. Watari, F. Tateishi, and H. Ishikawa, "Preparation and characteristics of the TlBr-TlI fiber for a high power CO_2 laser beam," *J. Appl. Phys.*, Vol. 60, pp. 3035–3039 (1986).
24. D. Bunimovich, S. Shalem, and A. Katzir, "Effects of thermal treatment on the infrared transmission of polycrystalline silver halide fibers," *Appl. Opt.*, Vol. 36, pp. 285–290 (1997).
25. S. S. Alimpiev, V. G. Artjushenko, L. N. Butvina, S. K. Vartapetov, E. M. Dianov, Y. G. Kolesnikov, V. I. Konov, A. O. Nabatov, S. M. Nikiforov, and M. M. Mirakjan, "Polycrystalline IR fibres for laser scalpels," *Int. J. Quantum Electron.*, Vol. 3, pp. 333–344 (1988).
26. K. Takahashi, N. Yoshida, and M. Yokota, "Optical fibers for transmitting high-power CO_2 laser beam," *Sumitomo Electric Tech. Rev.*, Vol. 23, pp. 203–210 (1984).
27. M. Saito, M. Takizawa, and M. Miyagi, "Infrared optical fibers with vapor-deposited cladding layer," *J. Lightwave Technol.*, Vol. 7, pp. 158–162 (1989).
28. V. G. Artjushenko, A. A. Lerman, E. G. Litvinenko, A. O. Nabatov, V. I. Konov, R. I. Kuznetsov, V. G. Plotnichenko, I. L. Pylnov, V. A. Shtein-Margolina, A. A. Urusovskaja, V. V. Vojtsekhovsky, N. D. Zaharov, W. Neuberger, and K. Moran, "Mechanisms of optical losses in polycrystalline fibers," Proc. SPIE, Vol. 1591, pp. 83–89 (1992).
29. V. Artjushenko (private communication).
30. S. Sakuragi, "Polycrystalline KRS-5 infrared fibers for power transmission," Proc. SPIE, Vol. 320, pp. 2–9 (1982).

31. J. A. Harrington and A. G. Standlee, "Attenuation at 10.6 μm in loaded and unloaded polycrystalline KRS-5 fibers," *Appl. Opt.*, Vol. 22, pp. 3073–3078 (1983).
32. V. G. Artjushenko, L. N. Butvina, V. V. Vojtsekhovsky, E. M. Dianov, and J. G. Kolesnikov, "Mechanisms of optical losses in polycrystalline KRS-5 fibers," *J. Lightwave Technol.*, Vol. LT-4, pp. 461–465 (1986).
33. J. A. Wysocki, R. G. Wilson, A. G. Standlee, A. C. Pastor, R. N. Schwartz, A. R. Williams, Guan-Dao Lei, and L. Kevan, "Aging effects in bulk and fiber TlBr-TlI," *J. Appl. Phys.*, Vol. 63, pp. 4365–4371 (1988).
34. L. Kupper, H. M. Heise, and L. N. Butvina, "Novel developments in mid-IR fiber-optic spectroscopy for analytical applications," *J. Mol. Struct.*, Vol. 563/564, pp. 173–181 (2001).
35. L. Grigorjeva, D. Millers, E. Kotomin, R. Eglitis, and A. A. Lerman, "Optical properties of silver halide fibres: ageing studies," *J. Phys. D.*, Vol. 29, pp. 578–583 (1996).
36. N. Barkay, A. Levite, F. Moser, and A. Katzir, "Mechanical properties of mixed silver-halide crystals and polycrystalline optical fibers," *J. Appl. Phys.*, Vol. 64, pp. 5256–5258 (1988).
37. N. Barkay and A. Katzir, "Elasticity of mixed silver-halide polycrystalline optical fibers," *J. Appl. Phys.*, Vol. 74, pp. 2980–2982 (1993).
38. A. German, N. Barkay, and A. Katzir, "High-cycle fatigue of silver halide infrared fibers," *Appl. Opt.*, Vol. 33, pp. 2734–2736 (1994).
39. N. Barkay and A. Katzir, "Transmission of infrared laser radiation through silver-halide optical fibers during repeated plastic deformation," *J. Lightwave Technol.*, Vol. 11, pp. 1889–1894 (1993).
40. D. H. Jundt, M. M. Fejer, and R. L. Byer, "Characterization of single-crystal sapphire fibers for optical power delivery systems," *Appl. Phys. Lett.*, Vol. 55, pp. 2170–2172 (1989).
41. R. Nubling and J. A. Harrington, "Optical properties of LHPG sapphire fibers," Proc. SPIE, Vol. 2677, pp. 49–54 (1996).
42. R. S. F. Chang, S. Sengupta, G. J. Dixon, L. B. Shaw, and N. Djeu, "Growth of small laser crystals for study of energy kinetics and spectroscopy," Proc. SPIE, Vol. 1104, pp. 244–250 (1989).
43. A. K. Pryshlak, J. R. Dugan, and J. J. Fitzgibbon, "Advancements in sapphire optical fibers for the delivery of erbium laser energy and IR sensor applications," Proc. SPIE, Vol. 2677, pp. 35–42 (1996).
44. R. S. Feigelson, "Pulling optical fibers," *J. Cryst. Growth*, Vol. 79, pp. 669–680 (1986).
45. T. J. Bridges, J. S. Hasiak, and A. R. Strnad, "Single-crystal AgBr infrared optical fibers," *Opt. Lett.*, Vol. 5, pp. 85–87 (1980).
46. Y. Okamura, Y. Mimura, Y. Komazawa, and C. Ota, "CsI crystalline fiber for infrared transmission," *Jap. J. Appl. Phys.*, Vol. 19, pp. L649–L651 (1980).
47. Y. Mimura, Y. Okamura, and C. Ota, "Single-crystal CsBr infrared fibers," *J. Appl. Phys.*, Vol. 53, pp. 5491–5497 (1982).

48. M. Saifi, B. Dubois, and E. M. Vogel, "Growth of tetragonal $BaTiO_3$ single crystal fibers," *J. Mater. Res.*, Vol. 1, pp. 452–456 (1986).
49. Y. Mimura, Y. Okamura, Y. Komazawa, and C. Ota, "Growth of fiber crystals for infrared optical waveguides," *Jap. J. Appl. Phys.*, Vol. 19, pp. L269–L272 (1980).
50. J. A. Harrington, A. G. Standlee, A. C. Pastor, and L. G. DeShazer, "Single-crystal infrared fibers fabricated by travelling-zone melting," Proc. SPIE, Vol. 484, pp. 124–127 (1984).
51. R. S. F. Chang, V. Phomsakha, and N. Djeu, "Recent advances in sapphire fibers," Proc. SPIE, Vol. 2396, pp. 48–53 (1995).
52. D. B. Gasson and B. Cockayne, "Oxide crystal growth using gas lasers," *J. Matr. Sci.*, Vol. 5, pp. 100–104 (1970).
53. J. S. Haggerty, W. P. Menashi, and J. F. Wenekkus, U.S. Patent No. 3,944,640 (Issued 3-16-1976).
54. H. E. LaBelle, "EFG, the invention and application to sapphire growth," *J. Cryst. Growth*, Vol. 50, pp. 8–17 (1980).
55. J. T. Pollock, "Filamentary sapphire: Part 3. The growth of void-free sapphire filament at rates up to 3.0 cm/min," *J. Matr. Sci.*, Vol. 7, pp. 787–792 (1972).
56. R. Nubling and J. A. Harrington, "Optical properties of single-crystal sapphire fibers," *Appl. Opt.*, Vol. 36, pp. 5934–5940 (1997).
57. M. M. Fejer, J. L. Nightingale, G. A. Magel, and R. L. Byer, "Laser-heated miniature pedestal growth apparatus for single-crystal optical fibers," *Rev. Sci. Instrum.*, Vol. 55, pp. 1791–1796 (1984).
58. K. M. Kim, A. B. Dreeben, and A. Schujko, "Maximum stable zone length in float-zone growth of small diameter sapphire and silicon crystals," *J. Appl. Phys.*, Vol. 50, pp. 4472–4474 (1979).
59. V. Phomsakha, R. S. F. Chang, and N. Djeu, "Novel implementation of laser heated pedestal growth for the rapid drawing of sapphire fibers," *Rev. Sci. Instrum.*, Vol. 65, pp. 3860–3861 (1994).
60. J. H. Sharp, R. Illingworth, and I. S. Ruddock, "Graded-index characteristics in single-crystal fibers," *Opt. Lett.*, Vol. 23, pp. 109–110 (1998).
61. K. Eigenmann and H. H. Gunthard, "Valence state, redox reactions and biparticle formation of Fe and Ti doped sapphire," *Chem. Phys. Lett.*, Vol. 13, pp. 58–61 (1972).
62. H. F. Wu, A. J. Perrotta, and R. S. Feigelson, "Mechanical characterization of the single crystal alpha-Al_2O_3 fibers grown by laser-heated pedestal technique," *Light Metal Age*, Vol. 49, pp. 97–98 (1991).

Chapter 7

Hollow-core Fiber Optics

7.0 Introduction

The first optical-frequency hollow waveguides were simple parallel-plate guides. The early pioneering work in this field was done by Garmire et al.[1] Her group made a rectangular waveguide using aluminum strips separated by about 0.5 mm with bronze shim stock. This structure is shown in Fig. 3.2. Even when the aluminum was not well polished, these guides worked surprisingly well, with losses at 10.6 μm well below 1 dB/m. As early as the late 1970s, Garmire demonstrated the high power-handling capability of an air-core guide by delivering more than 1 kW of CO_2 laser power through this simple structure. These rectangular waveguides, however, never gained much popularity, primarily because their overall dimensions (about 0.5×10 mm) were quite large in comparison to circular cross-section guides and also because the rectangular guides cannot be bent uniformly in any direction. As a result, hollow circular waveguides with diameters of 1 mm or less fabricated using metal, glass, or plastic tubing are the most common guide today.

In general, hollow waveguides are attractive for laser power delivery because of the inherent advantage of their air core. Hollow waveguides not only enjoy the advantage of high laser power thresholds but also low insertion loss, no end reflection, ruggedness, and small beam divergence. A disadvantage, however, is a loss on bending that varies as $1/R$, where R is the bending radius. In addition, the losses for these guides vary as $1/a^3$, where a is the radius of the bore. Unfortunately, this means that the flexibility of the very small bore ($\lesssim$250 μm) guides is somewhat negated by their higher loss. However, the loss can be made arbitrarily small for a sufficiently large core. The bore size and bending radius dependence of all hollow waveguides is a characteristic of these guides not shared by solid-core fibers. Initially these waveguides were developed for medical and industrial applications involving the delivery of CO_2 laser radiation, but more recently they have been used to transmit incoherent light for broadband spectroscopic and radiometric applications.[2] They are today one of the best alternatives for power delivery in IR laser surgery and industrial laser delivery systems, with losses as low as 0.1 dB/m and transmitted cw laser powers as high as 2.7 kW.[3]

Hollow-core waveguides may be grouped into two categories: (1) those whose inner core materials have refractive indices greater than 1 (leaky guides) and

(2) those whose inner wall material has a refractive index less than 1 (attenuated total reflectance, i.e., ATR, guides). These two structures for circular cross-section guides are illustrated in Fig. 7.1. Leaky or $n > 1$ waveguides have metallic and dielectric films deposited on the inside of metallic, plastic, or glass tubing. ATR guides are made from dielectric materials with refractive indices less than 1 in the wavelength region of interest.[4,5] Therefore, $n < 1$ guides are fiberlike in that the air-core index ($n \approx 1$) is greater than the clad index. Hollow sapphire fibers operating at 10.6 μm ($n = 0.67$) are an example of this class of hollow guide.[4]

7.1 Structure of hollow waveguides

7.1.1 Parallel-plate hollow waveguides

The first hollow-core waveguides were fabricated by Garmire et al.[1] from parallel plates of aluminum sheeting used for photographic lighting. The plates were cut into strips 1 cm wide and then held apart by brass shim stock spacers as shown in Figs. 3.2 and 7.2(a). Uncoated and nonanodized aluminum was chosen as the reflecting surface because it was convenient even though aluminum is not the best

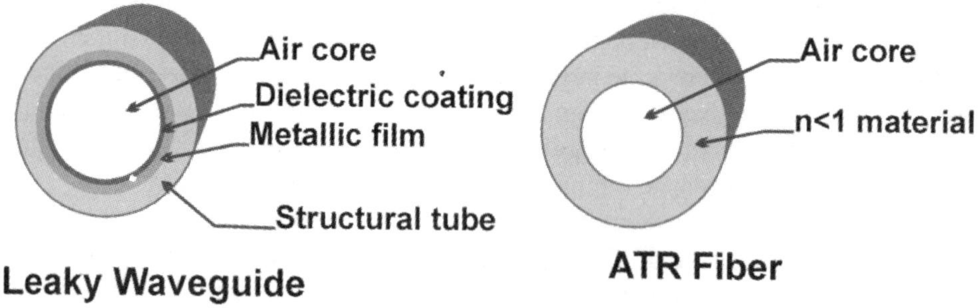

Figure 7.1 Two types of hollow waveguides are the leaky guide, composed of metallic and dielectric coatings, and the $n < 1$ waveguide in which the air core ($n = 1$) is greater than the inner wall material ($n < 1$). [Copyright 2000 from Ref. 2. Reproduced by permission of Taylor & Francis, Inc., http://www.routledge-ny.com.]

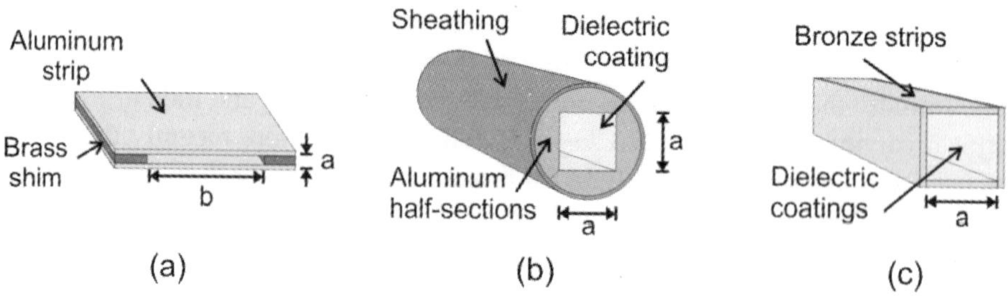

Figure 7.2 (a) Cross section of a hollow rectangular waveguide made by Garmire and/or Kubo et al. The spacers for the Garmire waveguides were made using brass shim stock and for Kubo guide they were made of Teflon and square cross-section waveguides made by Laakmann (b) and by Miyagi group (c).

reflector in the infrared region. Kubo et al.,[6,7] following the work of Garmire, used aluminum foil and Teflon strips to fashion a similar rectangular waveguide. The cross sections of both waveguides are shown in Fig. 7.2(a). Using aluminum foil and plastic instead of Al sheet stock made the guides somewhat more flexible. Both groups found that using polished aluminum improved the transmission somewhat, but not enough to warrant the effort necessary to achieve a polished surface. As may be seen in Figs. 3.2 and 7.2(a), the rectangular cross section of the guides for both researchers were rather large, certainly from the viewpoint of a viable fiber delivery system for invasive surgical procedures. Additionally, the rectangular shape restricts bending to the obvious easy directions. In addition to the additional loss on bending, there is further loss when these guides are twisted. Twist losses are not a concern for the circular cross-section guides.

There have been several attempts to reduce the size and increase the flexibility of the rectangular guides by going to a square cross section. Levy and Laakmann[8] fabricated the rather novel square waveguide shown in Fig. 7.2(b). This waveguide is assembled from two semicircular halves with inner right-angle square cross sections. Each half-section is coated with an optimized thin film of ThF_4 or ZnSe prior to final assembly. Machida et al.[9] devised a clever scheme of forming a 1 mm × 1 mm cross-section guide from phosphor bronze strips. The metal strips, shown in Fig. 7.2(c), were coated with an optimized thin layer of PbF_2 prior to assembly into the guide. Either two or all four inner walls of the guide were coated. The smaller overall dimensions allowed these square guides to be more easily bent in any direction compared to the rectangular guides.

7.1.2 Circular hollow waveguides

Circular waveguides have the advantage of a cross section considerably smaller than that of rectangular waveguides. Furthermore, they can be bent in any direction. This is in part why today circular guides have replaced parallel-plate fibers as the hollow-core guide of choice. In particular, the hollow silica glass guides resemble solid-core silica fibers in that they have the flexibility and feel of standard silica fibers. The cross-sectional structure of the two categories of circular guides is shown in Fig. 7.1. Most circular guides involve one or more dielectric coatings deposited over a metallic coating as shown in Fig. 7.1(a). The ATR guides that have a refractive index of less than 1 over a select wavelength range are either made from special oxide glasses drawn into tubing or from SC sapphire tubing grown by the EFG process (see Sec. 6.2.2).

By far the most actively studied circular guides are the leaky waveguides involving metal and dielectric coatings deposited inside a structural tube. A variety of tubing has been used to form the structural support for the optical films, including tubing made from polymers, glass, and metal. The early structural tubes were stainless-steel hypodermic needle tubing. Stainless tubing is one type of hardened tubing that is flexible but will not kink under normal bending conditions. But the hardness and springy character of stainless steel tubing can be a disadvantage because it is less flexible than glass or plastic tubing. Plastic tubing seems almost ideal in that it is very flexible and inexpensive. Croitoru et al.[10,11] employed polyethylene and Teflon tubing as the structural foundation for their Ag/AgI coatings. The problem with these plastic materials is that they tend to be too soft and the inside surfaces are somewhat rough. The softness leads to deformation of the circular cross section on bending, and the roughness increases the scattering losses. Better results are obtained when harder polymers like polycarbonate tubing are used. Polycarbonate is also much smoother than polyethylene or Teflon tubing. An excellent choice of support tubing seems to be silica glass tubing. It is inherently strong, smooth, and uniform. Waveguides based on hollow silica are today the lowest-loss guides and the only ones commercially available.

The optical properties of the leaky waveguides are dictated for the most part by choice of metallic and dielectric films deposited on the inside of the tubing. The simplest way of viewing the situation is by analogy with thin-film coatings used to enhance the reflectivity of mirrors. The easiest means to improve the reflectivity of a metal mirror in the IR is to add a single dielectric layer. For example, the best reflector at 10 μm is silver. If a single-layer, quarter-wavelength coating of a suitable dielectric is deposited over the silver, the reflectivity can be increased by about a factor of 8 compared to a silver-only film. Successive quarter-wavelength dielectric layers of alternating high- and low-index materials can increase the reflectivity even further. Generally, the ideal number of layers of thin-film coatings with either high/low or low/high index layers is 1, 3, 5, etc. Such a dielectric stack is shown in Fig. 3.6. Furthermore, each high-index layer will have the same thickness and each low-index material the same but different thickness. Normally, it is common that most films have an optical thickness of $\lambda/4$. Non–quarter-wavelength-thick layers are possible and even desirable in some cases, but these are considered nonstandard configurations. Little work has been done on this type of structure because it is sufficiently difficult just to deposit $\lambda/4$ layers; thus there has been virtually no effort to go beyond this standard approach. The theory of single and multilayer thin films for hollow guides is covered in Sec. 3.2.2.

7.2 Hollow waveguide fabrication

Hollow-core waveguides have been fabricated using an enormous variety of techniques. The parallel-plate guides can be simply made by separating the reflecting metal strips with a metal or plastic spacer. The circular guides are more complicated to produce. Some of the methods used to fabricate these guides include physical vapor deposition (PVD) of silver and dielectric layers on metallic substrates,[12] sputtering of metallic, dielectric, and semiconductor films on a leachable mandrel followed by electroplating,[13] and liquid phase formation of coatings inside plastic tubing[14] and glass tubing.[15] Except in the case of some special $n < 1$ glass tubing, none of the methods used to date is capable of producing continuous lengths of hollow guides as in glass fiber drawing. In fact, most of the hollow guides made do not exceed 3 m in length.

7.2.1 Rectangular and square hollow waveguide fabrication

Garmire et al.[1] fabricated some of the earliest parallel-plate waveguides using a bright aluminum sheet held apart by brass shim stock spacers. These guides are shown in Fig. 3.2. In the early guides, simple clamps held the Al-shim stock sandwich together. Later, the guides were permanently fixed by gluing together the Al and brass spacers. The longest lengths were about 1.4 m. In general, Garmire did not polish the aluminum to improve the reflectivity, as the stock aluminum sheet worked well enough.

Kubo et al.[7] improved somewhat on these basic Al plate guides by using Al foil and plastic spacers, as shown in Fig. 7.2(a). Kubo's guides were overcoated with a plastic coating, and they were quite flexible as a result of the use of Al foil rather than Al-sheet stock. Their guides were used to transmit both IR and UV laser radiation.

One of the chief problems with the rectangular waveguides is their large cross section. As may be seen from the dimensions given in Fig. 7.2(a), the long dimension is approximately 1 cm. This large size inhibits the acceptance of these guides in essentially all applications except for high-power laser delivery. Even then there has not been a rush to use the rectangular guides, as circular guides are both smaller in cross section and flexible in any plane. Furthermore, circular guides have been shown to transmit almost 3 kW of CO_2 laser power. To overcome the problem of the large rectangular size, much smaller square cross-section guides have been constructed. Machida et al.[9] made a square cross-section guide using phosphor bronze strips with PbF_2 coatings vacuum-deposited on the strips. The guides were 1 mm $\times$ 1 mm $\times$ 1 m in length. Either two or all four inner walls of the guides were coated prior to fabrication of the guide. The coated strips were permanently assembled into a square waveguide by passing the four strips through a heated soldering solution. The guides could be bent to a 30-cm radius.

The square guides shown in Fig. 7.2(b) are the work of Laakmann and her colleagues at Laakmann Electro-Optics (now part of Lumenis, Inc.). The unique feature of these guides is that they are made in two halves, which are then put together to make the final square geometry. The structural member of the guide is one-half of a stainless-steel tube. The inside of the guide is made from a circular wire that has a polished V-groove machined into the wire, as shown in Fig. 7.2(b). Each half of the guide is then placed in a vacuum chamber and a $\lambda/4$-thick layer of ThF_4 or ZnSe is deposited in the groove. Inserting two stainless-steel tube halves into a plastic or metal sleeve makes the final guide. Clearly this is a rather laborious process, and it is not a common approach today. Nevertheless, the measured losses for these guides are below 1 dB/m.

7.2.2 Circular hollow waveguide fabrication

7.2.2.1 Waveguides based on metal tubing

Professor Miyagi and his coworkers at Tohoku University[13,16] have pioneered the development of metal-tubing waveguides. The structural tubing in their guides is nickel. The fabrication process used by Miyagi et al.[13] involves three steps. In the first step, shown in Fig. 7.3(a), a pipe or mandrel typically made of aluminum is placed into a sputtering chamber, and a dielectric layer followed by a metallic film is sputtered onto the outer surface of the pipe. Next, the coated pipe in Fig. 7.3(b) is placed into an electroplating tank, where a thick nickel layer is deposited on top of the sputtered layers. Finally, in Fig. 7.3(c), the pipe is etched away with NaOH, leaving the structure shown in Fig. 7.4. The earliest metal-tube guides were composed of only a Ge coating. Thereafter, metallic coatings of Ag were used as the base over which Ge, ZnSe, and ZnS were deposited. The addition of a dielectric layer of the correct optical thickness greatly reduced the losses at 10.6 μm. For example, the losses for the Ge/Ni guides are predicted to be about a factor of two and six times greater than Ge/Ag/Ni and ZnSe/Ag/Ni coated guides, respectively. For this reason, all waveguides made today have a metallic and dielectric layer for increased reflectivity.

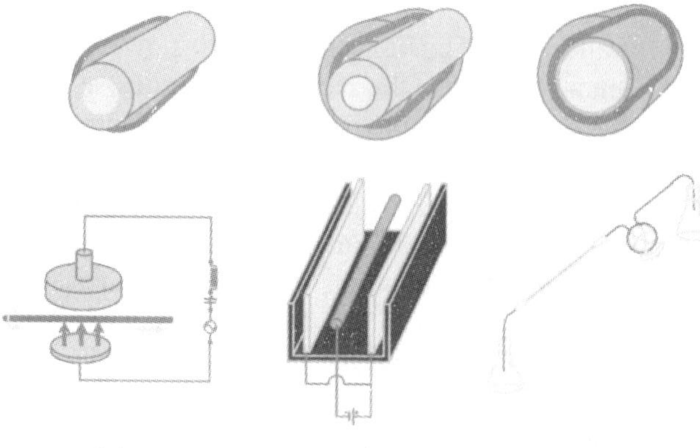

(a) (b) (c)

Figure 7.3 Process used by Miyagi et al. to form a metal-tube waveguide with inner metallic and dielectric coatings. The structural tube is nickel that has been deposited electrically.

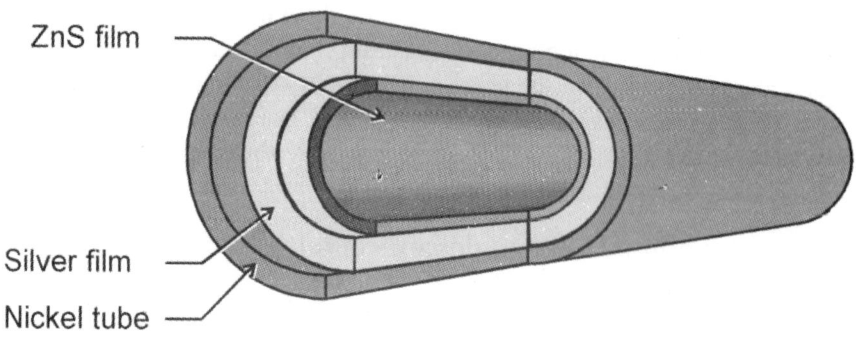

Figure 7.4 Final hollow metal-tube-type waveguide fabricated using by Miyagi's group at Tohoku University using the procedure outlined in Fig. 7.3.

More recently, Komachi et al.[17] used a very similar over-coating technique to produce Ge/Ag, Ge/Au, and Ge/Cu guides. Their method involved using an etchable glass tube as the mandrel. This tube was continuously fed into a vacuum chamber, where a 0.1- to 0.5-μm thick layer of Ge was deposited on the outer surface of the tube by electron-beam evaporation. Next, the metallic layer was deposited over the Ge again using the electron-beam techniques. Finally, a 200-μm-thick layer of Ni was electroplated over the Ag, Au, or Cu coating. Then the glass tube was removed using a nitric acid etching solution. Their final waveguides had bore sizes of 0.5, 1.0, and 1.5 mm. Komachi et al. obtained the best results for the Ge/Ag guides because good-quality Au and Cu coatings were difficult to deposit.

The waveguides developed by Morrow et al.[18] are constructed from a silver tube. Instead of depositing a metallic layer inside a hollow mandrel, they begin with an extruded silver tube and then deposit a silver halide film on the inside of the tube, as shown in Fig. 7.5. The first step involves chemically polishing the extruded Ag tubing using an acid solution of chromic oxide; this results in a smoother inner surface and, therefore, lower loss. Next, a single layer of either AgBr or AgCl was deposited on the inner Ag wall using either a liquid or gas phase reaction with Ag. That is, a thin layer of Ag was depleted to form the Ag-halide film. The Ag-halide layer was formed from the liquid phase by flowing bromine or iodine liquids through the polished tubing. Gaseous methods involved passing nitrogen through bromine or iodine solutions and then through the Ag tubing. The process continued until the correct $\lambda/4$-thick dielectric layer was formed. The advantage of this method is that the structural tube is made of silver, so that there is no need to deposit an Ag coating on the inside of another tube. The disadvantage is that extruded Ag tubing is not smooth, and the roughness present even after polishing can lead to higher loss and to the transmission of higher-order modes. Typical bore sizes ranged from 400 to 1500 μm, and lengths as long 3 m have been made. Generally, the final silver tube is placed in a plastic tube for protection.

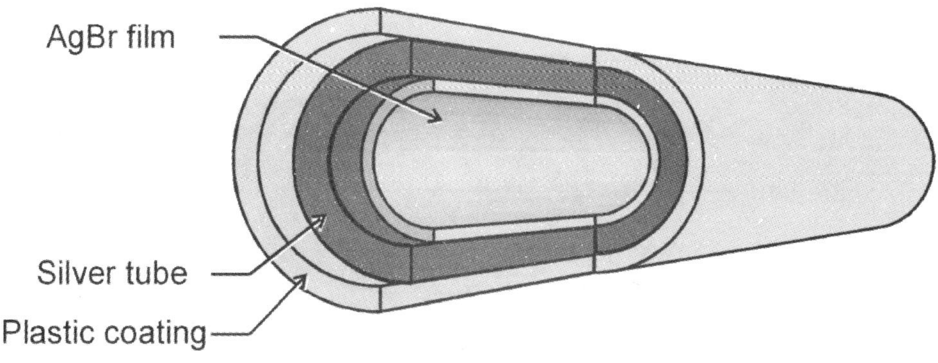

Figure 7.5 Metal-tube waveguide made by Morrow et al. in which the entire tube is made of silver and then a dielectric layer of AgBr is deposited on the inside.[2] [Reproduced with permission.]

Levy has developed a rather unusual approach to forming a metal-tube guide. His method involves coating a thin metallic strip with Ag plus a dielectric film, and then rolling the strip and inserting it into a stainless-steel tube.[12] The details of the waveguide are shown in Fig. 7.6. The thin, flexible molybdenum ribbon is placed in a vacuum chamber and an Ag film is deposited, followed by a thin film of PbF_2. The PbF_2 film has the correct optical thickness for optimal reflectivity. The metal strip is then pulled through a succession of wire-drawing dies to form the ribbon into a tightly rolled cylinder. The ribbon is then inserted into stainless-steel hypodermic tubing as shown in Fig. 7.6. The advantage of this guide is that the coatings are very uniform because they are deposited using conventional physical vapor deposition techniques. The small seam that remains in the ribbon after rolling does not seem to adversely affect the loss or the laser-power handling capability of the guides.

7.2.2.2 Waveguides based on plastic tubing

Some of the earliest hollow waveguides were made using plastic tubing. These guides are naturally very flexible and inexpensive to make. Prof. Croitoru and his group at Tel Aviv University did the pioneering work in this area.[14,19,20] Using polyethylene and Teflon tubing as the structural member, they first deposited an Ag film on the inside of the tubing. Next they formed a dielectric layer of AgI using wet or liquid-phase chemistry techniques to convert some Ag to AgI.[20] The liquid-phase chemistry methods are essentially the same as those described in the next section for glass tubing. Initially, Croitoru used rather large bore tubing (about 2-mm bore), but more recently guides with bore sizes of about 1000 μm have been fabricated. Figure 7.7 shows the composition of the guide. In general, the losses are higher for these plastic guides compared to the metal or glass tube guides, because the rougher inner surface of the plastic tubing leads to increased scattering losses. Also, all plastic guides have poor laser-power handling capability as a result of their low softening temperature.

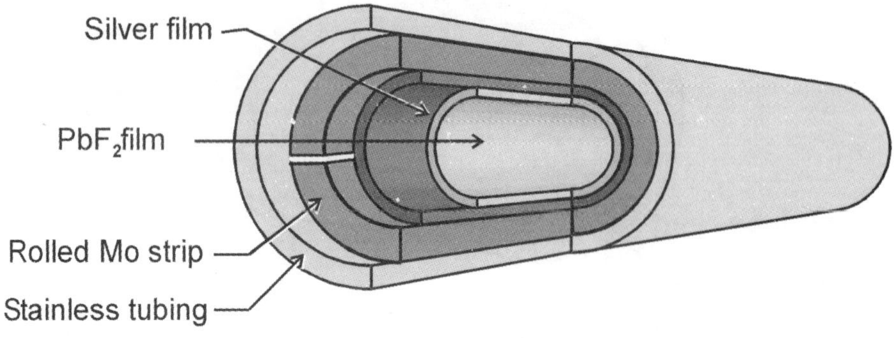

Figure 7.6 Metal-tube waveguide made by Levy and Laakmann in which a thin Mo strip is coated and then rolled to fit into a stainless steel tube.[2]

Harrington and his group at Rutgers have used similar liquid-phase chemistry methods to deposit Ag/AgI films inside polycarbonate tubing.[21] Extruded polycarbonate tubing in lengths of 2 m and bore sizes from 500 μm to 2 mm was used to make the guides. The structure of these guides is similar to that shown in Fig. 7.8 for glass tubing, but polycarbonate is substituted for the glass and plastic coating. The advantage of polycarbonate tubing over polyethylene or Teflon tubing is that it is much smoother. Other polymer tubing used by George and Harrington[21] to make Ag/AgI waveguides is butyrate, polyimide, and PMMA. In all cases, it was essential to first sensitize the tubing using SnCl or $PdCl_2$ solutions to improve the adhesion of the Ag films. Generally, the Ag films do not adhere as well to polycarbonate as they do to glass or metal tubing. But if the plastic is first sensitized, there is much better adhesion.

7.2.2.3 Waveguides based on glass tubing

One of the most popular hollow waveguides today is the hollow glass waveguide (HGW) developed by Harrington's group at Rutgers University.[2] The advantage

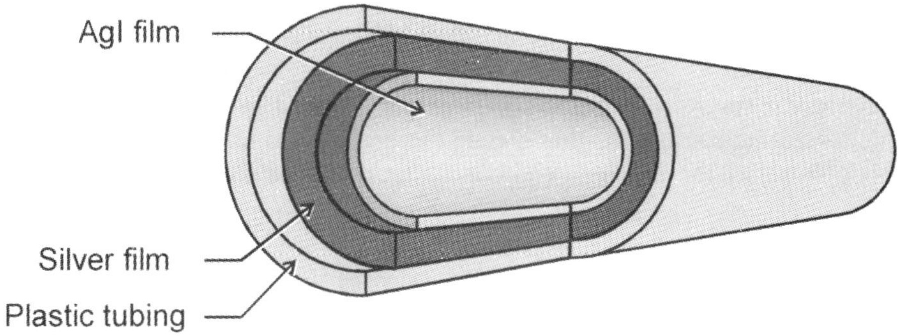

Figure 7.7 Structure and bending loss for hollow plastic waveguide made by Croitoru's group at Tel Aviv University.[2]

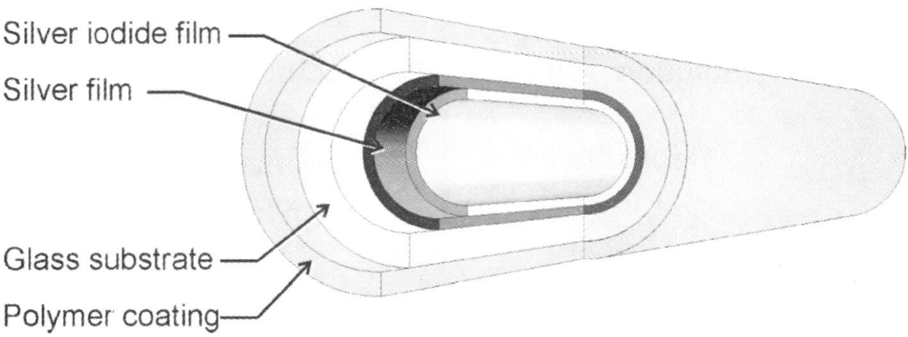

Figure 7.8 Structure of the HGWs showing the metallic and dielectric films deposited inside silica glass tubing using the liquid-phase chemistry techniques shown in Fig. 7.9.[2]

of the hollow glass structure over other hollow structures is that it is simple in design, flexible, and, most important, it has a very smooth inner surface. HGWs have a metallic layer of Ag on the inside of silica glass tubing and then a dielectric layer of AgI over the metal film identical to that used to make the hollow plastic guides. Figure 7.8 shows a cross section of the structure of the HGWs. The fabrication of HGWs begins with silica tubing, which has a polymer (UV acrylate or polyimide) coating on the outside surface. A liquid-phase chemistry technique, shown in Fig. 7.9, is used to deposit the Ag and AgI films inside the glass tubing.[22] This technique is similar to that used by Croitoru and his coworkers[14] to deposit metal and dielectric layers on the inside of plastic tubing. The first step involves depositing a silver film using standard Ag plating technology. Next, a very uniform dielectric layer of AgI is formed through an iodization process in which some of the Ag is converted to AgI. As shown in Fig. 7.9, a peristaltic or suction pump can be used to force the silver and reducing solutions through the silica tubing. Generally, the Ag film is between 0.5 and 2 μm thick, and it is deposited slowly over a period of about 1 hour. Immediately after the silver is deposited, an iodine solution is pumped through the tubing and, through a subtraction process, a layer of AgI is formed. By controlling the concentration of the iodine solution and reaction time, an AgI film of the correct optical thickness can be deposited. Using these methods, HGWs with bore sizes ranging from 100 to 1200 μm and lengths as long as 13 m have been made.

There are several key process parameters involved in the deposition of the metallic and dielectric films that dictate the quality and uniformity of the films. Among these are the flow rate and concentration of the solutions used. Because

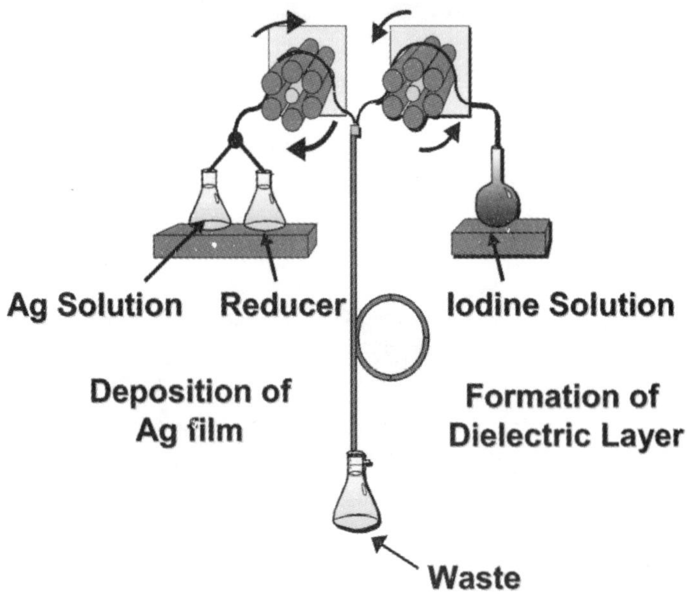

Figure 7.9 Schematic of the experimental setup for depositing the Ag metallic and AgI dielectric films inside silica tubing to form the HGWs.[2]

the AgI film is formed from the Ag layer, it is essential to have a sufficiently thick silver film; otherwise the Ag layer will be too thin for the formation of a sufficiently thick AgI film over Ag. However, thick Ag layers are rougher than thin Ag films. Matsuura et al. have studied the chemical kinetics involved in the deposition of Ag, AgI, and Au films inside silica tubing.[22] They were able to relate film-thickness to reaction times and also to make an independent measurement of the film thickness using gravimetric analysis. For example, they obtained a correlation between the iodination time and the thickness of the AgI layer for different concentrations of iodine solutions. With this information they were able to readily deposit coatings with the appropriate optical thickness. Their film-thickness measurements using gravimetric analysis agreed very well with the thickness calculated using Eq. (3.18). The Au films were also of high quality. The difficulty with Au films is that it is not easy to deposit a dielectric film on the Au layer using a subtractive process analogous to AgI. This is because Au is not very reactive.

Rabii and Harrington[23] have studied the uniformity of the AgI films. Since the deposition generally involves solution flow in only one direction, as shown in Fig. 7.9, it is reasonable to assume that the AgI layer will be thicker at the input end compared to the output end. These investigators measured the thickness variation in a 6-m-long HGW made using the normal solution flow parameters. When they examined the film thickness variation they found that there was a 14.5% end-to-end variation in the thickness of the AgI film. Rabii and Harrington found that the thickness variation could be lessened by increasing the flow rate by a factor of three to 7 ml/min. Using this higher flow rate, they were able to reduce the thickness variation of the film from 14% to 2% for a 6-m-long fiber. A consequence of a nonuniform dielectric film is that the losses increase in the 2- to 12-μm region. Further details of the optical implications of these nonuniform films are given in Sec. 7.3.4.2.

It is important that the surface roughness of both the metallic and dielectric films be as small as possible to reduce scattering losses. Rabii et al. studied the roughness of these films as a function of silvering time.[24] They found that the grain size and therefore the roughness of the Ag film increases as $t^{1/2}$, where t is the silvering time. A $t^{1/2}$ dependence is characteristic of a diffusion-controlled process. In Fig. 7.10, topographical AFM pictures are shown of the surface of Ag films made using the shortest (5 min) and longest (80 min) silvering times. As the silvering time increases, the RMS surface roughness increases from 12.3 nm to 71.8 nm for the shortest and longest times, respectively. Rabii et al. further related the roughness to an increase in attenuation. These optical measurements are discussed in detail in Sec. 7.3.4.2. In general, they found that the absorption coefficient increases as the square of the roughness. Therefore, a thin, smooth Ag film produces the lowest loss. However, the Ag must be thick enough so that light does not propagate through a partially reflecting silver layer.

Another distinctly different approach to forming a single-dielectric layer on an Ag film is that used by Miyagi and his coworkers at Tohoku University.[25–27] They deposited a variety of different polymer films on the inner surface using liquid-phase techniques similar to those used to deposit AgI. The primary advantage

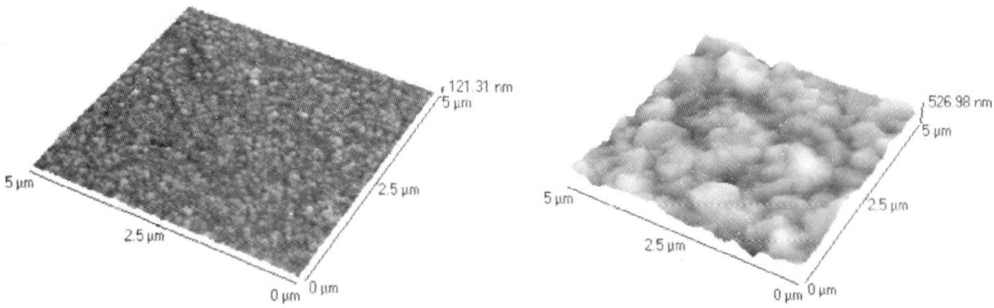

Figure 7.10 Topographical AFM scans showing the surface roughness of Ag films deposited on glass substrates for 5 and 80 min. The top image is for a 5-min deposition time while the lower image is for 80 min. The dimensions on these AFM scans have been corrected from the original Rabii et al. Ref. 24. [Reprinted with permission.]

to this method is that they deposit a dielectric film that does not depend on the subtraction of the metallic layer to form the dielectric, as is the case for the AgI deposition technique. This enables them to deposit their films over smoother Ag surfaces, because the Ag film does not need to be as thick as it does for the Ag/AgI combination. The polymer films studied by Miyagi's group include polyimide,[28] fluorocarbon polymer (FCP),[29] and cyclic olefin polymer (COP).[25,30] In Fig. 7.11, their setup is shown for the deposition of Ag and polymer films. Wang et al.[26] noted, as did Rabii et al., that the roughness of the Ag film increased with silvering time. The disadvantage of polymer films is that they often have rich infrared absorption spectra, and they are generally not as low loss as an ionic dielectric material. For example, polyimide has a number of absorption bands between 5 and 10 μm and a lesser number between 2 and 4 μm. These absorption bands are due to various hydrocarbon absorptions. Nevertheless, it is possible to find polymeric materials that have relatively transparent regions at some common IR laser wavelengths. The best polymer film to date seems to be COP, because it has the fewest absorption bands and it is the best of the polymer films at CO_2 laser wavelengths.

Deposition of the polymer films is accomplished by dissolving the polymer in a suitable solvent and then pumping the mixture through the Ag-coated silica, as shown in Fig. 7.11. At times the polymer films are heat cured, but more often they are merely cured at room temperature by flowing dry N_2 through the tubing. Wang et al.[31] give an extensive analysis of their FCP films with respect to film thickness and process parameters. They also found thickness variations in the FCP films as a function of coating velocity just as Rabii et al. found with AgI films. From their study they were able to obtain the optimal deposition parameters. The polyimide and FCP materials have strong absorption bands beyond about 5 μm, so their use is restricted to the shorter IR wavelengths and Er:YAG laser transmission at 2.94 μm. The COP material is unusual in that it may be used at the longer wavelengths even out beyond 10 μm. Abe et al.[30] have obtained low loss at CO_2 laser wavelengths using a Ag/COP combination with COP coating thicknesses of about 0.77 μm. It should be noted that polymers generally have rather low refractive indices (for example, $n = 1.53$ at 10.6 μm for COP). Also, while these film do give reasonably

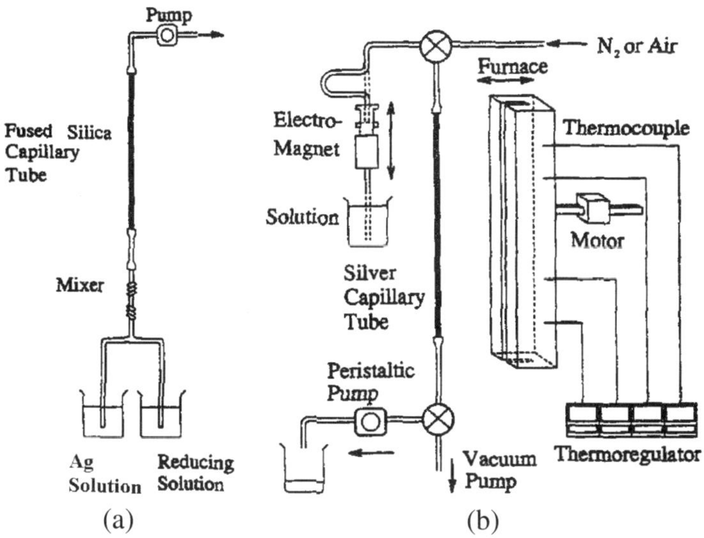

Figure 7.11 (a) Liquid-phase chemisty process used by Miyagi, et al. to first deposit an Ag film inside silica tubing (b) and then to deposit and thermally cure a thin polymer layer over the Ag.[29]

low loss, they are not suitable for ultrahigh power laser delivery as they do not have the temperature capability of the metal halide or Ge-coated waveguides.

7.3 Optical properties of hollow waveguides

The losses in hollow waveguides depend critically on bore size, bending radius, wavelength, mode, and polarization. As discussed in Chapter 3, the attenuation coefficient varies as $1/a^3$ and as $1/R$, where a is the bore radius and $1/R$ is the curvature. Furthermore, the loss also depends on λ^2. This may be seen in Eqs. (3.7) and (3.8) for the rectangular waveguides and Eq. (3.13) for circular waveguides. It may also be observed from these same equations that the loss is strongly dependent on the propagating mode varying as the mode parameter squared.

7.3.1 Metallic-tube, rectangular, and square waveguides

The earliest rectangular waveguides were those studied by Garmire et al.[1] Their guides are composed of simple aluminum strips spaced a distance a apart, as shown in Fig. 3.2. They studied the losses for these guides using CO_2 laser energy polarized parallel (TE) to the aluminum strips. The straight losses for different waveguide heights a are given in Fig. 7.12. The measured losses in Fig. 7.12 can be compared with those calculated using Eq. (3.7). From the data it is seen that the loss does indeed vary as $1/a^3$ as predicted by Eq. (3.7). However, Garmire noted that the loss calculated using Eq. (3.7), with $n^* = 27 - i70$ for a high-quality Al surface, is much lower than that measured. In fact, the calculated loss for a spacing of 400 μm is approximately 0.04 dB/m compared to a measured value of about

1 dB/m. The discrepancy between these two values is due to the non-ideal surface of the Al strips. In particular, Garmire attributes essentially all the loss to increased scattering from surface irregularities. Nevertheless, the losses for the guides are quite low at 10.6 μm. Specifically, they are below 0.5 dB/m for waveguide heights of 600 μm or greater. The additional loss due to bending for two waveguides with $a = 500$ μm and $a = 250$ μm is shown in Fig. 7.13. The guides were bent 180° in the easy direction, and there were no twists. The data for both waveguide heights is essentially the same, and from Fig. 7.13 the predicted linear relationship between loss and curvature is evident. These bending losses are again higher than expected,

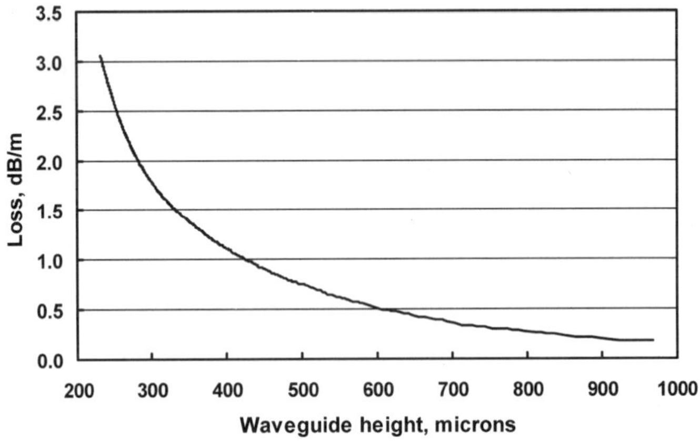

Figure 7.12 Measured losses for rectangular waveguide made by Garmire et al. as a function of spacing, a, between the aluminum plates.[1] [Copyright 2003 IEEE.]

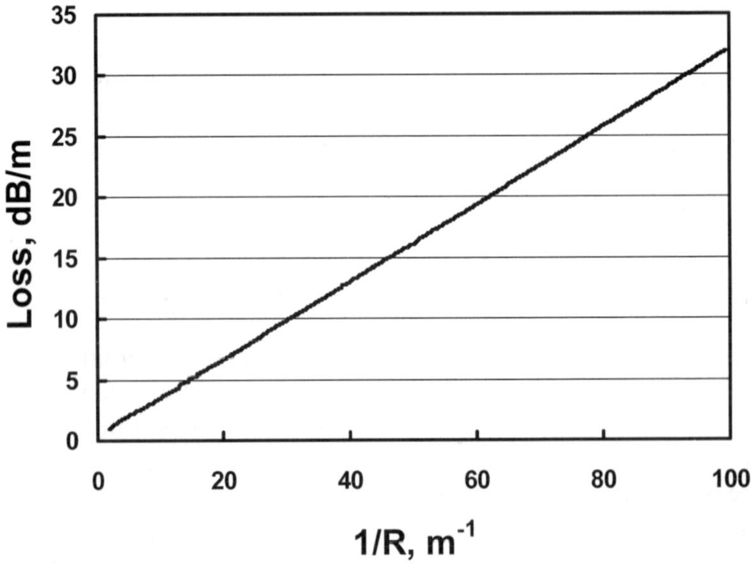

Figure 7.13 Bending losses for the rectangular waveguides shown in Fig. 7.12, with plate spacings of $a = 250$ and 500 μm. There is little difference in loss between these two spacings.[1] [Copyright 2003 IEEE.]

by about the same amount as for the straight waveguide. Again, Garmire attributes this discrepancy between theory and experiment to surface roughness.

The losses for a similarly constructed waveguide by Kubo and his coworkers[7] at Kinki University in Japan are somewhat lower than those of Garmire. For their guides, shown in Fig. 7.2(b) with a height of 500 μm, the best loss for a straight, 1-m-long guide measured at 10.6 μm was 0.46 dB/m. The results for the same guide measured using an Er:YAG laser at 2.94 μm were 1.9 dB/m. It should also be noted that Kubo's waveguides have also been used quite successfully in the ultraviolet region. For example, the losses for the same Al strip waveguides studied in the infrared have losses of about 5 dB/m at the KrF excimer laser wavelength of 248 nm.[6] The losses are high as a result of the increased scattering at shorter wavelengths.

The square cross-section waveguides made by Miyagi et al., described in Sec. 7.2.1, also have low loss at 10.6 μm. These researchers fabricated a 1 mm × 1 mm × 1 m–long guide with a PbF_2 coating over the phosphor bronze metal walls. This is the structure shown in Fig. 7.2(b). The straight losses for these guides are approximately 0.1 dB/m measured using a CO_2 laser.[32]

7.3.2 Metallic-tube, circular waveguides

The most extensive studies on metallic, circular cross-section waveguides have been done by Miyagi and his colleagues at Tohoku University, Japan.[33] Their pioneering research fabricating the metallic/dielectric structures, described in Sec. 7.2.2.1, led the way for many of today's circular waveguides. Their earliest work involved the deposition of Ge coatings inside a Ni tube used for structural support.[34] Later improvements involved coating the Ge films over Ag deposited inside the Ni tubing and, finally, the use of two-six materials such as ZnS and ZnSe over Ag films. The fabrication methods for these coatings are shown in Fig. 7.3. In Fig. 7.14 typical loss curves are shown for some of their best coatings.[35] The data

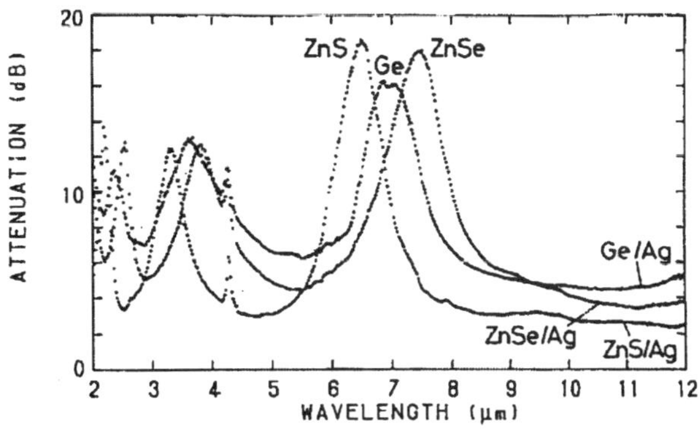

Figure 7.14 Loss for several different dielectric coating materials deposited over Ag films using the process developed by Miyagi's group at Tohoku University. These waveguides were made using the procedure outlined in Fig. 7.3.[35] [Reproduced with permission.]

shown is for 1000-μm-bore guides 1 m in length. The advantage of these coating materials is that there is a broad spectral response. In general, the exact bandpass and region of minimum loss can be tailored to meet the application. This is accomplished by adjusting the thickness of the dielectric films. For the data in Fig. 7.14, the thickness of each dielectric film is chosen to optimize the transmission for the 10.6-μm wavelength of the CO_2 laser. For this case, the thickness of both the ZnS and ZnSe films is about 0.7 to 0.8 μm.

There are several key points in their data. First, note that the lowest loss at 10.6 μm is for the two-six coating materials, with the ZnS/Ag combination having the lowest loss. The broad peaks in the spectra are due to thin-film interference effects and not impurities. Finally, the loss for the guides shown in Fig. 7.14 seems quite high when compared to the straight losses measured using a CO_2 laser. The CO_2 laser results are 0.3 dB/m and 0.45 dB/m for ZnS/Ag and ZnSe/Ag, respectively. The reason that the spectral losses are higher than the 10.6 μm results is that incoherent light from the spectrometer is launched with a much lower f/# than that used for the laser measurements. This has the effect of launching higher-order modes, which are much more lossy than the lowest-order HE_{11} mode (see Sec. 3.2 and Eq. (3.13) for details on the dependence of loss on the mode parameter). If incoherent light is in fact launched into the guide with the correct f/#, then the losses for the coherent (laser) and incoherent (spectrometer) light will be the same. Normally, however, incoherent light is focused into the guides with a low f/# because the overall throughput energy will be higher than if a high f/# or low-order mode launch is used. It should also be noted that the spectral data in Fig. 7.14 does not exhibit the λ^2 dependence on the absorption coefficient mentioned at the beginning of Sec. 7.3 or as seen in Eq. (3.13). The primary reason for this is that the scattering losses increase as $1/\lambda^2$, and this tends to offset the predicted λ^2 dependence.

The bending losses for the two-six coated waveguides are shown in Fig. 7.15. For these measurements the first 20 cm of the guide was left straight and the re-

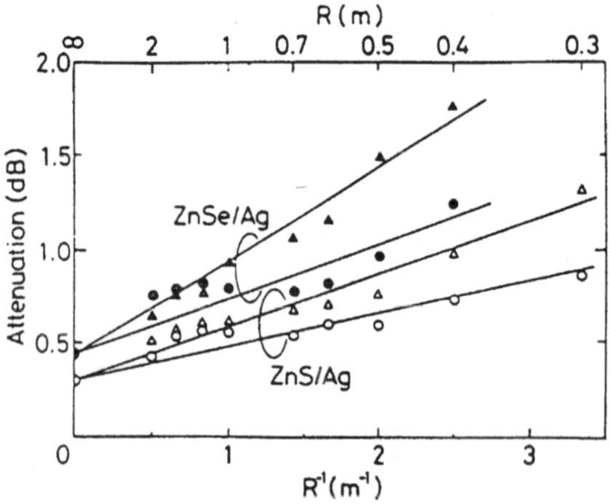

Figure 7.15 Bending losses for 1000-μm bore, 1-m long waveguides made by Miyagi and his group. These are some of the same guides shown in Fig. 7.14.[35] [Reproduced with permission.]

maining 80 cm was bent to a uniform radius *R*. Therefore, the amount of fiber under bend remains constant, and as *R* decreases there will be more loops in the fiber. The measurements were made using a CO_2 laser. From the data in Fig. 7.15 it can be seen that the loss increases linearly with curvature, as predicted by Eq. (3.2). Furthermore, the loss depends on the polarization of the laser beam with respect to the plane of bending. Losses are always lowest for s-pol light or light polarized perpendicular to the bending plane (see Sec. 3.2.1). The bending loss for either s-pol or p-pol (parallel to the plane of bending) is also seen to be lower for the ZnS/Ag–coated waveguides. As one might expect, bending losses are lower for the lowest-loss guides. In fact, a closer look at the bending data reveals that the slope of the bending loss curves is less for the lower-loss fiber. This point is discussed further in Sec. 7.3.4. Finally, note that the actual amount of bending in the data of Matsuura and Miyagi is not great. The maximum curvature of 3.3 is a bend radius of 30 cm; that is, the waveguide would be bent to a rather large 60-cm-diameter loop.[35] At this diameter the loss has approximately doubled over the straight loss.

The silver-tube waveguides developed by Morrow et al.[18] have a somewhat rougher inner wall than the waveguides of Miyagi described above. Nevertheless, the losses can still be rather low. The bending loss at 10.6 μm for one of Morrow's guides with a 1000-μm bore and 1 m in length is shown in Fig. 7.16. The losses for the straight waveguide are quite low. Unfortunately, the guides tend to mix modes and, therefore, the beam output is rather multimode. This aspect will be examined in greater detail in Sec. 7.3.4 when a comparison is made to the cleaner output of the hollow glass waveguides. The poor mode quality of the silver-tube guides results in large part to the rougher inner surface of the extruded silver tubing.

The waveguides of Levy, based on rolling a PbF_2-coated metal strip inserted into a stainless-steel tube, also exhibit low loss at 10.6 μm. The bending loss for one of Levy's guides is shown in Fig. 7.17.[2] This waveguide has a 750-μm bore and it is 1 m in length. As in the case of the guides made by Miyagi and Morrow, the maximum amount of bending is limited. In this case they are bent only to a 50-cm loop, and at this bend diameter the bending loss is essentially double the straight loss. The primary reason that these metal tube guides are not bent to smaller radii is that they tend to kink when bent too strongly.

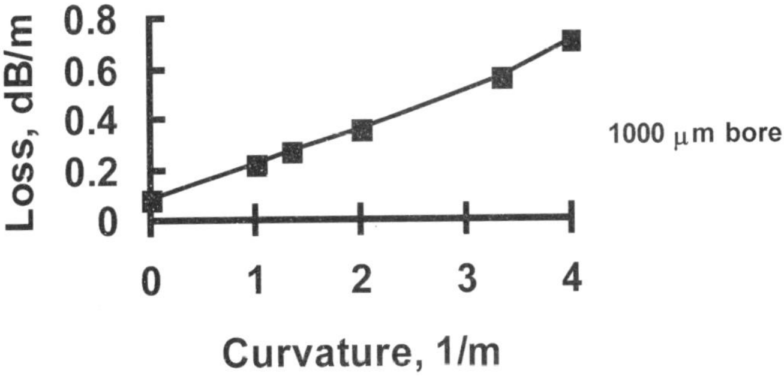

Figure 7.16 Bending loss for hollow guide made from silver tubing with a AgBr film deposited on the inside surface.[2]

7.3.3 Plastic-tube waveguides

Hollow waveguides made from plastic tubing are generally quite flexible but suffer somewhat from higher scattering losses resulting from the inherently rougher inner surface compared to glass. Nevertheless, they were some of the first guides made with silver and dielectric coatings. Croitoru and his group at Tel Aviv University made the first plastic waveguides.[20,36] They used polyethylene and Teflon tubing as the structural member. Their fabrication methods are described in Sec. 7.2.2.2.

Initially, Croitoru used large-bore plastic tubing with bore sizes in excess of 1000 μm. More recently they have fabricated guides with bore sizes a little less than 1000 μm.[36] In Fig. 7.18, bending loss data is given for their 1000-μm-bore waveguide measured at 10.6 μm. The losses are higher than those measured for the metallic tubes. This is primarily due to increased scattering losses. In an independent measurement, more than 65 W of CO_2 laser power was transmitted through an 1850-μm-bore guide for several minutes.

As discussed in Sec. 7.2.2.2, plastic waveguides have also been made using polycarbonate tubing as well as other optical polymer tubing. The advantage of the polycarbonate tubing over polyethylene or Teflon tubing is that it has a much smoother inner surface. It is also a tougher, more rigid plastic than the other two, which means that it is less likely to deform on bending. The lowest loss has been measured for the polycarbonate guides with a Ag/AgI coating. The measured

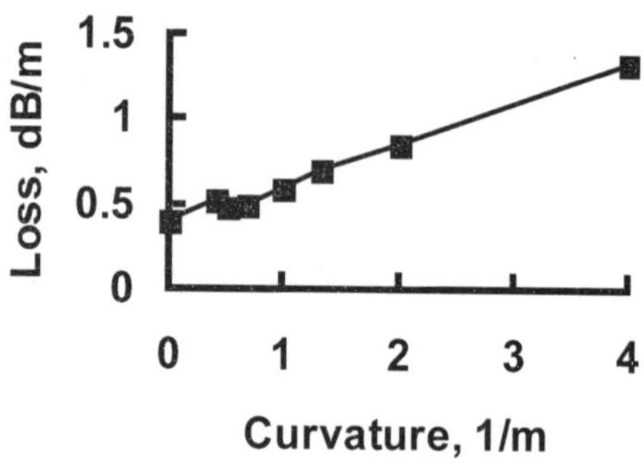

Figure 7.17 Bending loss for hollow guides made by Levy et al.[2]

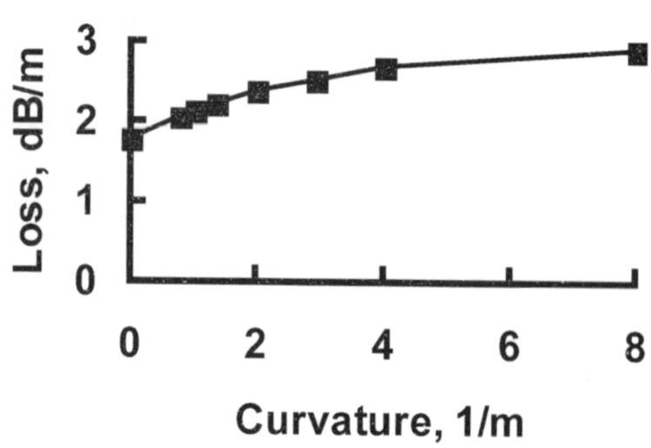

Figure 7.18 Loss for hollow plastic waveguides made by Croitoru et al.[2]

straight loss at 10.6 μm for these polycarbonate waveguides is shown in Fig. 7.19. Note that the loss falls below 0.1 dB/m for bore sizes greater than about 1200 μm. The lowest measured loss is 0.012 dB/m for a 2000-μm-bore guide.[21] This is the lowest reported loss for a hollow waveguide, yet it is still above the theoretical limit, as indicated by the solid line in Fig. 7.19. The primary reason for the low loss is the larger bore size of 2000 μm. If this bore size had been made using silica tubing the waveguide would have been essentially inflexible, or at least limited to rather large bending radii. Large bore sizes are a particular advantage for the plastic guides.

The bending losses for an 840-μm-bore polycarbonate guide at 10.6 μm are shown in Fig. 7.20.[21] This is one of the smaller-bore plastic guides, yet the bending loss is rather small compared to the bending losses for the polyethylene or Teflon guides. Furthermore, notice in Fig. 7.20 that the tubing is bent into a rather small radius. The minimum radius is 10 cm, or about an 8-inch-diameter circle. It is possible to bend polycarbonate tubing to this small a diameter and retain its shape because polycarbonate tubing is stiffer than the other polymers and, therefore, less likely to go out of round upon bending.

The output mode profile is important for applications requiring good beam quality and/or preservation of the lowest-order Gaussian-type mode. In general, the metal waveguides do not preserve a single mode launched into the guide. The modal properties of the metal and plastic waveguides have been studied in some detail, but none preserve the mode as well as the small-bore hollow silica guides. As a means of introducing the subject of mode mixing, it is worthwhile to look first at the hollow polycarbonate guides, as they illustrate well the general aspects of mode mixing and the effect of roughness and bore size on the mode coupling. The smallest-bore waveguides exhibit the least amount of mode mixing, as would be predicted by the general dependence of α on the mode parameter u_{lm} in

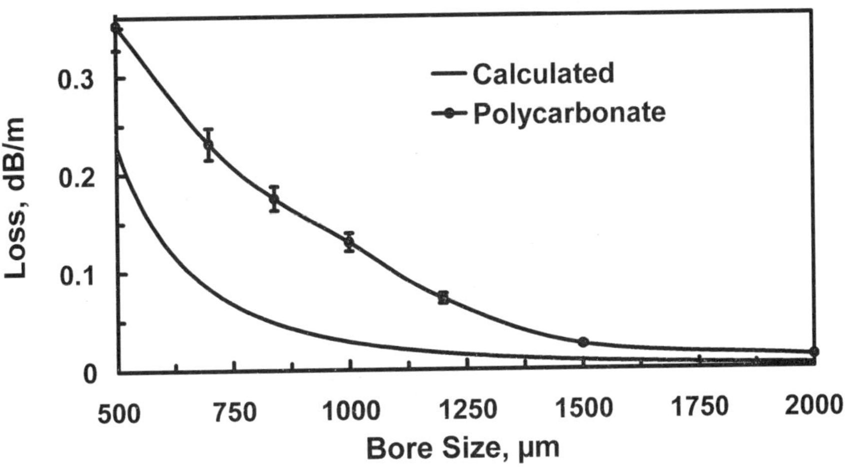

Figure 7.19 Measured and calculated losses at 10.6 μm for hollow polycarbonate waveguides with Ag/AgI coatings. Note the very low measured loss of 0.012 dB/m for the 2000-μm-bore waveguide.[21]

Eq. (3.19). That is, the loss for higher-order modes is relatively higher in small-bore waveguides compared to large bore sizes. This effect may be seen qualitatively in the mode profile data in Fig. 7.21 for two quite different bore sizes. Figure 7.21(a) shows the mode profile of 500-μm tubing, while (b) shows it for the 1000-μm-size

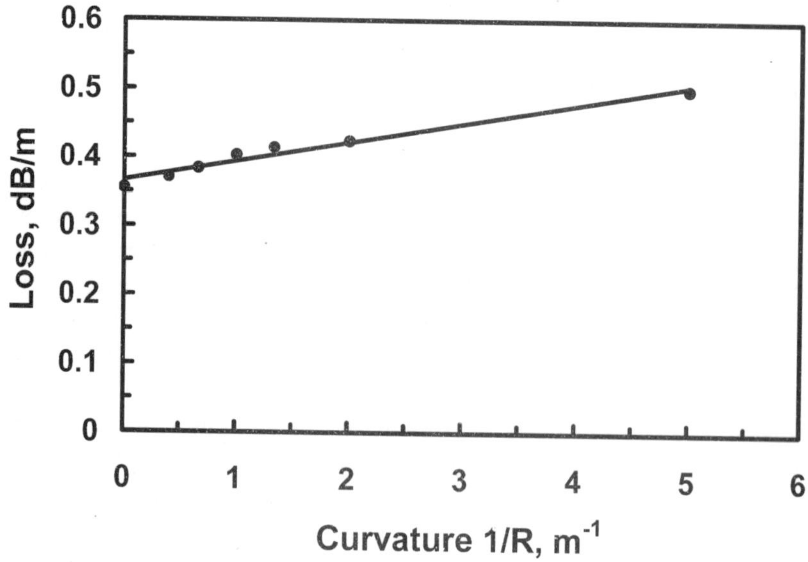

Figure 7.20 Bending losses for the 840-μm-bore hollow polycarbonate waveguides.

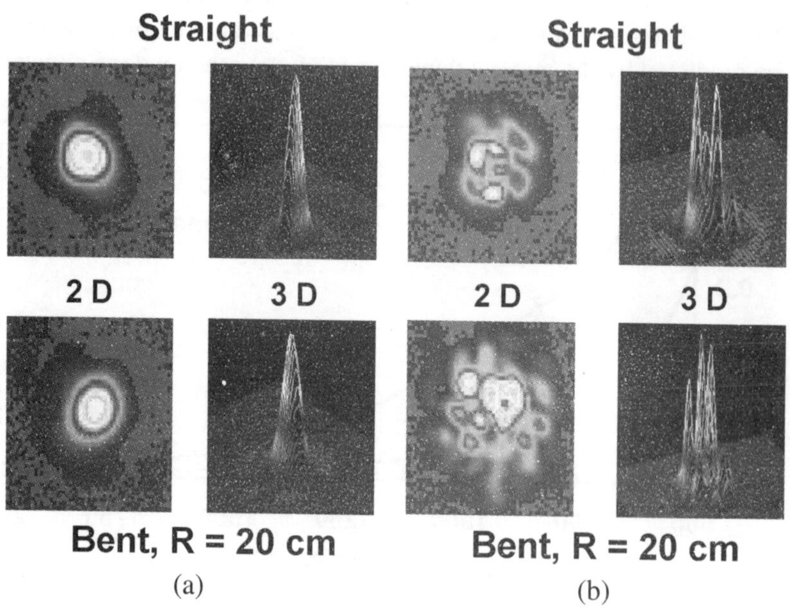

Figure 7.21 CO_2 laser output beam profiles from a (a) 500-μm and (b) 1000-μm bore hollow polycarbonate waveguide.[21]

waveguide. The 500-μm-bore waveguide remains largely single mode even when bent to a radius of 20 cm. The 1000-μm-bore tubing shows a distinctly multimode pattern even when straight, and bending further mixes the modes. The dramatic difference in modal properties between the two sizes is largely due to scattering from surface roughness, which leads to coupling of energy into the higher-order modes. In the next section on hollow glass waveguides it will be seen that the mode mixing is not as severe owing to the inherently smoother surface of drawn glass versus extruded plastic tubing.

Other plastic tubing has been used to make waveguides. In most cases, Ag/AgI films were deposited in the same manner as applied to the polycarbonate or Teflon tubing. Waveguides made using 2000-μm-bore PEEK tubing had a loss of 0.19 dB/m at 10.6 μm on straight tubing.[21] Other polymer tubing studied included 2000-μm-bore polyimide and 3000-μm-bore butyrate tubing with 10.6-μm losses at 0.74 dB/m and 0.08 dB/m, respectively. The polyimide tubing had a very thin wall that deforms easily; therefore, it is not suitable for most applications, as it is too delicate. One method used to improve the structural integrity of polyimide tubing is to place heat-shrinkable tubing over the polyimide and then heat the tubing for a tight tubing-to-tubing bond. Unfortunately, the act of shrinking can itself lead to ripples in the polyimide, which in turn leads to scattering loss.

7.3.4 Hollow glass waveguides (HGWs)

As discussed in Sec. 7.2.2.3, HGWs composed of metallic and dielectric films are easily fabricated and simple in structure. Furthermore, HGWs have some of the lowest losses measured, especially at 10.6 μm. In Chapter 3, the theoretical spectral response of a typical hollow waveguide was discussed and illustrated for a 700-μm-bore guide with Ag/AgI films (see Fig. 3.7). The measured spectral response for two different, 700-μm bore, 1-m long HGWs is given in Fig. 7.22. One guide has a

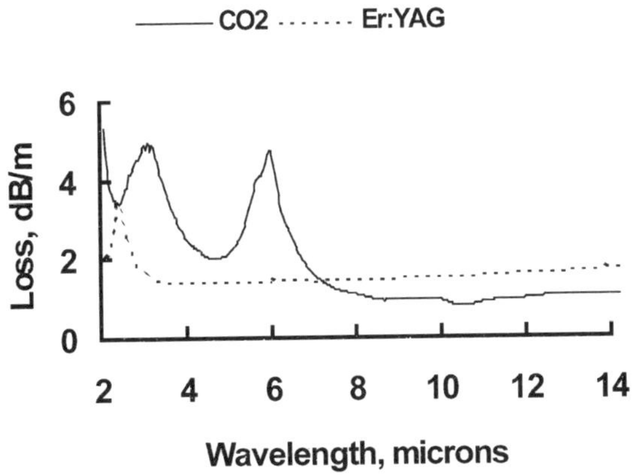

Figure 7.22 Typical spectra of HGWs with Ag/AgI coatings. The guide with the thinner AgI layer (dotted line) is suitable for Er:YAG laser transmission at 3 μm, while the thicker AgI layer (solid curve) has the lowest loss at CO_2 laser wavelengths.[37] [Reprinted with permission.]

0.3-μm-thick AgI film and the other a 0.8-μm film. Harrington's group at Rutgers University fabricated both guides.[2,37] The spectral response of the HGW with the thinner AgI layer is appropriate for the shorter IR wavelengths as, for example, for transmission of Er:YAG laser energy. The thicker AgI film gives the lowest loss at 10.6 μm. Clearly, for broadband IR applications it is desirable to have a thin AgI layer. This is because the thinner the dielectric layer, the shorter the wavelength of the first interference peak, as predicted by Eq. (3.18). This effect is seen in the data of Fig. 7.22. It would seem, therefore, that merely making even a thinner dielectric layer would yield a waveguide that would transmit well into the visible region of the spectrum. It is possible in principle to tailor the optical response to achieve short-wavelength transmission, but this is very difficult using AgI or, for that matter, Ag films. The reason is that AgI (purple color) does not transmit well in the visible region, and Ag has a plasma resonance that leads to excessive absorption below about 500 nm. It is possible, however, to prepare HGWs that work well in the visible and near-IR spectrum. Abe et al.,[25] using visibly transparent polymer films in place of AgI, have accomplished this in some recent work. The optical spectra for these films are discussed later in this section. Finally, note again that the losses are high for guides measured spectroscopically. This is due to the low f/# launch resulting in transmission of many higher-order modes. Also note that the loss data does not show a λ^2 dependence.

The strongly bore-size-dependent loss for straight HGWs is shown for two guides in Fig. 7.23.[37] This data was taken using CO_2 and Er:YAG lasers and the

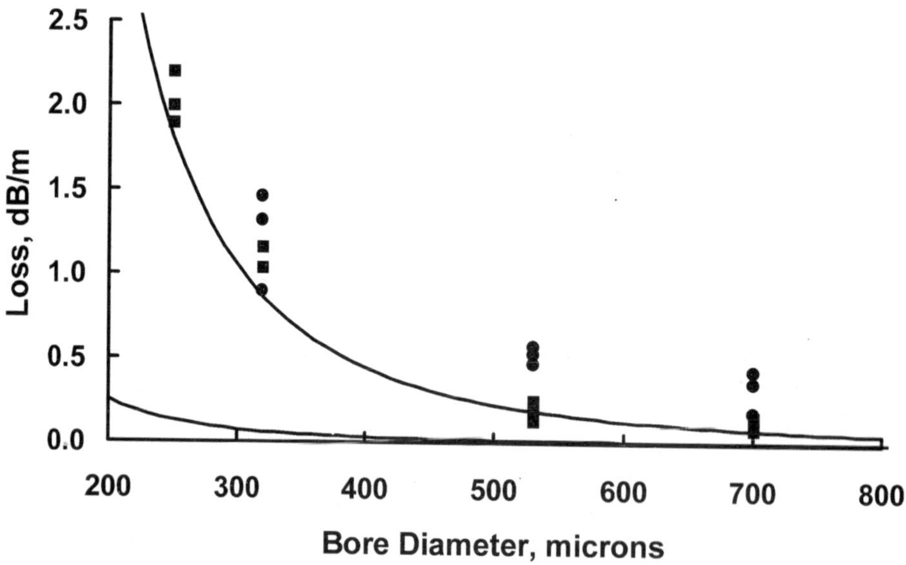

Figure 7.23 Measured and calculated losses for straight HGWs with Ag/AgI coatings. The solid line near the data points is the calculated loss for 10.6 μm and it can be seen that there is good agreement with the measured values. The lower solid curve is the calculated loss for 2.94 μm where there is poor agreement with theory.[37] [Reprinted with permission.]

guides were optimized for minimal loss at 10 and 3 μm, respectively. The solid curves are theoretical calculations of the losses for the lowest-order HE_{11} mode showing again the $1/a^3$ dependence predicted by MS theory. At the CO_2 laser wavelengths, the calculated losses agree quite well with those measured. However, at 3 μm the calculated losses are much lower than the measured values. This is a result of increased scattering losses at the shorter wavelengths and the multimode character of the Er:YAG laser used in the measurements.

The increased loss on bending HGWs is shown for two 530-μm-bore guides in Fig. 7.24.[37] This data shows that the loss increases as the curvature, or $1/R$, increases. All data was taken with a constant length of waveguide under bending. A curvature of 20 m^{-1} represents a bend diameter of only 10 cm! This is sufficiently small for most applications. The data in Fig. 7.24 also reveals that there is a difference in the slope of the bending loss at 3 and 10 μm. In general, the slope, or increase in bending loss, is least for guides with the lowest loss. Greater detail on bending losses in general is given later in this section.

One of the most successful dielectric coatings other than the Ag halides are the polymer films deposited by Matsuura and Miyagi's group at Tohoku University.[25,26] As previously discussed in Sec. 7.2.2.3, these films are applied over metallic Ag that has been deposited inside the silica tubing using conventional liquid-phase chemistry methods. A variety of polymer films have been studied including polyimide,[38] COP,[39] and FCP (a fluorocarbon polymer known as Teflon AF and as CYTOP).[31] The problem with any polymer is that the absorption in the IR region is generally not as low as it is for crystalline films. Furthermore, the low melting point of polymers means that they have limited use in the delivery of high-

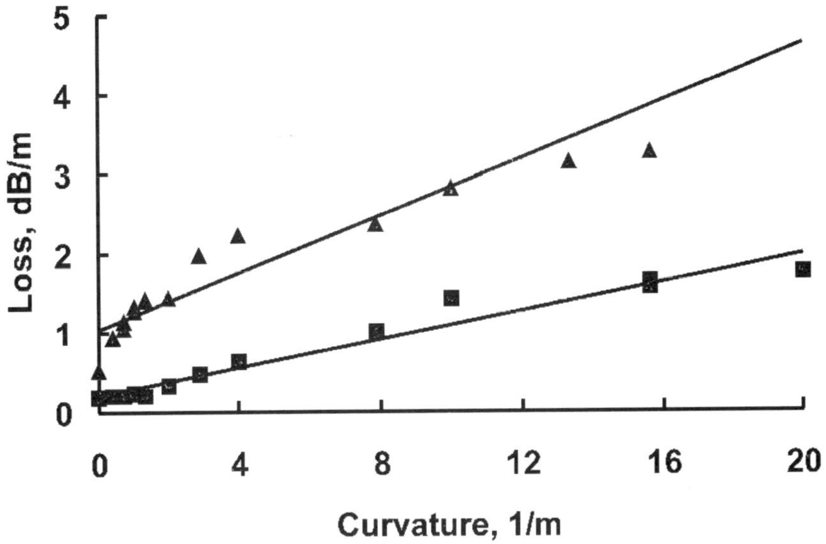

Figure 7.24 Bending losses for two Ag/AgI, 530-μm-bore HGWs. (▲) Data taken at the Er:YAG laser wavelength of 2.94 μm and (■) data taken at CO_2 laser wavelength of 10.6 μm.[2]

average-power laser energy. It is also difficult to find polymers that transmit well at the longer IR wavelengths, for example, at CO_2 laser wavelengths.

Polyimide and FCP films are best suited for operation below about 4 μm. In particular, they have been shown to work well with Er:YAG lasers at 2.94 μm and also for transmitting Nd:YAG energy at 1.06 μm. Kato et al. found that their Ag/polyimide coated, 700-μm-bore waveguide had a loss of about 1 dB/m at 2.94 μm.[38] The Ag/FCP-coated 700-μm-bore waveguide had a loss of about 0.2 dB/m at the same wavelength.[31] Beyond 4 μm the absorption increases due to the presence of numerous C—H bonded molecules. The polymer of choice seems to be COP. This polymer has the unique ability to transmit beyond 4 μm well into the 10 μm region. The spectral loss data of Abe et al. for an Ag/COP HGW is shown in Fig. 7.25(a).[30] In their data there is also a comparison with the Ag-only guide. While there is some absorption due to COP around 7 μm, the losses are

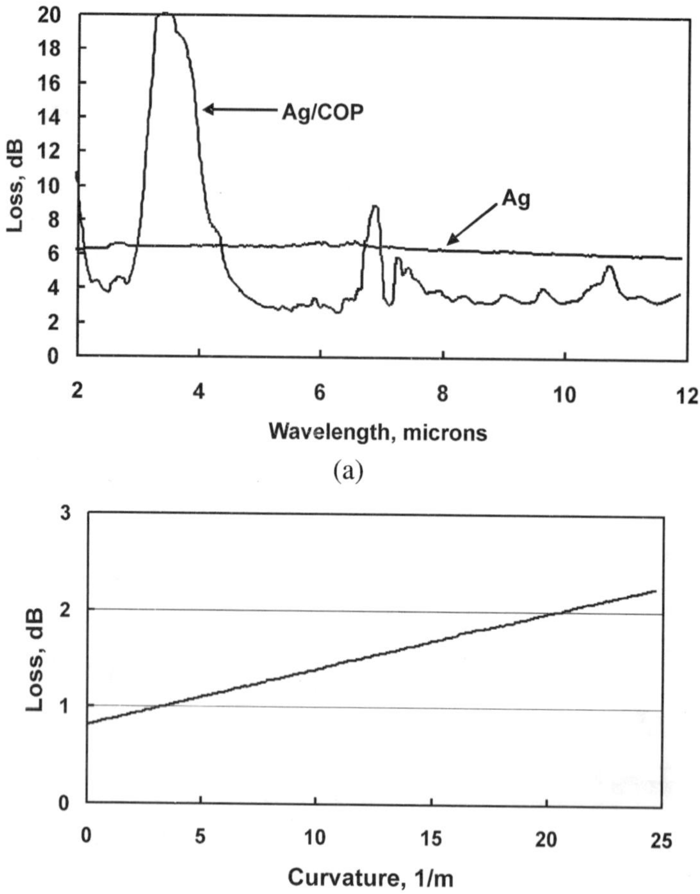

(a)

(b)

Figure 7.25 (a) Spectral response for a 700-μm-bore, 1-m-long Ag/COP-coated HGW compared to a Ag-only guide. (b) Bending losses for a similar Ag/COP-coated HGW 1-m long but with a 250-μm bore.[30] [Reprinted with permission.]

reasonably low at 10 μm. Specifically, they measure a loss of 0.6 dB/m for their 700-μm-bore, 1.8-m-long waveguide at 10.6 μm. COP is also quite useful at the shorter wavelengths. In Fig. 7.25(b), the bending loss data of Abe et al.[30] is shown for a 250-μm-bore waveguide measured using an Er:YAG laser. The straight loss for this guide was a low 0.8 dB/m. This is an excellent result for this short wavelength. This low loss can be attributed in part to the very smooth surface of both the polymer and Ag films.

7.3.4.1 Modal properties of HGWs

The output beam profile of HGWs depends heavily on the bore size and on the quality and launch conditions of the input laser. In principle, HGWs are low-order mode guides because the higher-order modes are attenuated by the factor $(u_{lm})^2$ (see Eq. (3.17)). In practice, however, mode distortion can occur even with a TEM_{00} input beam. There are several ways in which higher-order modes may be generated in these guides. The simplest method is to launch a higher-order mode into the guide with a low f/# launch. This situation is commonly encountered when using an Er:YAG laser, as these lasers are often multimode. Another method is to bend the guide or apply a radial force to the outside of the guide. Other factors influencing the generation of higher-order modes are the surface roughness, the thickness of the glass wall, and the coupling efficiency.

The modal purity of an output beam is determined by the overlap integral between a theoretical profile of the lowest-order HE_{11} mode and the measured profile. This can be very difficult to evaluate when many modes are present, but for small-bore waveguides with only a few propagating modes the process is tractable. Matsuura et al.[40] calculated the output beam profile for the HE_{11} mode using the well-known Kirchhoff–Huygens equation, assuming that the bore size, $2a$, is very much larger than the wavelength, or $2a \gg \lambda$. A transverse component of the field at the end of the guide is given by

$$E(r, \theta, z) = \frac{ik_o}{2\pi} \iint E(\rho, \phi, 0) \frac{e^{-k_o R}}{R^2} \left(1 - i\frac{1}{k_o R}\right) dS, \tag{7.1}$$

where k_o is the wavenumber in a vacuum, and R is the distance between a point light source at the aperture and the observation point on the z-axis. Assuming transmission of only the HE_{11} mode and linear polarization uniform in the circular direction, then the amplitude E for $k_o R \gg 1$ is

$$E = \frac{ik_o z}{2\pi} \int_0^{2\pi} \int_0^{a} J_o\left(u_o \frac{\rho}{a}\right) \frac{e^{-ik_o R}}{R^2} \rho d\rho d\phi. \tag{7.2}$$

The modal purity is obtained from an evaluation of the overlap integral, Eq. (7.2), calculated at a distance R from the end of the guide and the measured profile. The excellent modal purity for a 250-μm-bore waveguide is shown in Fig. 7.26. The distance from the output end was 86 cm, and the modal purity was calculated to be as high as 97%.

The modal purity decreases as the bore size increases and the guides are bent. The data taken by Matsuura et al.[40] in Fig. 7.27 clearly demonstrates this effect. From Fig. 7.27 it may be seen that the small, 250-μm-bore guide remains at least 90% pure even when bent to a radius of 2.5 cm and with seven loops in the fiber. For the larger-bore guides, the HE_{11} purity decreases rapidly with bending. This effect may be seen qualitatively in Fig. 7.28(a) and (b), where the spatial profile of a bent 250-μm-bore guide is compared to a bent 530-μm-bore guide.[40]

The waveguide loss depends critically on the number and type of modes propagating in the guide. The discussion in Sec. 3.2.2 and Eq. (3.19) enables one to calculate the loss for the different modes. Nubling et al.[41] has done this for the

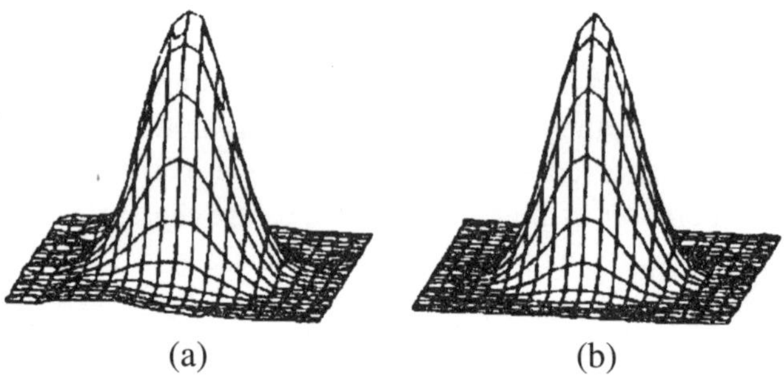

Figure 7.26 Beam profile from a straight, 250-μm bore HGW 86 mm from the output end. (a) Measured and (b) calculated beam profiles.[40] [Reprinted with permission.]

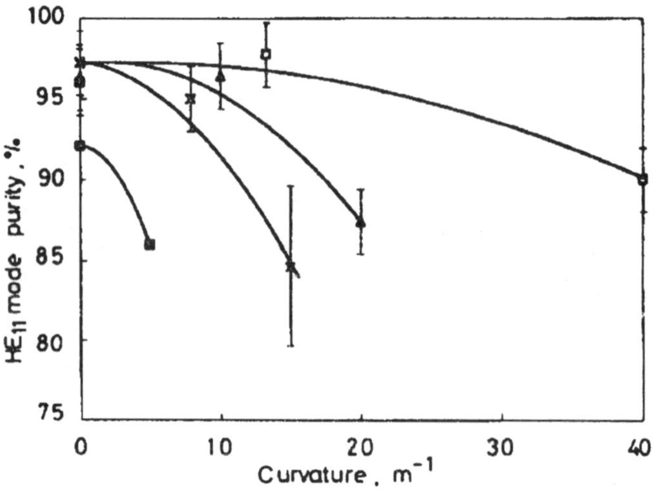

Figure 7.27 Modal purity of 250 μm (□); 320 μm (▲); 530 μm (×); 700 μm (■) waveguides.[40] [Reprinted with permission.]

HE_{nm} modes using the appropriate equation in Eq. (3.19):

$$\alpha_{nm} = \left(\frac{u_{nm}}{2\pi}\right)^2 \frac{\lambda^2}{a^3}\left(\frac{n}{n^2+\kappa^2}\right)\left[\frac{1}{2}\left(1+\frac{n_d^2}{\sqrt{n_d^2-1}}\right)^2\right]. \tag{7.3}$$

Figure 7.29 shows the theoretical attenuation coefficients for the first four HE_{1m} modes in a Ag/AgI hollow glass waveguide as a function of bore diameter. These theoretical curves show the $1/a^3$ dependence of loss as well as the dramatic increase in loss for the higher-order modes. Note that the attenuation of the higher-order modes is particularly severe for the small-bore waveguides. Because the higher-order modes are so rapidly attenuated, it becomes possible to deliver a near-

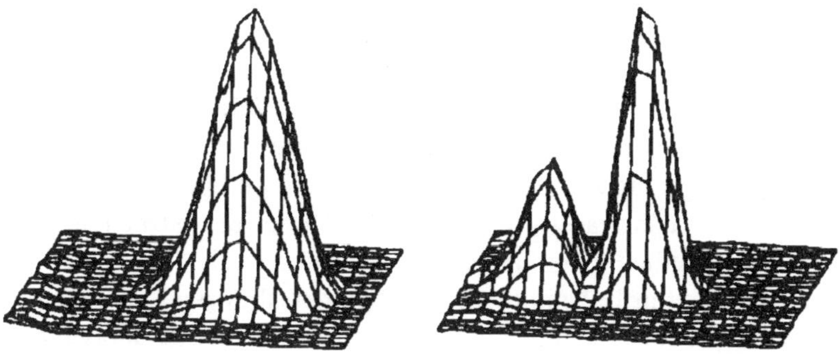

Figure 7.28 (a) Measured output beam profile for 250-μm bore bent to a 25-mm radius and (b) beam profile for 530-μm bore bent to a 69-mm radius.[40] [Reprinted with permission.]

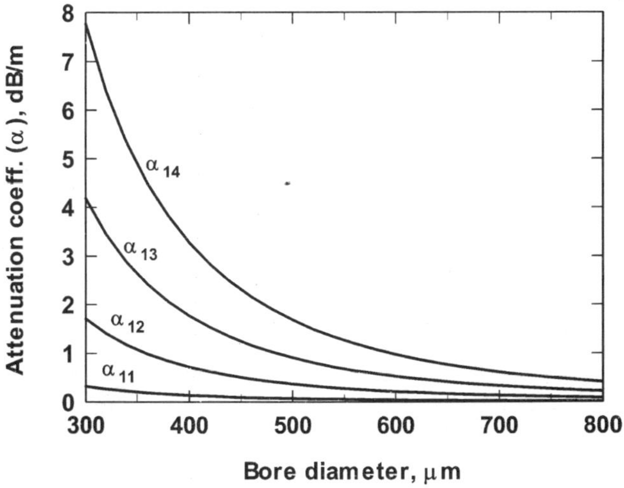

Figure 7.29 Calculated attenuation coefficients for the first four HE_{1m} modes in hollow glass waveguides.[41]

single-mode beam through the small-bore waveguides. The larger-bore waveguides will transmit the higher-order modes over longer lengths, but the total loss will be substantially less than for the small-bore waveguides.

An interesting feature of the modal properties of HGWs is their ability to effectively filter higher-order modes from a multimode laser beam. This is a result of the larger attenuation of higher-order modes. An example of this type of modal filtering is shown in Fig. 7.30 for a 700-μm-bore waveguide.[42] A multimode CO_2 laser beam [Fig. 7.30(a)] is used as a source for transmission through the 1.5-m-long guide. The output, seen qualitatively in the beam profile data shown in Fig. 7.30(b), contains fewer modes. It is important to realize that the filtering of propagating laser radiation comes at the expense of power. In this example, the attenuation is ~0.8 dB/m. This measured loss is substantially higher than the 0.25 dB/m that would be observed for the same waveguide measured using a TEM_{00} CO_2 laser input beam.

Bending the waveguides can often lead to the generation of higher-order modes, especially for large bore sizes.[43] This is often undesirable because many applications require a near-single-mode output. Furthermore, it has also been observed, somewhat surprisingly, that HGWs packaged in standard fiber optic furcation tubing with connectors can lead to mode mixing. The effects of bending on mode mixing have been studied quantitatively by Matsuura et al.[44] Instead of inducing mode mixing by arbitrarily bending the waveguides, they devised a simple set of experiments to reproduce the mode generation effect merely by applying a known stress to one side of the waveguide. This simulates the distortion that the tubing would experience in bending. Specifically, they applied a known point stress to the

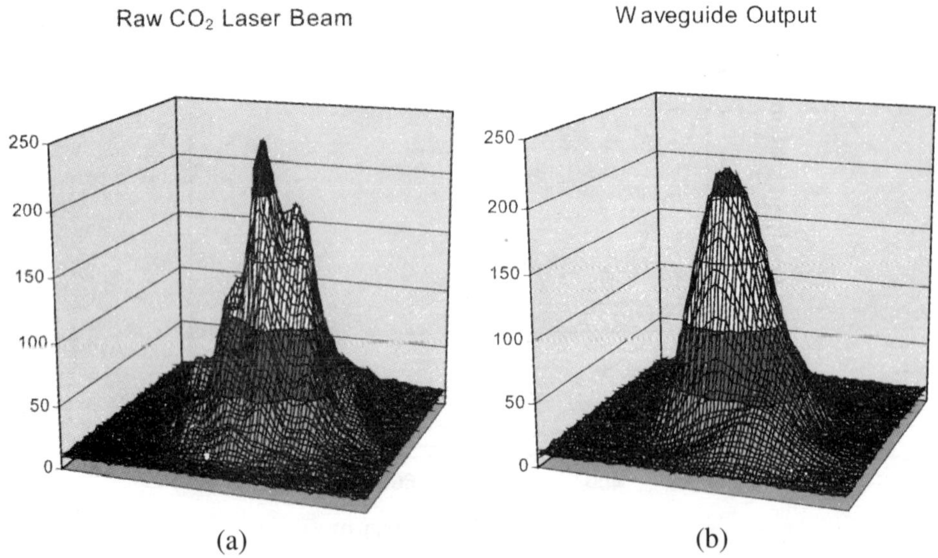

Figure 7.30 Illustration of mode filtering by HGWs; multimode raw CO_2 laser beam (a) filtered into a purer mode (b) as the beam exits a 700-μm-bore waveguide.

side of 530-μm-bore guides, each with different glass wall thickness. The results of these measurements at 10.6 μm are given in Fig. 7.31. From their data, it is clear that the thicker the glass wall, the less loss is generated. That is, the extent of mode mixing depends on the amount of deformation of the glass. Large deformation for the thinnest wall tubing generates the greatest number of high-order modes and, therefore, the highest additional loss. This is confirmed qualitatively from the output mode profiles shown in Fig. 7.32. If it is important to minimize mode mixing,

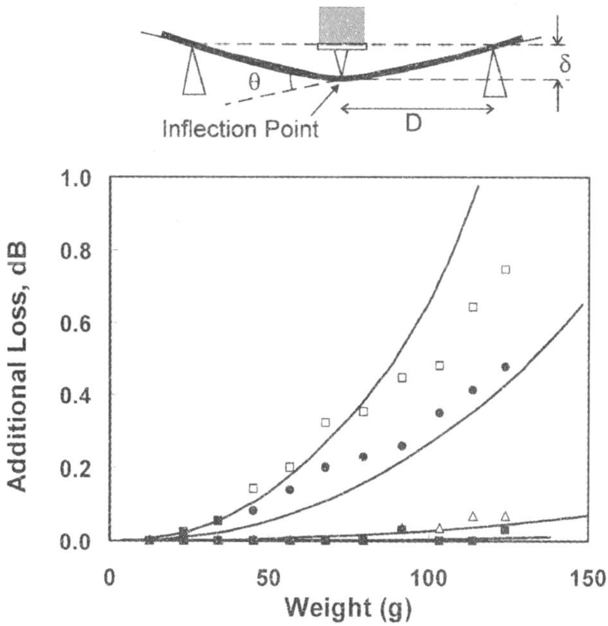

Figure 7.31 Experimental setup to measure the additional losses due to pressure on hollow glass waveguides with various wall thicknesses of (□) 68 μm, (●) 97 μm, (△) 135 μm, and (■) 210 μm. The solid lines are the calculated losses for each thickness.[44] [Reprinted with permission.]

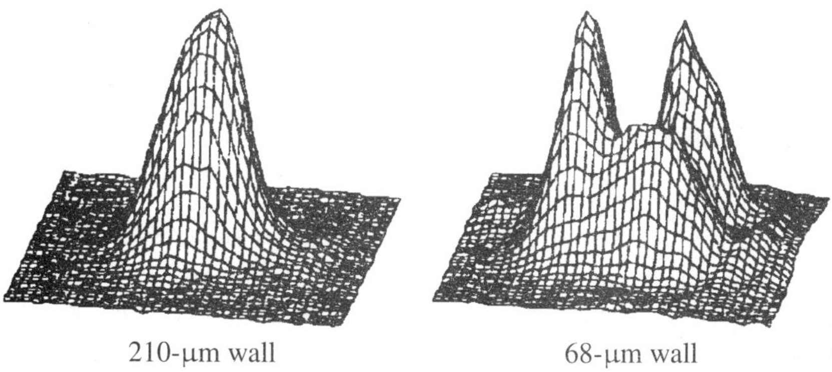

Figure 7.32 Measured output beam profiles of HGWs when a 45-g weight is placed on the waveguides as shown in Fig. 7.31.[44] [Reprinted with permission.]

Table 7.1 Full-angle beam divergence out of hollow waveguides in mrad.

	Bore diameter					
	250 μm	500 μm	750 μm	1000 μm	1500 μm	2000 μm
HE_{11}	65	32	22	16	11	8
HE_{12}	149	74	50	37	25	19
HE_{13}	234	117	78	58	39	29
HE_{14}	318	159	106	80	53	40
HE_{15}	403	202	134	101	67	50
HE_{16}	488	244	163	122	81	61
HE_{17}	573	286	191	143	95	72

the best approach is to keep the glass wall thickness greater than about 200 μm. The downside to this is that the overall OD of the tubing increases, which means that the minimum-bending radius increases. An alternative method to reduce mode mixing is to choose a smaller bore size, with the concomitant increase in loss.

The output-beam divergence of a HGW depends critically on the modes propagating in the guide, the wavelength, and the bore size. In principle the output beam divergence can give an indication of the number of high-order modes propagating in the waveguide. The HE_{1m} modes will couple to free-space modes with a half-angle beam divergence, θ, given by

$$\theta_m \cong \sin(\theta_m) = \frac{u_{1m}\lambda}{2\pi a}. \tag{7.4}$$

So for example, the HE_{11} mode will have a predicted full-angle, 2θ, beam divergence of ~22 mrad out of a 750-μm-bore hollow waveguide at 10.6 μm. A summary of the full-angle beam divergence of the first few HE_{1m} modes for various waveguide at 10.6 μm for different bore sizes is shown in Table 7.1. When the guides are propagating the lowest order HE_{11} mode, it is clear from the data in Table 7.1 that the beam divergence is quite small. For example, the 750-μm-bore guide has a full-angle divergence that is less than 2° for the HE_{11} mode. In practice, however, the beam divergence is usually larger than that for the lowest-order mode because higher-order modes are propagating. Harrington and his coworkers have found that the actual full-angle divergence is often at least 4° for most guides. This would translate to an NA_{eff} of about 0.035.

7.3.4.2 Scattering, surface roughness, and film uniformity of HGWs

Scattering can lead to additional loss in both solid- and hollow-core fibers. In solid-core fibers, scattering most often results from inhomogeneities and defects in the core and cladding materials. In hollow guides, however, the primary transmission medium is air, which scatters much less than a solid material. Thus, scattering in HGWs results from surface roughness of the substrate tubing and the thin-film coatings. Surface roughness leads to mode mixing and, therefore, higher loss. Takatani et al.,[45] Danilov et al.,[46] Alaluf et al., Inberg et al.,[47] and Rabii et al.[42] have studied the effects of surface roughness on scattering losses in hollow waveguides. In general, there is a decrease in the reflectivity from that of an ideal smooth surface, $R(\theta)$, to $R'(\theta)$, given by

$$R'(\theta) = R(\theta) \cdot \exp\left[-\left(\frac{4\pi \cdot n_i \cdot \sigma \cdot \sin\theta}{\lambda} \right)^2 \right], \tag{7.5}$$

where n_i is the index of the core ($n_i \approx 1$), θ is the incident angle with respect to the wall of the guide, and σ is the RMS surface roughness. This is a common expression for scattering from rough surfaces, whether from films in a hollow structure or other optical components like mirrors and scattering filters. The key feature of Eq. (7.5) is that the scattering loss increases with increasing roughness and decreasing wavelength. Assuming that the total loss of the guide is proportional to the reflection loss per bounce, then the length-independent attenuation coefficient for an air-core guide is proportional to

$$\alpha \sim -\ln\left(\frac{R'(\theta)}{R(\theta)} \right) = -\left(\frac{4\pi \cdot \sigma \cdot \sin\theta}{\lambda} \right)^2. \tag{7.6}$$

The surface roughness, σ, is quite small for the pristine silica tubing or even the polycarbonate tubing. Generally, $\sigma_{\mathrm{RMS}} \approx 1$ nm for both materials. The difference is that the polycarbonate tubing can have striations and other defects in certain regions that contribute to the scattering loss. Normally the first film deposited inside the tubing is Ag. The roughness of the Ag film has been shown by Rabii et al.[24] to increase with silvering time. Specifically, Rabii et al. showed that $\sigma \sim t^{1/2}$, where t is the silvering time. This increasing roughness with deposition time is easily seen in the AFM pictures shown in Fig. 7.33 for Ag on silica.[24] The actual roughness of the Ag films is given in Fig. 7.34. From the data in Fig. 7.34 it can be seen that not only does the roughness increase approximately fivefold as the silvering time increases from 5 to 80 min, but also that the trendline through the data gives $\sigma \sim t^{1/2}$. It would seem, therefore, that it is desirable to use the shortest silvering times to produce the smoothest Ag films and the lowest loss guides. However, the deposition of AgI is a subtraction process in which Ag is converted to AgI. This means that to produce HGWs operating at the longest wavelengths such as 10 μm, the thickness of the Ag film must be sufficiently thick to permit the deposition of AgI films with a thickness of about 0.8 μm. In practice this requires silvering times of a minimum of 60 min. It has been found that the roughness of the final Ag/AgI coating is depends largely on the roughness of the starting Ag layer itself. Specifically, it has been found that after the formation of the AgI layer the final Ag/AgI coating roughness is increased by at most 10% over the Ag layer alone.

The effect of roughness on loss may been seen quite dramatically in measurements of loss at 2.94 μm using an Er:YAG laser. In Fig. 7.35, the loss for four different, 700-μm-bore waveguides, each made with a different silvering time, are

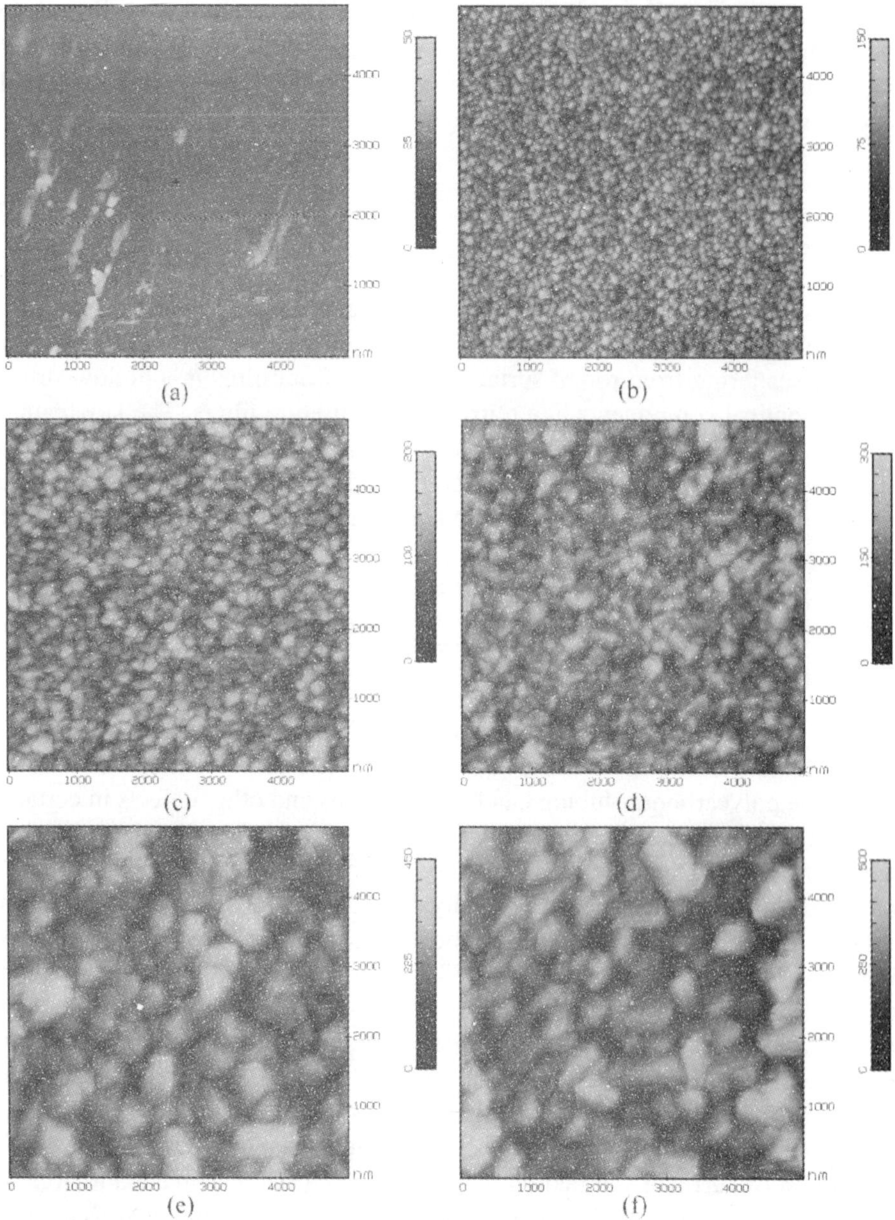

Figure 7.33 AFM images of silvering process: (a) uncoated slide, (b) 5-min, (c) 10-min, (d) 20-min, (e) 40-min, (f) 80-min silvering time.[24] [Reprinted with permission.]

shown.[24] The data shows that the losses increase linearly for silvering times greater than about 20 min. This linear behavior is predicted by Eq. (7.6) using the fact that $\sigma \sim t^{1/2}$. That is, Eq. (7.6) becomes

$$\alpha \sim= -\left(\frac{4\pi \cdot \sin\theta}{\lambda}\right)^2 t, \tag{7.7}$$

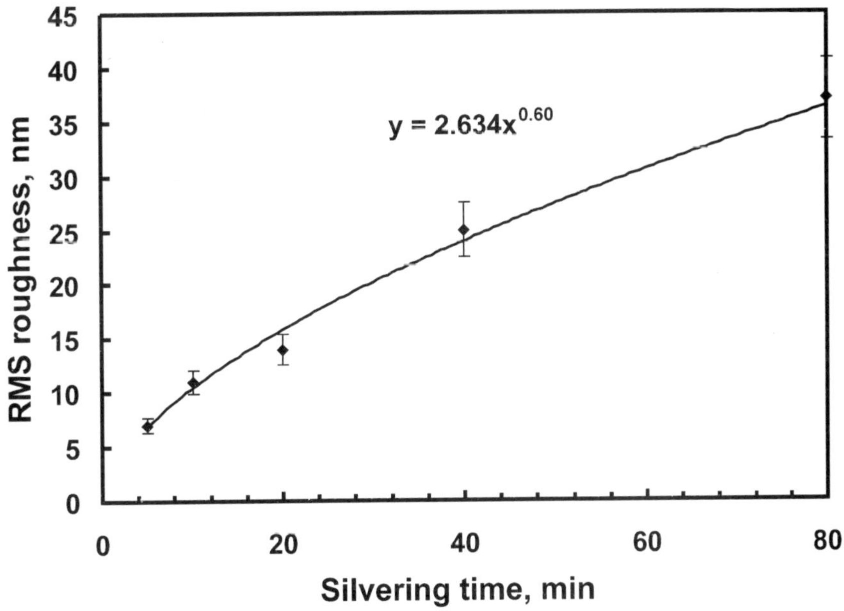

Figure 7.34 Increase in surface roughness of Ag film as the silvering time increases.[24] [Reprinted with permission.]

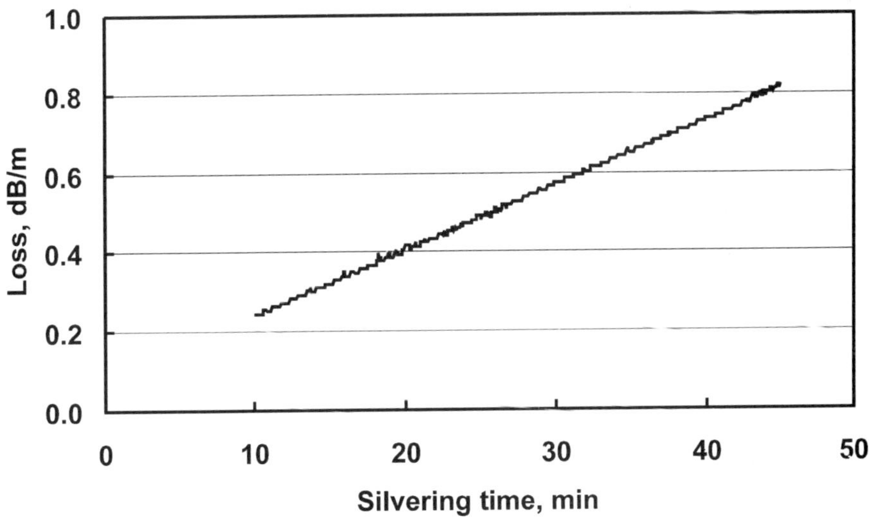

Figure 7.35 Increase in loss with increasing silvering time for the Ag/AgI HGWs. The loss after only a 10-min silvering time is not included in the trendline, as this loss is quite high because the thickness of the silver layer is too thin to support the required AgI layer thickness.[24] [Reprinted with permission.]

or $\alpha \sim t$. The shortest silvering time of 10 min produces a very high loss. This is because the Ag layer is too thin to support both an AgI film of thickness ~0.3 μm and a sufficiently thick Ag layer. That is, the Ag is too thin to provide the requisite reflection, and some of the light goes through the film into the highly absorbing silica substrate tubing. In fact, it is possible to completely convert the Ag film to a stand-alone AgI film. This might be a useful approach in the design of multilayer dielectric coatings, but here it leads to excessive absorption. However, thin, smooth Ag films are the ideal approach if the dielectric film can be deposited independent of the Ag layer; that is, without resorting to subtraction of the metallic layer to produce the dielectric layer. This is the case for the polymer films deposited over Ag by Matsuura et al.[39] Their Ag/polymer guides have very low loss, especially at shorter wavelengths.[48]

Surface roughness can also lead to mode mixing and, therefore, higher loss. Inberg et al.,[36] Gregory and Harrington,[49] and Ben-David et al.[50] have studied the effects of roughness on waveguide loss. Inberg et al. calculated the losses for Ag/AgI waveguides made using both silica and Teflon tubing. For silica tubing they used $\sigma = 10$ nm, whereas for the Teflon tubing they used both $\sigma = 25$ and 100 nm. Their calculated output spectra were compared to the measured output when a near-Gaussian CO_2 laser was used as the input source. Furthermore, they also studied different bore sizes. Their spatial profiles and calculations reveal a very rich mode spectrum. As expected, the output profiles with the greatest number of modal lobes were those for guides with the roughest surfaces (Teflon) and the largest bore.

In some related work, Takatani et al.[45] have studied the losses in pure silica glass hollow guides that have been intentionally roughened. The basic idea is to measure the transmission of pristine hollow silica (no coatings) as a baseline and then roughen the inside surface of the tubing using different abrasive powders. Their experiments did show the expected loss increase with increased roughness, but they also found an unusual result at the long wavelengths just past the Reststrahl peak around 10 μm. In this region the losses actually decreased for rough surfaces. They account for this effect by mathematically replacing the roughened surface by a thin film of varying thickness.

The uniformity of the metallic and dielectric layers can affect the losses as well as the spectral response of the waveguides. In Chapter 3, the classic theoretical spectrum of a single-layer AgI film over Ag was calculated and the results presented in Fig. 3.7. One sees from this data the interference effects for a perfectly uniform dielectric film with optimal thickness given by

$$d_p = \frac{m \cdot \lambda_p^{(m)}}{4 \cdot \sqrt{n_1^2 - 1}}, \tag{7.8}$$

where m is the order of the interference maxima, λ_p is the wavelength of the mth absorption peak, and n_1 is the refractive index of the dielectric film. Equation (7.8) is similar to Eq. (3.18), but Eq. (7.8) is written in terms of the wavelength of the interference peaks rather than the minimum in absorption. Equation (7.8) is a handy means of determining the thickness of the dielectric film from the absorption spectrum. When the dielectric film is nonuniform along the length of the guide, then the interference patterns tend to overlap and broaden. Given that the guides are fabricated by a liquid-phase process in which the liquid generally flows in one direction, it would not be surprising if the films are thicker at the input end of the guide compared to the output end. This can result from the fact that the plating solutions become more dilute as material is deposited on the tubing wall. Rabii et al.[51] have made a thorough study of this phenomenon. In general, they found that for guides 5- to 6-m in length that the end-to-end film thickness could vary by more than 10%. This effect is evident in their data (see Fig. 7.36) for a 2-m-long waveguide compared to a short, 5-cm length taken from one end of the same guide. As expected, the attenuation increases as the length increases, but the characteristics of the signature interference peaks also change. The spectral response of a long waveguide possesses interference peaks with large widths compared to the peaks for a short waveguide. In general, short HGWs (<1 m) have interference peaks resembling those of the theoretical spectrum, while the longer guides may have absorption peaks several times as broad.

Rabii et al.[51] measured the nonuniformity in the films by sectioning a long (6-m) waveguide and calculating the film thickness using Eq. (7.8). In general, they found that the first interference band (peak at 6 μm in Fig. 7.36) always occurred at longer wavelengths for segments taken at the input end of the fiber, compared to segments taken at the output end. That is, for liquid flowing in only one direction from the input to the output end, the thickness of the films was greater at the input end, and therefore the interference peak occurs at longer wavelengths. They attribute this feature to processing parameters. In one experiment in which they used a more-or-less standard liquid-phase processing method, they found an end-to-end AgI film thickness variation of 14.5% over a 6-m length of waveguide. When the iodine flow rate was increased by a factor of 4, the end-to-end variation

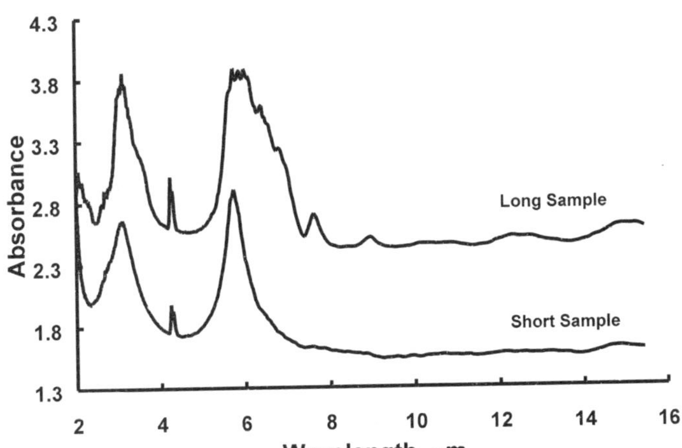

Figure 7.36 Spectral response for a long (2-m)-long Ag/AgI waveguide compared to a short (5-cm) long sample taken from one end. The broader interference peaks in the long sample reflect the end-to-end nonuniformity of the AgI film thickness.[51]

was found to be only 1.3% over a 5-m length of waveguide. They explain this result in terms of the kinetics of the AgI film formation process. By increasing the iodine flow rate, they reduce the quantity of iodine per unit length per time removed from solution. Thus, increasing the flow rate and reaction time leads to a more uniform longitudinal concentration profile at the AgI surface, which leads to a more uniform film. Finally, the effect of film nonuniformity on transmission was also considered. When the film is nonuniform, the minimum loss calculated using Eq. (3.18) is not achieved because there is a spread in the film thickness. Rabii et al. estimate that the minimum loss at 10.6 μm can increase by about 15% for a film variation of 14%.

7.3.4.3 Bending loss in HGWs

Bending losses for hollow guides are discussed in general in Sec. 3.2. From the discussion in Sec. 7.3.4, the data in Fig. 7.24, and elsewhere, there is a clear increase in the loss on bending that varies as $1/R$. When the bending loss, α_R, is added to the straight loss, α_∞, the total loss is obtained. From the linear dependence of α_R on the curvature (see Fig. 7.24), $\alpha_R = m/R$, where m is the proportionality constant or the bending slope. The value of m varies for different waveguides and wavelengths. Although little has been published on what determines m, it seems to be a function of bore size, roughness of the metallic and dielectric films, and mode parameter. From bending data in the literature, m has been found to vary between 0.13 and 0.60 dB for hollow waveguides of various sizes.[52,53]

For guides with a constant length under bend, it is convenient to write the total loss of the waveguide as the sum of the straight and bending losses,

$$Loss_{total}(\text{dB}) = \alpha_\infty L + \left(\frac{m}{R}\right) L_R = \alpha_\infty L + \alpha_R L_R = \alpha_\infty L + 2\pi m N, \qquad (7.9)$$

where L is the total length, L_R is the length under bend, and N is the number of loops in the waveguide. One way to interpret Eq. (7.9) is that the total loss for a given bore size depends only on the number of loops, increasing by $2\pi m$ per loop. Measurements of bending loss are typically made by varying the bending radius while maintaining a constant length under bend. The bending slope is then obtained by plotting the measured loss, $Loss_{total}$, against curvature, $1/R$.

An alternative approach to bending loss is to consider the loss for a guide bent to a constant bend angle, θ. That is, letting $L_R = R\theta$ and substituting into Eq. (7.9) gives

$$Loss_{total}(\text{dB}) = \alpha_\infty L + \frac{m}{R} \cdot R\theta = \alpha_\infty L + m\theta. \qquad (7.10)$$

Equation (7.9) implies that the loss is independent of the radius of curvature and directly proportional to bending slope, m.

The bending loss for different bore-size HGWs under constant bending length measured at 10.6 μm is shown in Fig. 7.37(a). The source CO_2 laser was linearly polarized perpendicular to the bending plane (minimal loss configuration). For the constant-bend length measurement, the last 100 cm of a 125-long HGW was bent to different radii. The straight losses and bending slopes obtained from Fig. 7.37(a) are summarized in Table 7.2. From this data it is seen that the value of m is quite low and almost constant for the different bore sizes. Generally, the lowest-loss guides have the lowest m values. Higher values of m are obtained when a multimode laser source is used. The loss, $Loss_{2\pi}$, for a single loop, i.e., $\theta = 2\pi$, calculated from the values in Table 7.2, is also given in the table.

For the loss measurements with a constant bend angle, the HGW was bent into a complete loop and placed between two plates like that shown in Fig. 2.13 for strength testing. Using a stepper motor to drive the faceplates together, the loop was flattened as the plate separation, D, was reduced. While maintaining the bend angle ($\theta = 2\pi$), the length under bend and bending radius were both decreased as D was reduced from 17 cm to the point of waveguide failure. The guide in this configuration has an almost uniform curvature with $R = D/2$. The data in Fig. 7.37(b) shows the results for $Loss_{2\pi}$ for the same three fibers measured in Fig. 7.37(a), but this time bent to one loop, i.e., under constant bending angle. When the radius of the loop is not too small, the total loss should be constant, as may be seen from Eq. (7.10). This is the case shown in Fig. 7.37(b). Furthermore, the measured loss for this case compares well with the $Loss_{2\pi}$ calculated using Eq. (7.10) with m and α_∞ from the previous constant length measurements (Fig. 7.37(a) and Table 7.2). These losses are given in the last column in Table 7.2. However, as the bending radius is reduced, the measured loss begins to exceed $Loss_{2\pi}$ due to the propagation of higher-order modes.

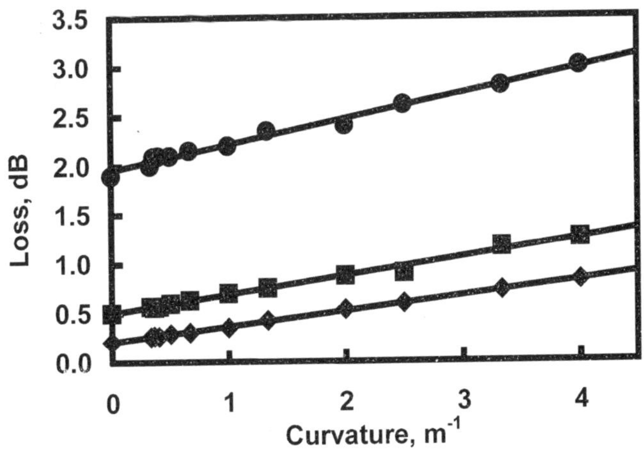

Figure 7.37(a) Constant bend length (CBL) losses for (●) 320-, (■) 530-, and (◆) 700-μm-bore HGWs.[53]

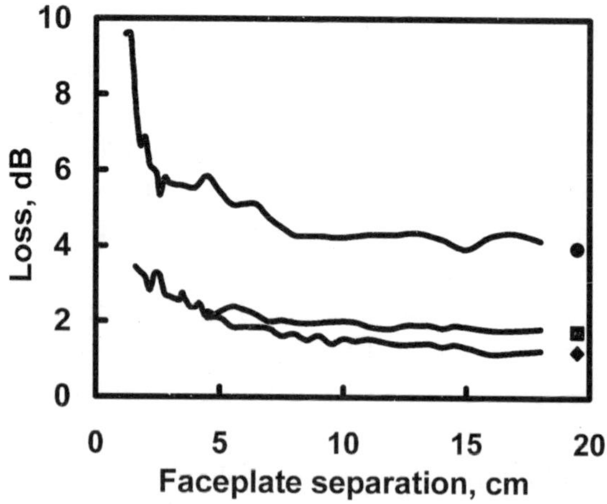

Figure 7.37(b) Small radius, constant bend angle, $\theta = 2\pi$, loss for (●) 320- (■) 530-, and (◆) 700-μm bore HGWs at 10.6 μm from which the bending slope, m, is calculated as a function of curvature, $1/R$. The ●, ■, and ◆ on the right represent the values obtained from the CBL measurement.[53]

Table 7.2 Measured losses for HGWs at 10.6 μm; m is the value of the slope of the loss (dB/m) vs. curvature (m^{-1}) plots.

Bore radius, μm	Straight loss, dB/m	Bending slope m, dB	Loss for one loop, dB
700	0.21	0.15	1.20
530	0.53	0.18	1.79
320	1.94	0.25	4.00

7.3.4.4 Coupling light into HGWs

The waveguide loss depends on the nature and number of modes propagating in the guide. It is straightforward to calculate the loss α_{nm} for each mode using Eq. (3.19) and, as explained earlier in this chapter, the loss increases dramatically for the higher-order modes. Therefore, it is often desirable to launch only the lowest-order HE_{11} mode into the guide to achieve the lowest loss. Several authors have considered the problem of coupling efficiency into hollow guides, and their results are important for the design of optical systems using these guides.

When a Gaussian beam is focused on-axis into a hollow waveguide, only the HE_{1m} modes are excited. Assuming the usual form for a Gaussian beam focused to a $1/e^2$ waist equal to ω, or,

$$E(r) = E_o \exp\left(\frac{-r^2}{\omega^2}\right), \tag{7.11}$$

where $0 \leq r < a$. Nubling and Harrington[41] calculate the power-coupling efficiency, η_m, from the overlap integral,

$$\eta_m = \frac{\left[\int_0^a \exp\left(\frac{-r^2}{\omega^2}\right) J_o\left(u_m \frac{r}{a}\right) r dr\right]^2}{\int_0^\infty \exp\left(\frac{-2r^2}{\omega^2}\right) r dr \int_0^a J_o^2\left(u_m \frac{r}{a}\right) r dr}. \tag{7.12}$$

Equation (7.12) gives the amount of power coupled to the HE_{1m} mode for a given spot size to bore-size ratio, ω/a. These authors give the coupling results using Eq. (7.12) for different bore sizes. As an example, Fig. 7.38 shows the coupling efficiencies for a 530-μm bore guide calculated using Eq. (7.12) for the first seven HE_{1m} modes. The coupling efficiency to the lowest-loss HE_{11} mode is highest with a beam waist to bore radius ratio, $\omega/a \cong 0.64$. Under these conditions $\eta_1 = 98.1\%$ and $\eta_2 = 0.5\%$. This is the same result as that obtained by Abrams.[54] Also from Fig. 7.38 it may be seen that the total loss (dotted curve) drops precipitously for a higher f/# launch. Thus, correct coupling can have a dramatic impact on the performance of the guides. It is easy to understand from this data why the losses are so high for guides measured with low f/# launch conditions as encountered, for example, in most IR spectrometers. In fact, if the correct f/# or beam size is used, coupling of incoherent light into the waveguides is as efficient as for coherent light.

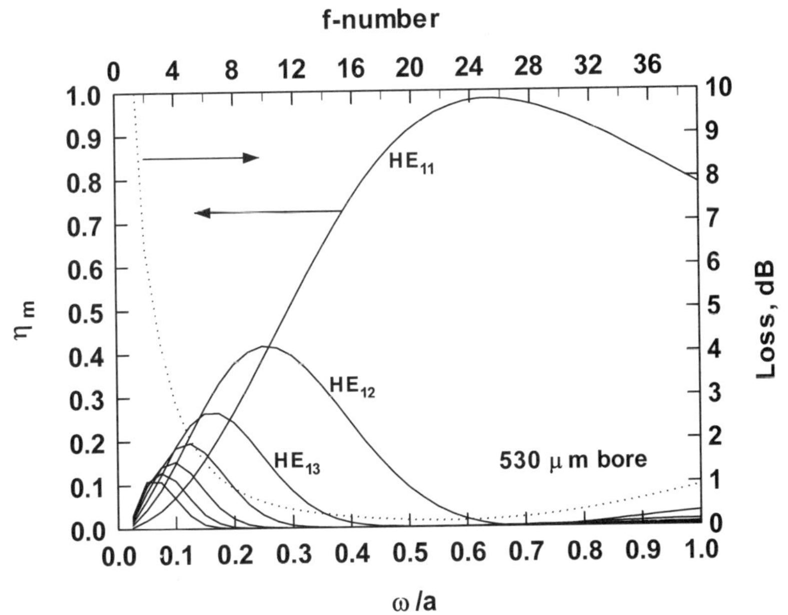

Figure 7.38 Calculated coupling coefficients (solid curves) and total loss (dotted curve) of a 1-m-long, 530-μm-bore, hollow-glass waveguide.[41]

The total loss for a waveguide depends on both the attenuation and coupling efficiency to each of the modes. Nubling and Harrington[41] discuss the optimal coupling conditions for different bore sizes. In particular, they present their coupling data in terms of the f/number rather than the spot-size ratio as given in Fig. 7.38. Data in terms of f/# is of some practical importance, because many end users are unable to easily measure spot sizes. The data to achieve minimum loss is given in Fig. 7.39 for three different bore sizes. Maximum transmission occurs at an f/# launch of ~f/15 ($\omega/a = 0.614$), ~f/22 ($\omega/a = 0.569$), and ~f/30 ($\omega/a = 0.544$) for 320, 530, and 700-μm-bore waveguides, respectively.

7.3.4.5 Miscellaneous HGW structures

A rather unusual, dual-core hollow guide was constructed by Rabii and Harrington.[55] The guide is dual-core because the waveguide has two separate regions for the propagation of light: a silica annulus for transmission of visible and near-IR light, and the hollow bore for transmission of wavelengths greater than 2 μm. This structure is shown in Fig. 7.40(a). The basic idea is to use the glass annulus as both a support member and as a fiber optic. To do this, Rabii and Harrington had to provide a cladding layer on the inside and outside of the silica annulus. Specifically, they deposited a fluorine-doped cladding layer on the inner wall and a low-index silicone polymer on the outer surface. The index profile of the dual-core guide is shown in Fig. 7.40(b). The inside bore of the tubing was then coated with the usual Ag/AgI coatings for IR transmission.

The bending losses for the dual-core waveguide are given in Fig. 7.41(a). When either Er:YAG laser or CO_2 laser light is used in the hollow core, the bending loss is

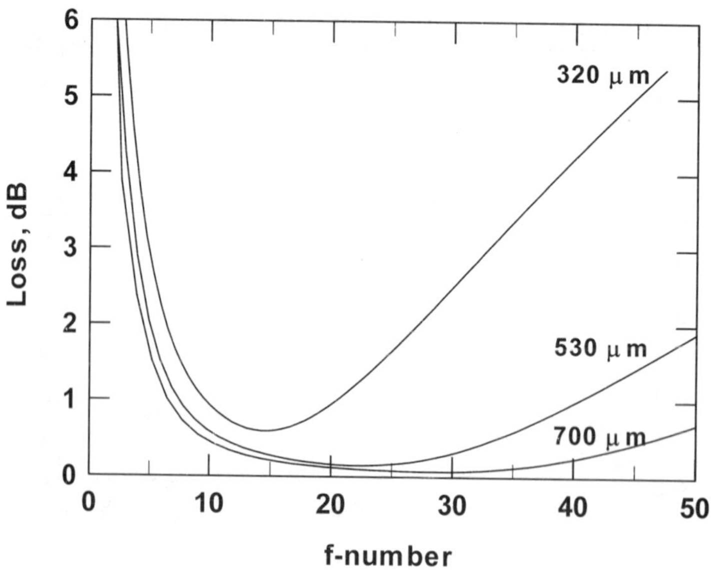

Figure 7.39 Minimum-loss coupling conditions for the 1-m-long, 320-μm-bore, 530-μm-bore, and 700-μm-bore hollow-glass waveguides.[41]

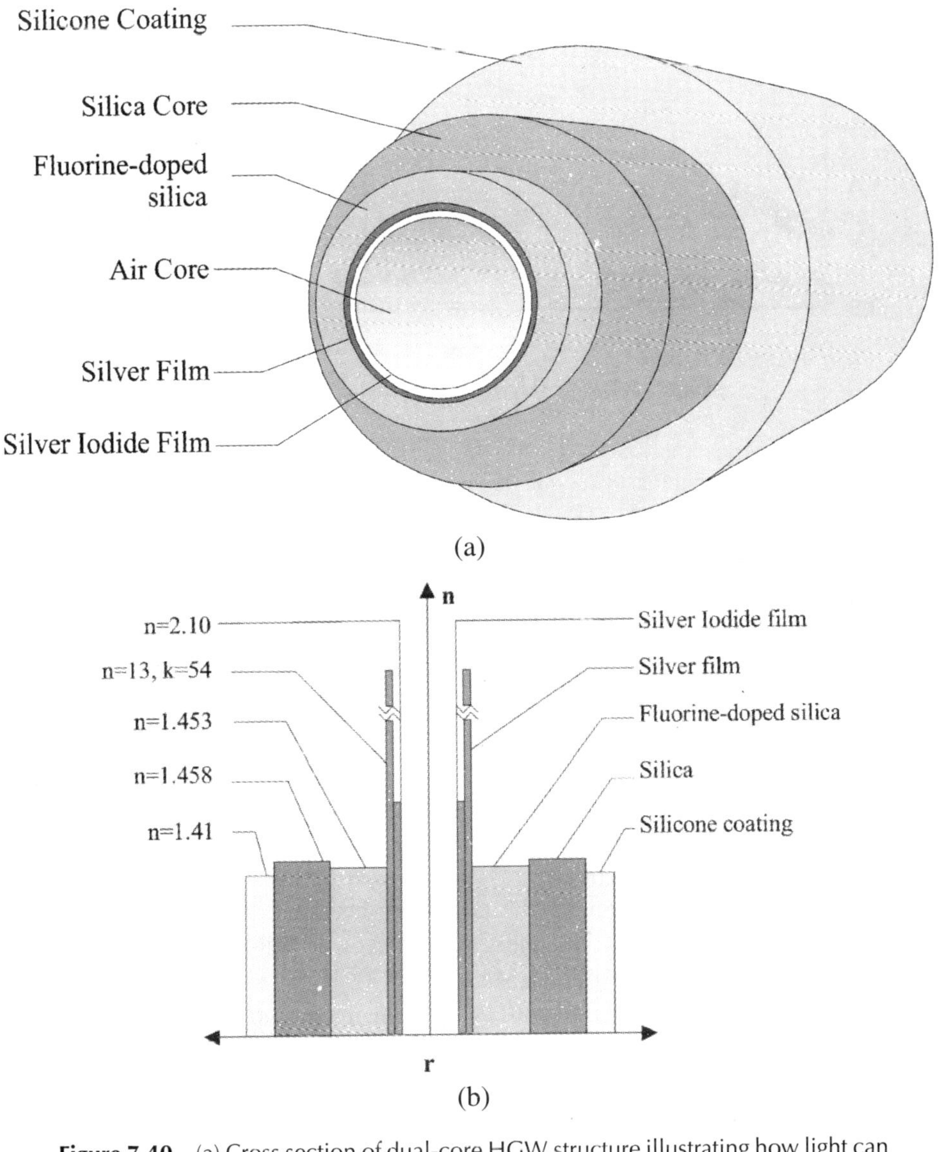

Figure 7.40 (a) Cross section of dual-core HGW structure illustrating how light can be transmitted in the annulus of waveguide. (b) Index profile of composite dual-core structure with Ag-AgI films.

seen to increase as $1/R$. However, the transmission of HeNe light at 632.8 nm in the clad annulus does not show a dependence on bending, as expected for solid-core fibers. It is also interesting to note that the near-field output profile of the HeNe beam is ring shaped, as shown in the spatial profile in Fig. 7.41(b). Therefore, in those applications requiring an aiming beam such as in laser surgery, the red HeNe laser-aiming beam will always remain concentric with the IR laser beam. Naturally, in the far field the red ring shape becomes diffuse unless the output end is reimaged. Rabii and Harrington obtained somewhat high loss values for

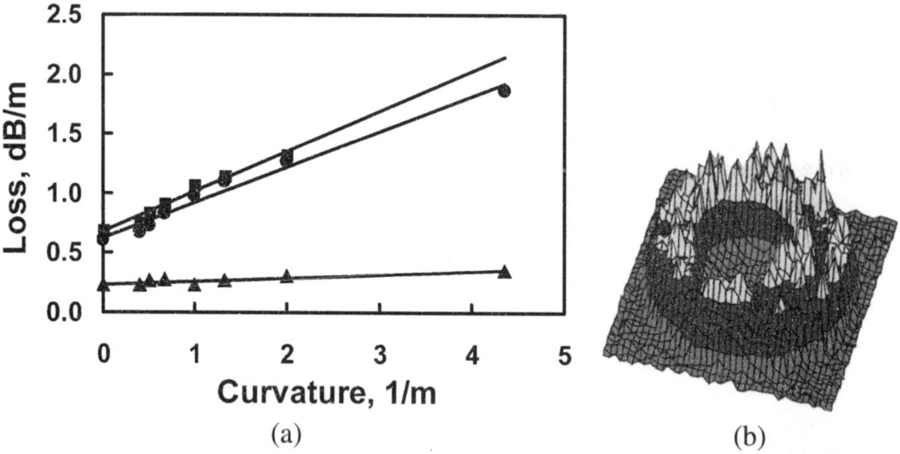

Figure 7.41 (a) Bending losses for dual-core HGW at three laser wavelengths; (■) Er:YAG at 2.94 μm; (●) CO_2 at 10.6 μm; and (▲) HeNe through the annulus at 632.8 nm. (b) Spatial profile of the HeNe laser beam near the output end of the HGW.[55] [Reprinted with permission.]

their 575-μm-bore dual-core structures at both 2.94 and 10.6 μm (approximately 0.6 dB/m at both wavelengths). They attribute this to variations in the bore diameter, which were abnormally high compared to the usual silica tubing without cladding.

A variety of configurations other than the conventional circular, straight-bore hollow guides have been studied. Tapered-bore HGWs have been placed at the input and output ends of straight guides to increase coupling efficiency and to condense light, respectively. Normally the tapers are less than 10 cm in length, and most are coated on the inside with the same Ag/AgI or Ag/polymer coatings applied to the long straight-bore guides. Matsuura et al.[56] made a careful study of the use of tapers at the input end of long HGWs as a means of improving the alignment tolerance between the input laser and the small-bore waveguide. That is, due to the small effective NA of hollow guides, these waveguides are especially sensitive to tilt and translation error of the input laser beam. By using an input taper to more or less "funnel" the light into the waveguide they were able to greatly reduce the misalignment error resulting from transverse displacement of the laser beam. Matsuura et al. made several tapers from Pyrex glass. The best taper turned out to be one with an input end tapered from 1400 to 700 μm with a length of 10 cm. This taper was coated with Ag/COP. The output end of the taper matched the 700-μm bore of the straight guide. The laser energy was focused into the taper with a 10-cm focal lens. Their results indicated that they had much less sensitivity to transverse displacement of the laser beam with the coupler, even though the coupler itself introduced a 0.5-dB loss. They also found that a shorter 5-cm-long taper was even lossier due to the higher bounce angles and, therefore, the higher-order modes launched into the guide. Waynant et al.[57] also studied the use of input tapers, but their glass tapers were uncoated.

Output tapers are a convenient means of concentrating the energy by reducing the spot size at the output end of a HGW. Often these tapers are short lengths of drawn glass tubing which may or may not be coated on the inside surface. Gibson and Harrington[58] studied a variety of tapers, including bent tapers. Tapers always introduce a loss for the simple reason that the angle that a light ray strikes the side of the taper will double on reflection. This means that rays bouncing through a taper will exit at angles considerably larger than the input angle. This leads to the generation of higher-order modes and, ultimately, high loss. Nevertheless, an output taper can often be a better choice than a lens placed at the output end of a long waveguide, as it is frequently inconvenient to place a lens at the output to reimage the light.

One of the most intriguing tapers is a long waveguide tapered uniformly and continuously from input to output end. Gibson and Harrington[59] drew silica tubing with tapers from 1000 μm to 530 μm and 700 μm to 530 μm with a length of about 1.5 m. The inside of the tapers were coated with Ag/AgI using the conventional wet-chemistry methods used on straight-bore guides. The losses for these long tapers measured using a CO_2 laser were 0.8 dB/m and 0.4 dB/m for the 1000- to 530-μm and 700- to 530-μm-bore guides, respectively. The same loss was measured irrespective of which end was used as the input end. These losses are higher than the loss for similarly sized nontapered HGWs. The higher losses of the tapered waveguides are likely due to the generation of higher-order modes. From a ray optic viewpoint, the increased loss results from an increase in bounce angle, which in turn leads to a greater number of bounces. The loss of the waveguide is proportional to the number of bounces and, therefore, the loss increases due to tapering. The bending losses for a tapered HGW show an unusual effect. The bending loss data measured with a CO_2 laser is shown in Fig. 7.42(a). The losses for the two

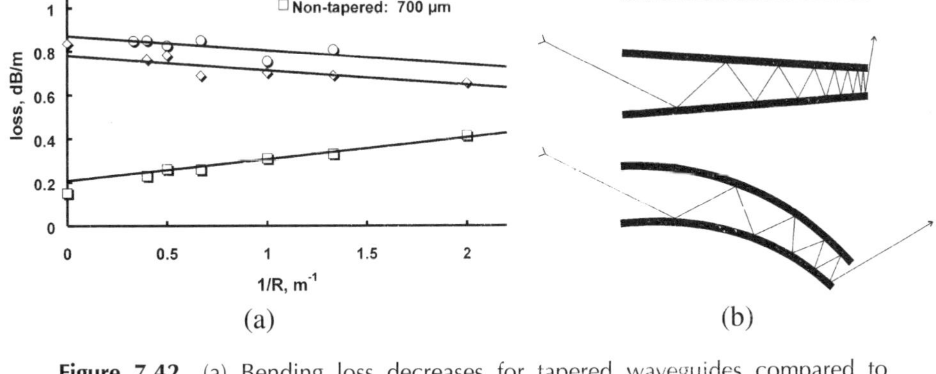

(a) (b)

Figure 7.42 (a) Bending loss decreases for tapered waveguides compared to straight-bore guides. Basically, decreasing the bounce angle decreases the number of bounces; hence the total loss of the tapered HGW decreases when bent. (b) Ray-optic approximations illustrate qualitatively this behavior for nontapered HGWs, tapered HGWs, and bent tapered HGWs.[59]

tapered guides are compared to those of straight-bore guides, and it can be seen that the bending loss *decreases* on bending for the tapered guides, in sharp contrast to the usual increase in loss as R is decreased. This reduction in loss may be qualitatively explained using ray optics, as shown schematically in Fig. 7.42(b). That is, as the tapered guide is bent, the bounce angles with respect to the wall decrease, meaning that there are fewer bounces and, thus, less loss. It also happens that as the tapered guide is bent, more rays strike only one side of the guide (grazing incidence, as shown in Fig. 3.3), which also leads to lower loss. While this is also true for nontapered guides, Gibson[60] points out that the whispering gallery (grazing incidence) modes are generated at smaller curvatures in tapered guides compared to nontapered ones. Zhou et al.[61] also studied tapered hollow waveguides and their losses.

Noncircular glass waveguides have been studied largely for their ability to preserve polarization. Rectangular and square cross-section metallic hollow waveguides are discussed in Sec. 7.1.1. Gibson and Harrington[58] studied an all-glass version of this waveguide. The two noncircular HGWs were fabricated from silica tubing, and they had a square (500 μm × 500 μm) and a rectangular (227 μm × 1253 μm) cross section. The coatings were Ag/AgI, and all measurements were made using a polarized CO_2 laser.

The polarization-preserving properties of the guides may be obtained by measuring the degree of polarization, V, defined as

$$V = \frac{P_{\text{max}} - P_{\text{min}}}{P_{\text{max}} + P_{\text{min}}}, \tag{7.13}$$

where P_{max} and P_{min} are the respective maximum and minimum powers determined by rotating an analyzer in front of the detector. The data for the raw polarized laser beam and for the square and rectangular guides is shown in Figs. 7.43(a)–(f). These polarization plots give the V values using the height and width at the center as P_{max} and P_{min}, respectively. The 530-μm circular HGW [see Fig. 7.43(d)], preserved only 26% of the input polarization, whereas the 500-μm square guide [see Fig. 7.43(b)] improved the polarization preservation since V increased to 69%. This value dropped to 54% when the square guide was rotated such that the plane of polarization was oriented 45° with respect to the walls. The rectangular guide maintained the incident polarization to a nearly perfect 96% when the long axis was aligned perpendicular to the plane of polarization, as shown in Fig. 7.43(e). They observed only small changes in V and some mode mixing on bending the guides.

It is often useful to alter the output, or distal, end by adding a disposable or bent tip. This is particularly useful in medical applications in which the end may either be contaminated or the tip needs to be bent to access a remote site, such as in laser dentistry. Laakman, then at Luxar, Inc., placed short, <2-cm tips made of alumina tubing at the end of their waveguides. Alumina or polycrystalline Al_2O_3 has an index of refraction less than 1, and, therefore, it will guide light well (see Sec. 7.3.5). When the alumina tip becomes contaminated it can be thrown away and

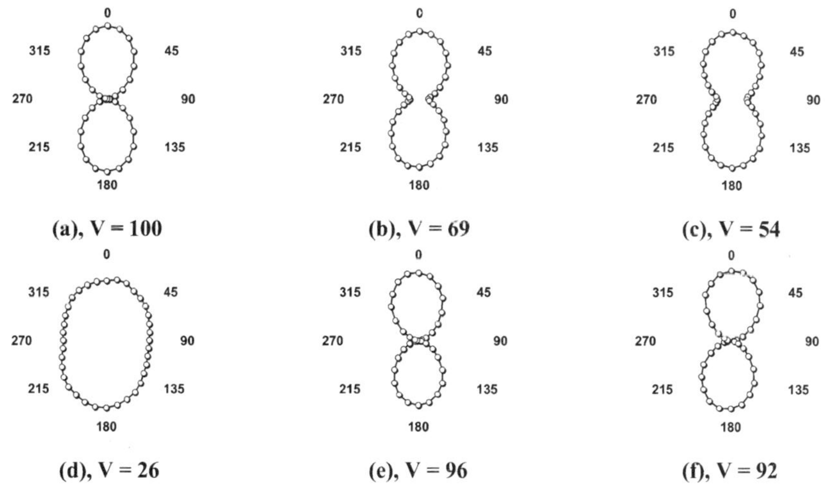

Figure 7.43 Degree of polarization plots for: (a) CO_2 laser input beam, (b) 500-μm square-bore HGW with walls perpendicular to the plane of polarization and (c) with walls at 45° to the plane of polarization, (d) 530-μm-bore circular HGW and (e) 227 × 1253-μm rectangular-bore HGW with the long axis perpendicular and (f) parallel to the plane of polarization. All HGWs are 125 cm in length.[58]

replaced with a new tip. In this way the trunk waveguide will not be affected by debris buildup on the interior walls. Gibson and Harrington[59] bent the last several centimeters of the distal end of a long waveguide by inserting the guide into a stainless-steel tube. In one example, the last 2 cm of different bore size HGWs were bent 90°, and the increase in loss due to the bent tip was found to be 0.56, 0.17, and 0.30 dB for the 320-, 530- and 700-μm-bore waveguides, respectively.

7.3.4.6 Multilayer-dielectric HGW structures

The losses in HGWs can be significantly reduced if the waveguide is fabricated with more than one dielectric layer. As discussed in Sec. 3.2.2, a stack composed of alternating layers of high- (n_H) and low- (n_L) index dielectric materials can greatly reduce the loss over the simple single-dielectric layer approach discussed so far. Miyagi and Kawakami[62] derived an expression for the loss as a function of the number of dielectric layers. Equation (3.17) is a general equation for the loss. In this equation, F_{film} is a complex term containing the values of n_H and n_L along with their thickness. In the simplest case, each dielectric layer is a quarter-wavelength thick. Equations (3.18) and (3.19) can be used to calculate the loss for a guide with only one $\lambda/4$ layer as, for example, the loss for the Ag/AgI structure shown in Fig. 3.7.

The thickness of a single-dielectric film is given by Eq. (3.18). For alternating layers of high and low index, the optimal film thickness is given by

$$d_i = \frac{\lambda}{4\sqrt{n_i^2 - 1}}, \qquad n_i = n_L, n_H. \tag{7.14}$$

A sketch of a dielectric stack showing the alternating layers is given in Fig. 7.44. Note that the last dielectric layer (next to air core) can have a different thickness than the other layers. This is required to achieve the lowest loss. Katagiri et al.[63] considered this case in detail as well as the advantages of first depositing a metallic layer on the glass wall and then adding the alternating dielectric layers. Using a thin metallic layer reduces the number of dielectric layers needed to achieve a low loss. The term F_{film} in Eq. (3.17) becomes, for multilayer structures with an odd number ($m = 2p + 1$) of dielectric layers,

$$F_{film}(odd) = \frac{1}{2}C^p\left[1 + \frac{n_L^2}{\sqrt{n_L^2 - 1}}\left(\frac{n_L}{n_H}\right)^{2p} C^{-p}\right]^2, \tag{7.15}$$

where p is the number of pairs, and $C = (n_L^2 - 1)/(n_H^2 - 1)$. For an even number ($m = 2p + 2$) of dielectric layers,

$$F_{film}(even) = \frac{1}{2}\frac{n_L^4}{n_L^2 - 1}\left(\frac{n_L}{n_L}\right)^{4p} C^{-p}\left[1 + \frac{n_L^2}{n_L^2\sqrt{n_H^2 - 1}}\left(\frac{n_H}{n_L}\right)^{2p} C^p\right]^2. \tag{7.16}$$

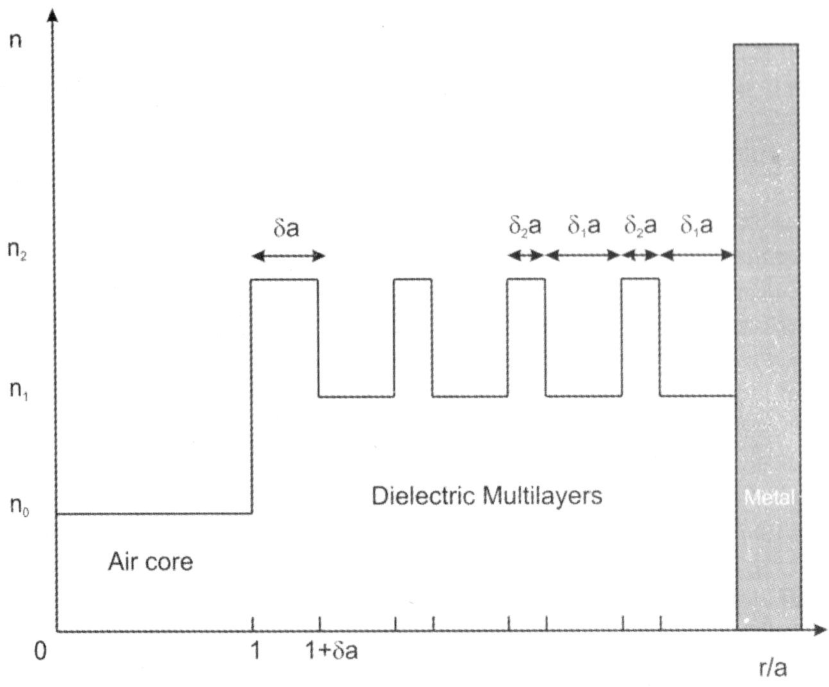

Figure 7.44 Refractive index profile of hollow waveguide with multiple inner dielectric layers.

Equations (7.15) and (7.16) are for the HE_{nm} modes when the thickness of the dielectric layers are as specified by Eq. (7.14) for all but the last dielectric layer. The thickness of this layer is discussed in the literature.[63]

As an illustration of the power of multiple dielectric films to reduce loss, Rabii[42] has used Eqs. (7.14)–(7.16) to calculate the loss reduction for different n_L/n_H combinations over a silver film. In Fig. 7.45, the amount of loss calculated for m layers is compared to the loss for an Ag-only waveguide (α_{metal}) at 10.6 μm. The ratio of $\alpha/\alpha_{\text{metal}}$ indicates that increasing the number of layers can markedly reduce loss. Furthermore, it can be seen from Fig. 7.45 that the loss is less when an odd number of layers is used compared to a comparable number of an even number of layers. Note in particular that the larger the index contrast, i.e., $\Delta n = n_H - n_L$, the greater the loss reduction. This is an important fact and one that is vital in the design of photonic bandgap hollow fibers. Photonic bandgap fibers are discussed in Chapter 11. In any case, it is clear that one would like to select materials with the largest index contrast in order to prepare waveguides with the least number of dielectric layers.

There has been only a limited amount of experimental effort to fabricate multilayer waveguides. Matsuura and Harrington[64,65] have used chemical vapor deposition (CVD) methods to deposit a three-layer stack over Ag. A 0.3-μm-thick Ag layer is deposited inside silica tubing using the conventional, wet-chemistry technology. Next, a three-layer dielectric stack is formed using CVD methods to de-

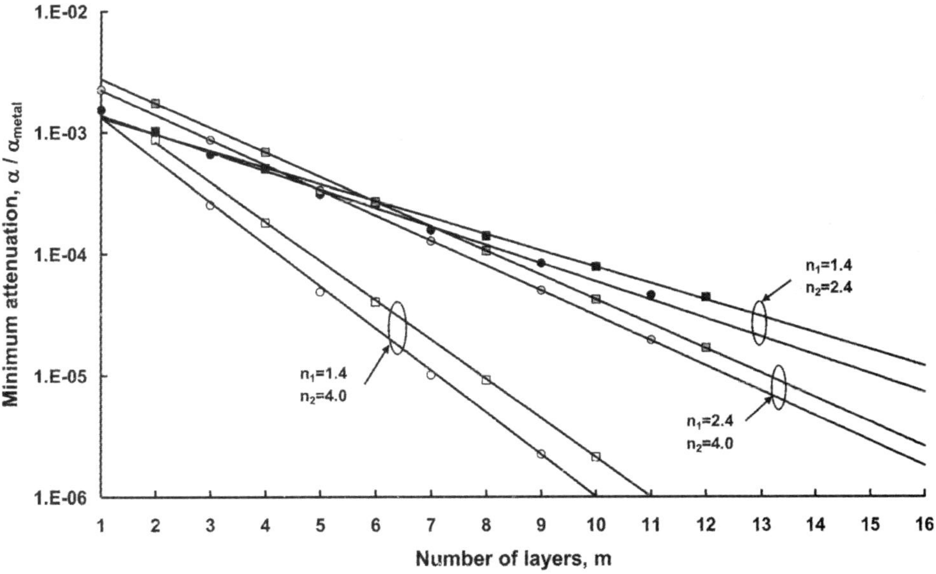

Figure 7.45 Loss reduction achieved through the use of multiple dielectric layers. Calculations based on structure shown in Fig. 7.44, for which a metallic layer of Ag is the first layer deposited on the inner wall of the glass tubing. (■) Even number of layers, and (●) odd number of layers.

posit $Al_2O_3/TiO_2/Al_2O_3$. These materials were chosen because they are relatively easy to deposit using CVD methods and they have good optical quality. There is also good index contrast between TiO_2 (2.5) and Al_2O_3 (1.71). While Al_2O_3 and TiO_2 are not transparent at wavelengths longer than about 9 μm, they are suitable materials for the fabrication of waveguides transmitting 3-μm Er:YAG and 5-μm CO laser wavelengths. Their results for a three-layer film optimized for operation at 3 μm are shown in Fig. 7.46.[64] These spectral loss measurements compare well with the calculated loss, but the authors note that the measured loss at 3 μm is almost the same as that for a single-layer Al_2O_3 layer over Ag. They attribute this discrepancy to the roughness of the metallic and dielectric films. More recently, Matsuura et al.[63] have used a liquid-phase process to deposit $SiO_2/Ta_2O_5/SiO_2$ over Ag. Again, the losses are not as low as those calculated.

7.3.5 Hollow waveguide structures with $n < 1$

A most unusual hollow waveguide is one whose inner wall material has a refractive index less than 1. As explained in Chapter 3, these ATR-type guides operate on the principle that the air core ($n = 1$) has a refractive index greater than the inner cladding wall so that $n_{clad} < n_{core}$ and the waveguide is fiberlike. The idea of an $n < 1$ structure originated with Hidaka et al.[66] in 1981. For most dielectric materials and even many metals, there is a wavelength region for which n becomes less than 1. This is generally near the region of anomalous dispersion. In this region, the optical constants n and k vary greatly. Unfortunately, k is usually not too small within the same wavelength region where $n < 1$. To be useful

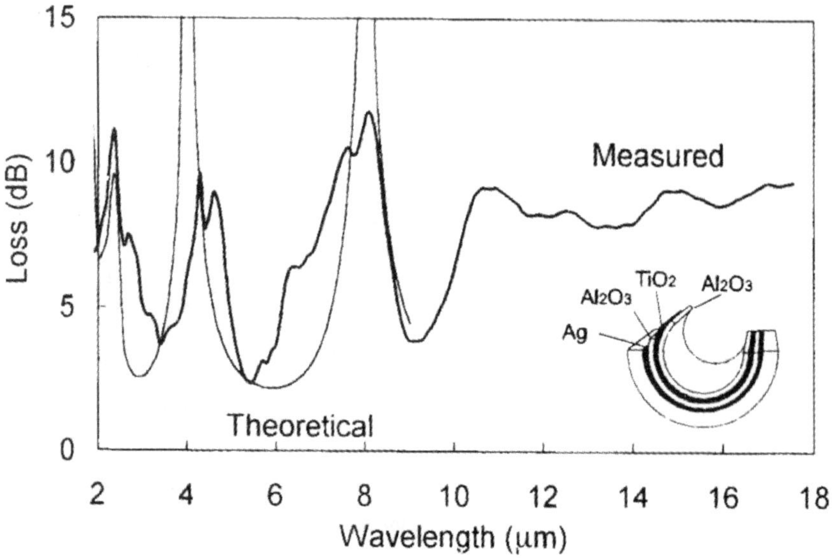

Figure 7.46 Spectral losses of the hollow waveguides with three-layer, $Al_2O_3/TiO_2/Al_2O_3$ coating. The thickness of the third Al_2O_3 layer is 0.32 μm.[64] [Reprinted with permission.]

for laser transmission, the ATR guides must have the region of anomalous dispersion fall within the laser wavelength range. The first $n < 1$ guides studied by Hidaka et al. focused on glass tubes made from lead- and germanium-doped silicates. By adding heavy ions to silica glass, they were able to shift the infrared edge to longer wavelengths so that the CO_2 laser wavelength occurred within the $n < 1$ region of anomalous dispersion.[5] Worrel[67,68] also studied $n < 1$ glasses for CO_2 laser transmission. In particular, he studied the germanate glasses. The losses in the hollow-glass, $n < 1$ guides, however, are very high due to a high k values, and an all-glass hollow waveguide structure with $n < 1$ has largely been abandoned.

A more promising $n < 1$ structure is hollow sapphire. It was first pointed out by Harrington and Gregory[4] that sapphire (Al_2O_3) has $n < 1$ from 10 to 16.7 μm and, in addition, it has a very small k value of 0.05 at 10.6 μm. This means that the theoretical loss predicted by MS theory is very low (less than 0.1 dB/m for a 1000-μm-bore tube) for this material. Single-crystal sapphire tubing is fabricated using the EFG process described in Sec. 6.2.2. Saphikon, Inc., in Milford, NH, has grown SC sapphire tubing in lengths as long as 2 m with bore sizes varying from 250 to 1070 μm. In Fig. 7.47(a) the losses measured by Gregory and Harrington are shown for five different bore sizes. These results are compared to the theoretical losses for both the lowest-order HE_{11} and next-higher-order HE_{12} modes. The measured losses are somewhat higher than those predicted by MS theory as a result of the roughness of sapphire's inner wall. Gregory and Harrington[49] showed that the surface roughness of hollow sapphire accounted for the increased attenuation over that predicted for the HE_{11} mode. In Fig. 7.47(b) the bending loss for the 530-μm-bore waveguide is shown. The curvature is not as great as it is for the HGWs because sapphire has a high modulus and, therefore, cannot be bent to small diameters. Hollow-sapphire delivery systems have been coupled to CO_2 lasers for use in gynecology and orthopedic surgery, and they have been packaged in a water-cooled jacket for the delivery of more than 1900 W of CO_2 laser power.[69]

7.4 Mechanical properties of hollow waveguides

Rabii and Harrington[70] have made a thorough study of the strengtn of HGWs. Since these guides are made of pure silica tubing, it is expected that their strength and Weibull modulus should be roughly similar to that of solid-core silica fibers. In the general discussion of strength and reliability of fibers in Sec. 2.4, it is seen that the ultimate failure of glass fibers is statistical in nature. That is, the maximum stress or strain that can be applied to a fiber depends on the statistical distribution and nature of the flaws in the fibers. A standard method of determining the strength of a fiber is to perform a two-point bending test as discussed in Sec. 2.5. Specifically, the maximum stress σ_{max} can be obtained from the geometry of the bend and the relation

$$\sigma_{max} = 1.198E\frac{2r}{D-d}, \tag{2.25}$$

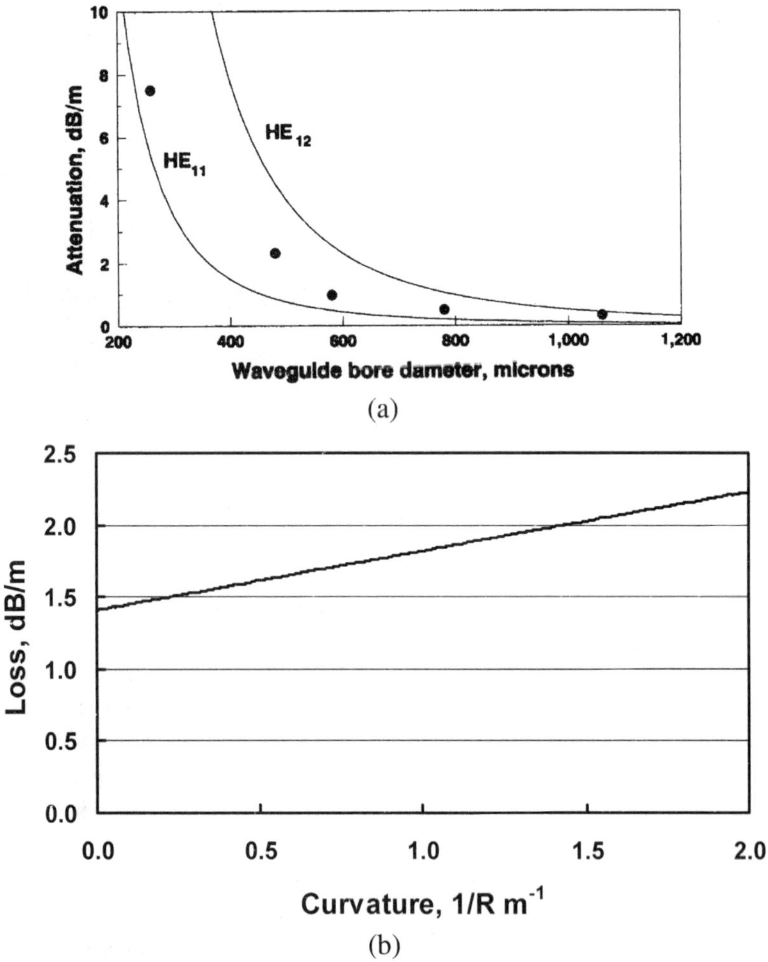

Figure 7.47 (a) Measured straight losses for different-bore-size hollow sapphire waveguides compared to the calculated loss for the two lowest-order modes. (b) Bending loss for 560-μm bore hollow sapphire waveguide. All data was taken at 10.6 μm.[4] [Reprinted with permission.]

where E is Young's modulus, D is the plate separation as shown in Fig. 2.13, d is the overall fiber diameter including coating, and r is the radius of the fiber.

Rabii and Harrington measured σ_{max} using two-point bending for each stage in the HGW process. In Fig. 7.48, Weibull plots are given for four different sample sets.[70] From the data in Fig. 7.48 it can be seen that the guide with the greatest strength is the raw, unprocessed tubing. The steep slope for this uncoated tubing indicates a high Weibull modulus [see the discussion around Eq. (2.27)]. In this case the Weibull modulus m is 96.9. This is very high and it is roughly equal to that for solid-core silica fiber. However, during the coating process the tubing is weakened, and m decreases. Specifically, the Ag-only and Ag/AgI guides have $m = 22.7$ and 18.5, respectively. This is indicative of a wider strength distribution. For example, the Ag-coated tubing samples (set no. 3) have a high mean-failure stress level of

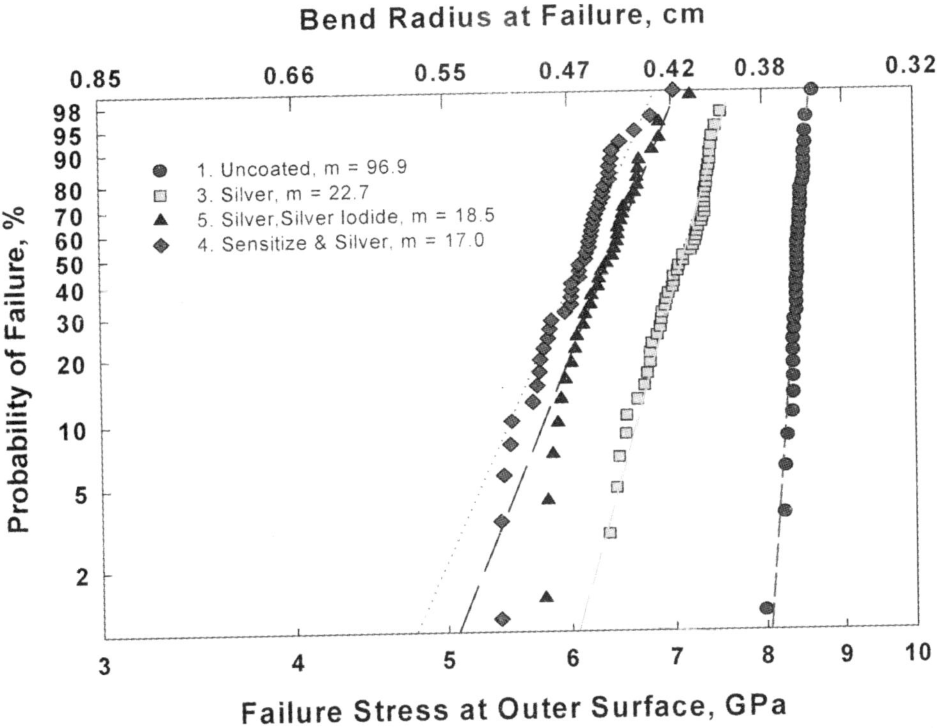

Figure 7.48 Weibull statistics for strength of waveguides after each processing stage. Constant faceplate speed $v = 1000$ μm/sec, $T \sim 20°C$, relative humidity ~25%. Weibull modulus, m, for each curve is given in the figure.[70]

6.99 GPa, but they are 17% weaker than the uncoated sample set. In short, the liquid-phase process weakens the guides. The weakening is due to the interaction of the solutions with the silica in the liquid-phase process. The silvering process involves the use of NH_4OH to raise the pH for optimum plating conditions. As a result, the final silvering solution has a pH of approximately 11.5. Silica surfaces are known to be susceptible to attack and dissolution by any pH solution, but the greatest effect is observed in basic solutions. The mechanism by which water corrodes silica surfaces involves the formation of silanol (Si—OH) units, which leads to small cracks opening from the surface and ultimately weakening the glass. Further fractography studies indicated that failure occurred at the inner surface of the tubing as a result of surface corrosion. Even though the tubing is somewhat weakened as a result of processing, the HGWs remain quite strong, with bending radii less than 2 cm for the 530-μm-bore tubing. Finally, the data in Fig. 7.48 may be used to estimate the minimum-bending radius for a certain probability of failure. Using the approximate expression in Eq. (2.29) and Young's modulus $E = 73.1$ GPa for silica, a minimum-bending radius may be calculated. For example, for a 50% failure probability, the bending radius at failure would be 0.46 cm and 1.21 cm for 320×450-μm and 530×700-μm-bore tubing, respectively.

References

1. E. Garmire, T. McMahon, and M. Bass, "Flexible infrared waveguides for high-power transmission," *IEEE J. Quantum Electron.*, Vol. QE-16, pp. 23–32 (1980).
2. J. Harrington, "A review of IR transmitting, hollow waveguides," *Fiber and Integrated Optics*, Vol. 19, pp. 211–227 (2000).
3. A. Hongo, K. Morosawa, K. Matsumoto, T. Shiota, and T. Hashimoto, "Transmission of kilowatt-class CO_2 laser light through dielectric-coated metallic hollow waveguides for material processing," *Appl. Opt.*, Vol. 31, pp. 5114–5120 (1992).
4. J. A. Harrington and C. C. Gregory, "Hollow sapphire fibers for the delivery of CO_2 laser energy," *Opt. Lett.*, Vol. 15, pp. 541–543 (1990).
5. T. Hidaka, J. Kumada, J. Shimada, and T. Morikawa, "GeO_2-ZnO-K_2O glass as the cladding material for 940 cm^{-1} CO_2 laser light transmitting hollow-core waveguide," *J. Appl. Phys.*, Vol. 53, pp. 5484–5490 (1982).
6. U. Kubo, "Medical applications of optical fibers," *Review of Laser Engineering*, Vol. 22, pp. 329–337 (1994).
7. U. Kubo, Y. Hashishin, and K. Okada, "Optical beam guides for medical CO_2 and excimer lasers," Proc. SPIE, Vol. 906, pp. 214–219 (1988).
8. M. B. Levy and K. D. Laakmann, "Flexible waveguide for CO_2 laser surgery," Proc. SPIE, Vol. 605, pp. 57–58 (1986).
9. H. Machida, Y. Matsuura, H. Ishikawa, and M. Miyagi, "Transmission properties of rectangular hollow waveguides for CO_2 laser light," *Appl. Opt.*, Vol. 31, pp. 7617–7622 (1992).
10. D. Mendlovic, E. Goldenberg, S. Ruschin, J. Dror, and N. Croitoru, "Ray model for transmission of metallic-dielectric hollow bent cylindrical waveguides," *Appl. Opt.*, Vol. 28, pp. 708–712 (1989).
11. N. Croitoru, D. Mendlovic, J. Dror, S. Ruschin, and E. Goldenberg, "Improved metallic tube infrared waveguides with inside dielectric coating," Proc. SPIE, Vol. 713, pp. 6–10 (1987).
12. K. D. Laakmann and M. B. Levy, U.S. Patent No. 5,005,944 (Issued 1991).
13. M. Miyagi, A. Hongo, Y. Aizawa, and S. Kawakami, "Fabrication of germanium-coated nickel hollow waveguides for infrared transmission," *Appl. Phys. Lett.*, Vol. 43, pp. 430–432 (1983).
14. N. Croitoru, J. Dror, and I. Gannot, "Characterization of hollow fibers for the transmission of infrared radiation," *Appl. Opt.*, Vol. 29, pp. 1805–1809 (1990).
15. T. Abel, J. Hirsch, and J. A. Harrington, "Hollow glass waveguides for broadband infrared transmission," *Opt. Lett.*, Vol. 19, pp. 1034–1036 (1994).
16. A. Hongo, K.-I. Morosawa, T. Shiota, Y. Matsuura, and M. Miyagi, "Transmission characteristics of germanium thin-film-coated metallic hollow waveguides for high-powered CO_2 laser light," *IEEE J. Quantum Electron.*, Vol. 26, pp. 1510–1515 (1990).
17. Y. Komachi, M. Wakaki, and G. Kanai, "Fabrication of hollow waveguides for CO_2 lasers," *Appl. Opt.*, Vol. 39, pp. 1555–1560 (2000).

18. P. Bhardwaj, O. J. Gregory, C. Morrow, G. Gu, and K. Burbank, "Performance of a dielectric-coated monolithic hollow metallic waveguide," *Mater. Lett.*, Vol. 16, pp. 150–156 (1993).
19. M. Alaluf, J. Dror, R. Dahan, and N. Croitoru, "Plastic hollow fibers as a selective infrared radiation transmitting medium," *J. Appl. Phys.*, Vol. 72, pp. 3878–3883 (1992).
20. R. Dahan, J. Dror, and N. Croitoru, "Characterization of chemically formed silver iodide layers for hollow infrared guides," *Mater. Res. Bull.*, Vol. 27, pp. 761–766 (1992).
21. R. George and J. A. Harrington, "Hollow plastic waveguides for sensor applications," Proc. SPIE, Vol. 4204, pp. 230–237 (2001).
22. K. Matsuura, Y. Matsuura, and J. A. Harrington, "Evaluation of gold, silver, and dielectric-coated hollow glass waveguides," *Opt. Eng.*, Vol. 35, pp. 3418–3421 (1996).
23. C. C. Gregory, "Hollow dielectric waveguides for carbon dioxide laser power delivery," PhD dissertation, Rutgers University (1992).
24. C. Rabii, D. Gibson, and J. A. Harrington, "Processing and characterization of silver films used to fabricate hollow glass waveguides," *Appl. Opt.*, Vol. 38, pp. 4486–4493 (1999).
25. Y. Abe, Y. W. Shi, Y. Matsuura, and M. Miyagi, "Flexible small-bore hollow fibers with an inner polymer coating," *Opt. Lett.*, Vol. 25, pp. 150–152 (2000).
26. Y. Wang, A. Hongo, Y. Kato, T. Shimomura, D. Miura, and M. Miyagi, "Thickness and uniformity of fluorocarbon polymer film dynamically coated inside silver hollow glass waveguides," *Appl. Opt.*, Vol. 36, pp. 2886–2892 (1997).
27. Y. W. Shi, Z. Y. Pan, Y. Matsuura, and M. Miyagi, "New and simple method for fabricating polymer-coated silver hollow fibers with large mechanical strength," *Optics and Laser Technology*, Vol. 32, pp. 273–275 (2000).
28. Y. Kato, M. Osawa, M. Miyagi, S. Abe, M. Aizawa, and S. Onodera, "Loss characteristics of polyimide-coated silver hollow glass wave-guides for the infrared," *Elect. Lett.*, Vol. 31, pp. 31–32 (1995).
29. J. A. Harrington, J. C. Harrington, C. C. Gregory, and S. Harman, "Properties of alkali halide optical fibers," Proc. SPIE, Vol. 906, pp. 176–182 (1988).
30. Y. Abe, Y. Matsuura, Y. W. Shi, Y. Wang, H. Uyama, and M. Miyagi, "Polymer-coated hollow fiber for CO_2 laser delivery," *Opt. Lett.*, Vol. 23, pp. 89–90 (1998).
31. Y. Wang, A. Hongo, Y. Kato, T. Shimomura, D. Miura, and M. Miyagi, "Thickness and uniformity of fluorocarbon polymer film dynamically coated inside silver hollow glass waveguides," *Appl. Opt.*, Vol. 36, pp. 2886–2892 (1997).
32. H. Machida, H. Ishikawa, and M. Miyagi, "Low-loss lead fluoride-coated square waveguide for CO_2 laser light transmission," *Elect. Lett.*, Vol. 27, pp. 2068–2070 (1991).
33. Y. Matsuura and M. Miyagi, "Er:YAG, CO, and CO_2 laser delivery by ZnS-coated Ag hollow waveguides," *Appl. Opt.*, Vol. 32, pp. 6598–6601 (1993).

34. M. Miyagi, Y. Shimada, A. Hongo, K. Sakamoto, and S. Nishida, "Fabrication and transmission properties of electrically deposited germanium-coated waveguides for infrared radiation," *J. Appl. Phys.*, Vol. 60, pp. 454–456 (1986).
35. Y. Matsuura and M. Miyagi, "Low-loss metallic hollow waveguides coated with durable and nontoxic ZnS," *Appl. Phys. Lett.*, Vol. 61, pp. 1622–1623 (1992).
36. A. Inberg, M. Ben David, M. Oksman, A. Katzir, and N. Croitoru, "Theoretical model and experimental studies of infrared radiation propagation in hollow plastic and glass waveguides," *Opt. Eng.*, Vol. 39, pp. 1316–1320 (2000).
37. Y. Matsuura, T. Abel, and J. A. Harrington, "Optical properties of small-bore hollow glass waveguides," *Appl. Opt.*, Vol. 34, pp. 6842–6847 (1995).
38. Y. Kato, M. Osawa, M. Miyagi, S. Abe, M. Aizawa, and S. Onodera, "Loss characteristics of polyimide-coated silver hollow glass wave-guides for the infrared," *Elect. Lett.*, Vol. 31, pp. 31–32 (1995).
39. Y. W. Shi, Y. Wang, Y. Abe, Y. Matsuura, M. Miyagi, S. Sato, M. Taniwaki, and H. Uyama, "Cyclic olefin polymer-coated silver hollow glass waveguides for the infrared," *Appl. Opt.*, Vol. 37, pp. 7758–7762 (1998).
40. Y. Matsuura, T. Abel, J. Hirsch, and J. A. Harrington, "Small-bore hollow waveguide for delivery of near singlemode IR laser radiation," *Elect. Lett.*, Vol. 30, pp. 1688–1690 (1994).
41. R. Nubling and J. A. Harrington, "Launch conditions and mode coupling in hollow glass waveguides," *Opt. Eng.*, Vol. 37, pp. 2454–2458 (1998).
42. C. Rabii, "Hollow glass waveguide for mid-infrared applications," PhD dissertation, Rutgers University (1998).
43. Z. Menachem, N. Croitoru, and J. Aboudi, "Improved mode model for infrared wave propagation in a toroidal dielectric waveguide and applications," *Opt. Eng.*, Vol. 41, pp. 2169–2180 (2002).
44. Y. Matsuura, C. Rabii, K. Matsuura, and J. Harrington, "Low-order multimode generation in hollow glass waveguides," *Elect. Lett.*, Vol. 32, pp. 1096–1098 (1996).
45. K. Takatani, Y. Matsuura, and M. Miyagi, "Theoretical and experimental investigations of loss behavior in the infrared in quartz hollow waveguides with rough surfaces," *Appl. Opt.*, Vol. 34, pp. 4352–4357 (1995).
46. O. B. Danilov, M. I. Zintchenko, Y. A. Rubinov, and E. N. Sosnov, "Transmission losses and mode-selection characteristics of a curved hollow dielectric waveguide with a rough surface," *J. Opt. Soc. Am. B*, Vol. 7, pp. 1785–1790 (1990).
47. A. M. Sagi-Dolev, J. Dror, A. Inberg, J. R. Ferncz, and N. Croitoru, "The influence of roughness on scattering of transmitted radiation through hollow waveguides," Proc. SPIE, Vol. 2677, pp. 72–80 (1996).
48. H. Jelinkova, J. Sulc, P. Cerny, Y. W. Shi, Y. Matsuura, and M. Miyagi, "High-power Nd:YAG laser picosecond pulse delivery by a polymer-coated silver hollow-glass waveguide," *Opt. Lett.*, Vol. 24, pp. 957–959 (1999).

49. C. C. Gregory and J. A. Harrington, "Attenuation, modal, polarization properties of $n < 1$, hollow dielectric waveguides," *Appl. Opt.*, Vol. 32, pp. 5302–5309 (1993).
50. M. Ben David, A. Inberg, I. Gannot, and N. Croitoru, "The effect of scattering on the transmission of infrared radiation through hollow waveguides," *J. Optoelectr. and Adv. Mat.*, Vol. 1, pp. 23–30 (1999).
51. C. Rabii and J. A. Harrington, "Measurement and control of thin film uniformity in hollow glass waveguides," *Opt. Eng.*, Vol. 38, pp. 2009–2015 (1999).
52. Y. Matsuura and M. Miyagi, "Bending losses and beam profiles of zinc selenide-coated silver waveguides for carbon dioxide laser light," *Appl. Opt.*, Vol. 31, pp. 6441–6445 (1992).
53. D. J. Haan and J. A. Harrington, "Hollow waveguides for gas sensing and near-IR applications," Proc. SPIE, Vol. 3596, pp. 43–49 (1999).
54. R. L. Abrams, "Coupling losses in hollow waveguide laser resonators," *IEEE J. Quantum Electron.*, Vol. QE-8, pp. 838–843 (1972).
55. C. D. Rabii and J. A. Harrington, "Optical properties of dual-core hollow waveguides," *Appl. Opt.*, Vol. 35, pp. 6249–6252 (1996).
56. Y. Matsuura, H. Hiraga, Y. Wang, Y. Kato, M. Miyagi, S. Abe, and S. Onodera, "Lensed-taper launching coupler for small-bore, infrared hollow fibers," *Appl. Opt.*, Vol. 36, pp. 7818–7821 (1997).
57. I. K. Ilev, R. Waynant, and M. A. Bonaguidi, "Attenuation measurement of infrared optical fibers by use of a hollow-taper-based coupling method," *Appl. Opt.*, Vol. 39, pp. 3192–3196 (2000).
58. D. Gibson and J. A. Harrington, "Tapered and noncircular hollow glass waveguides," Proc. SPIE, Vol. 3596, pp. 8–13 (1999).
59. D. Gibson and J. A. Harrington, "Transmission properties of hollow glass waveguides," Proc. SPIE, Vol. 3849, pp. 143–148 (2000).
60. D. Gibson, "Hollow glass waveguides: new variations," PhD dissertation, Rutgers University (2003).
61. J. Y. Zhou, Y. J. Deng, N. Ling, G. Lu, and Z. X. Li, "Laser transmission through a dielectric-coated hollow tapered metallic waveguide," *Opt. Commun.*, Vol. 142, pp. 30–33 (1997).
62. M. Miyagi and S. Kawakami, "Design theory of dielectric-coated circular metallic waveguides for infrared transmission," *J. Lightwave Tech.*, Vol. LT-2, pp. 116–126 (1984).
63. T. Katagiri, Y. Matsuura, and M. Miyagi, "Metal-covered photonic bandgap multilayer for infrared hollow waveguides," *Appl. Opt.*, Vol. 41, pp. 7603–7606 (2002).
64. Y. Matsuura and J. A. Harrington, "Hollow glass waveguides with three-layer dielectric coating fabricated by chemical vapor deposition," *J. Opt. Soc. Am. A*, Vol. 14, pp. 1255–1259 (1997).
65. Y. Matsuura and J. A. Harrington, "Infrared hollow glass waveguides fabricated by chemical vapor deposition," *Opt. Lett.*, Vol. 20, pp. 2078–2080 (1995).

66. T. Hidaka, T. Morikawa, and J. Shimada, "Hollow-core oxide-glass cladding optical fibers for middle-infrared region," *J. Appl. Phys.*, Vol. 52, pp. 4467–4471 (1981).
67. C. A. Worrell, "Infrared optical constants for CO_2 laser waveguide materials," *J. Matr. Sci.*, Vol. 21, pp. 781–787 (1986).
68. C. A. Worrell and V. Skarda, "CO_2 laser waveguides from germania-based glasses," *J. Phys. D (Appl. Phys.)*, Vol. 22, pp. 535–541 (1989).
69. R. K. Nubling and J. A. Harrington, "Hollow-waveguide delivery systems for high-power, industrial CO_2 lasers," *Appl. Opt.*, Vol. 34, pp. 372–380 (1996).
70. C. Rabii and J. Harrington, "Mechanical properties of hollow glass waveguides," *Opt. Eng.*, Vol. 38, pp. 1490–1499 (1999).

Chapter 8

IR Fiber Optic Sensor Applications

8.0 Introduction

A particularly attractive application for IR fiber optics is their use in sensor systems. In general, silica-based fiber optic sensors have been studied for many years and they now appear in a myriad of applications. These include simple on-off–type sensors, chemical and radiometric sensors, and the sophisticated fiber optic gyros. Not surprisingly, there have been far fewer analogous applications for IR fibers. The reason for this is that IR fibers do not have as low a loss as silica fiber and, therefore, they are usually not employed in lengths longer than about 3 m. Additionally, many of the most sensitive fiber sensors are those based on the interferometric measurement of very small phase shifts. Normally SM fibers are used, and frequently they must also preserve polarization. Other than some special fluoride glass fibers, IR fibers are not SM and, therefore, interferometric fiber sensors using IR fibers are generally not an option. In fact, there is probably little advantage to using an SM IR fiber over a conventional SM fiber in such applications even if a good SM IR fiber did exist. The real advantage of IR fibers in sensor applications is simply their ability to transmit long-wavelength information. Otherwise, silica fiber optics would be the sensor fiber of choice.

Fiber optic sensors are either intrinsic or extrinsic. Extrinsic fiber sensors are usually simple in design. In these applications the fibers may be thought of as merely light "wires" transmitting the source information to the detector. Thermometry is an example of this type of sensor. Intrinsic fiber sensors involve altering the optical properties of the fiber. An example of an intrinsic fiber sensor is the evanescent wave chemical sensor. These two concepts are sketched in Fig. 8.1.

There are two primary sensor applications for IR fibers. The first is their use in spectroscopy, or chemical sensing, and the second is in radiometry, or the measurement of temperature. Their unique advantage in chemical sensing is the ability of the fibers to transmit in the sensitive "fingerprint" region of the IR spectrum. In the region beyond about 2 μm, many materials have strong absorption bands characteristic of the material. A simple chart showing the region of absorption for a variety of molecules is given in Fig. 8.2. For example, from Fig. 8.2 it may be seen that it is possible to identify many C−H bonded solvents, drugs, petrochemicals, etc., by working in discrete regions of the IR spectrum. Clearly the specificity and

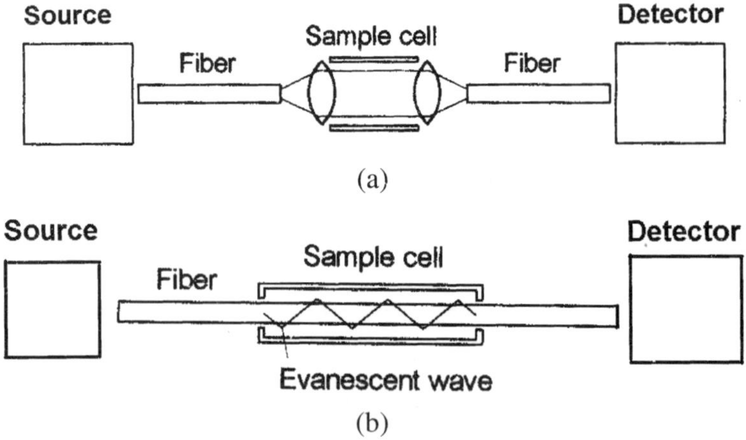

Figure 8.1 Simple examples of the two basic fiber optic sensors; (a) extrinsic sensor where the fibers merely link a source and detector, and (b) intrinsic sensor where the fiber itself is part of the sensor system.[2]

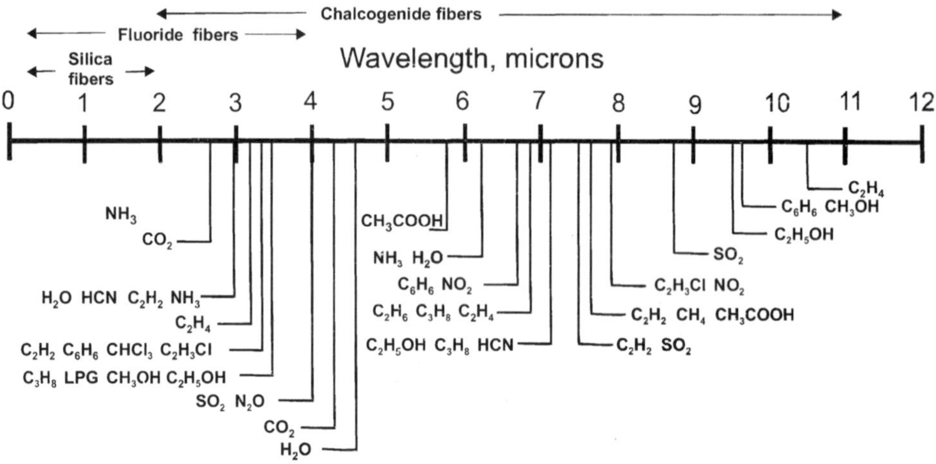

Figure 8.2 Absorption lines for chemical bonds, molecular ions, radicals, and compounds.

sensitivity is greater at infrared wavelengths; therefore, it is possible to configure chemical fiber sensors not only for broadband detection but also for the measurement of a specific compound using a tunable IR laser and detector. IR fibers are useful in temperature measurements because they can transmit long-wavelength, blackbody radiation to sensitive IR detectors. In particular, their broad spectral response beyond 2 μm can be very important when the fibers are used to detect low temperature. This may be seen from the blackbody radiation curves shown in Fig. 8.3. The Stefan-Boltzmann law says that the total power of a perfect blackbody (emissivity of 1), or the area under the curves in Fig. 8.3, is proportional to T^4, where T is the absolute temperature. Therefore, there is dramatically less radiated

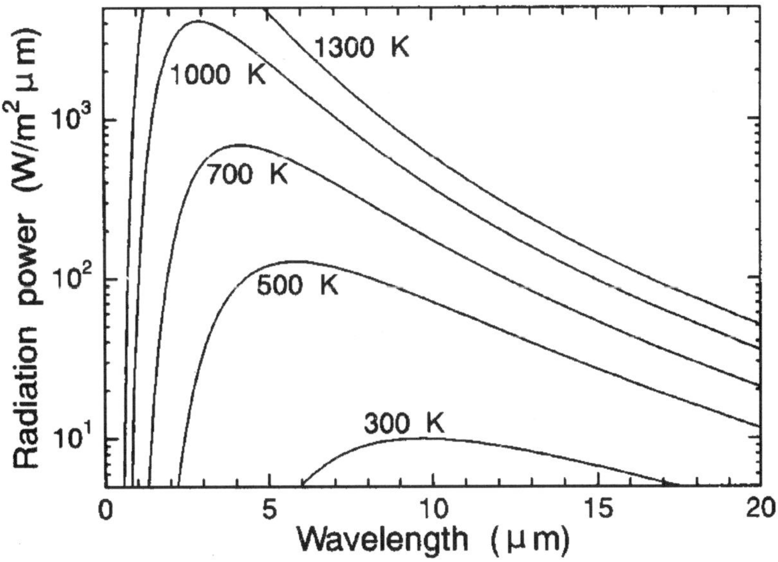

Figure 8.3 Blackbody emission at various temperatures. Note that the blackbody radiation for room temperature (~300 K) is peaked near 10 μm, where PC IR fibers and hollow waveguides work well.[2]

power for temperatures near room temperature. Moreover, the Wien displacement law indicates that the peak in the blackbody curve shifts to longer wavelengths as the temperature decreases. For example, the blackbody radiation near room temperature is peaked near 10 μm and, therefore, IR fibers are an excellent choice for biosensors used to measure body temperatures. The popular eardrum temperature sensors use a short hollow guide to transmit core body heat the short distance to a detector.

The evanescent wave sensor (EWS) is based on the simple idea that some small amount of energy can propagate beyond the core of the fiber into a surrounding medium to be analyzed. In Sec. 2.1.1 the basic principles of reflection and refraction were discussed. In the simplest case of a planar interface shown in Fig. 2.1 and Eq. (2.3), for all angles $\theta > \theta_c = \sin^{-1}(n_2/n_1)$ there is total internal reflection (TIR) and no energy is refracted out of the core material. However, as discussed in Sec. 2.1.3, the reflection is not perfect and it is possible to have some energy propagating beyond the core. The energy propagating in the medium beyond the core is called the evanescent wave energy. When one uses a wave optic approach and Maxwell's equations it can be shown that some energy will always be present in medium 2 or in the cladding region of a fiber even for TIR light energy. Furthermore, the amplitude of the evanescent field decays exponentially into medium 2. A measure of this amplitude decay is the penetration depth, d_p, defined as the distance required for the amplitude of the electric field to fall to $1/e$ of its value at the surface. The penetration depth is typically on the order of a wavelength except when $\theta \approx \theta_c$. Harrick[1] has written a detailed account on the subject of internal reflection and its use in spectroscopy. In his book he calculates d_p and finds

$$d_p = \frac{\lambda}{2\pi(\sin^2\theta - n_{21}^2)^{1/2}}, \quad (8.1)$$

where $n_{21} = n_2/n_1$ and $\theta > \theta_c$. From Eq. (8.1) it can be seen that as θ increases and approaches θ_c, the penetration depth increases, becoming infinite at $\theta = \theta_c$. Harrick used this analysis to perfect the technique of EWS, using pyramidal or trapezoidal plates for guiding the light. Light entering a trapezoid will multiply reflect from the lower and upper surfaces, as shown approximately in the sketch for a planar structure in Fig. 8.1(b). When a lower-index material is placed on the surface of the plate, the evanescent energy propagating in the material interacts with the material, leading to partial absorption or fluorescence. The spectral properties of the medium are then obtained, usually from their absorption properties. IR fibers would replace the trapezoidal plates and act as an EWS if the cladding of the fiber is removed or, in some cases, was never applied. There are several important advantages of a fiber over a trapezoidal plate. One is that the fiber core diameter, $2a$, can be made small, thus increasing the number of bounces and the points of interactions with the medium. The number of bounces, N, is given by

$$N = \left(\frac{L}{2a}\right)\cot\theta, \quad (8.2)$$

where θ is the bounce angle with respect to the normal to the surface as shown in Fig. 2.1, and L is the length of fiber surrounded by the medium. Another advantage of some IR fibers over silica fibers is their higher refractive index. From Eq. (8.1) it can be seen that a high-index fiber core, n_1, leads to a smaller d_p. A smaller penetration into the medium means that there will be less attenuation due to absorption. For example, a low index silica fiber coated with a highly absorbing liquid would absorb too much of the evanescent wave energy, leading to only a very small signal out of the fiber. In contrast, a high-index chalcogenide and Ag-halide fiber would confine more of the field to the core region and result in less absorption and a better fiber sensor. For these reasons much of the chemical sensing work with IR fibers has employed these two fibers. Saito and Kikuchi[2] have provided a good review of IR fiber sensors that includes many references. In particular, they give the IR fiber used and the sample analyzed for a large number of chemical species as well as for applications in thermometry.

Another sensor application for IR fibers is in thermal imaging. The basic idea is to form an image bundle using IR fibers, analogous to what has been done for many years with oxide glass fibers. Bundles of fibers that can transmit an image are called coherent bundles. Coherent fiber arrays are formed when the order of the fiber array is the same at both ends of the fiber. A generic coherent bundle is shown

Figure 8.4 Basic fiber optic image guide showing coherent alignment of fiber at both ends of the bundle.

in Fig. 8.4. Oxide-glass fiber image bundles have been fabricated since the early 1950s (see Sec. 2.0). Today these bundles find numerous applications from industrial borescopes to medical endoscopes. To achieve a high resolution it is important to have a small fiber core, a very thin clad, and a high fiber count. The resolution, R, in line-pairs/mm of a fiber image bundle is given as the range $1/d < R < 1/(2d)$, where d is the center-to-center spacing between fiber cores. For example, if the spacing $d = 20$ μm, $R_{max} = 50$ line pairs/mm and $R_{min} = 25$ line pairs/mm. Oxide fiber bundles can have a fiber count of more than 1 million and be packaged into endoscopic catheters with an overall diameter less than 1 mm. The coherent bundles are made using either a wrapping method, which involves making ribbons and stacking the ribbons in a close-pack configuration or by a leached-bundle technique. In the latter technique, individual fiber canes are bundled and drawn into a multifiber cane, which is then rebundled and redrawn again and so forth, until the final fiber count is reached. The final flexible fiber array is achieved by acid etching the outer acid-leachable glass layer that surrounds the core/clad fiber. The highest-quality fiber bundles are made using the leached-bundle technique.

While it might seem trivial to merely apply these same techniques to the fabrication of IR fiber coherent bundles, such is not the case. First, there has been no leached-bundle technology applied to IR fibers. In principle, it should be possible to find a suitable acid-leachable glass that can be used as an overclad for, say, the fluoride or chalcogenide glasses. This has just not been done to date. The approach that has been used for the glass fibers is a mandrel wrap. Chalcogenide and fluoride glass bundles have been made mostly for the purpose of thermal imaging. The problem using these fibers is that they are much more delicate than the oxide fibers and, therefore, the core sizes tend to be much larger than the 7- to 10-μm cores found in high-resolution oxide glass bundles. For example, chalcogenide glass fiber bundles have been fabricated with fiber counts as high as 8000 and core diameters normally around 75 μm. The 75-μm-diameter core greatly limits the resolution. More recently, Sanghera et al.[3] have used a fiber-stacking process to fabricate fiber bundles containing 6-μm-diameter cores. For most applications involving IR thermal imaging it is necessary to have a core size no larger than 25 μm. Another approach, loosely related to the leached-bundle method but without the leaching step, has been used for the Ag halide fibers. Katzir[4] and his students have made multiple extrusions of core/clad Ag halide fibers to form an image bundle. Their approach relies on the extrusion of a monofiber, which is then bundled with other monofibers fibers and extruded again. This process continues until they reach a high fiber count. The image quality of these fibers suffers a little because the Ag halides are quite malleable and the integrity of the core/clad interfaces tends to degrade after multiple extrusions. In general, coherent IR image bundles are an emerging technology that has promise but simply lags in development compared to the other sensor applications.

8.1 Chemical sensors

There are many attractive advantages to IR fiber chemical sensors: their inherent sensitivity in the fingerprint region of the IR spectrum, relatively high refractive index, and the short lengths necessary for most sensing elements. As mentioned, EWS has been done with these fibers for applications ranging from determination of alcohol content of beverages, to groundwater contamination, to the monitoring of epoxy cures. Most of these applications involve sensing of liquids such as solvents or other organic materials. Gases may be studied using hollow waveguides that are filled with the gas medium. This is a particularly sensitive method to measure the spectral response of a gas, as all of the light energy goes through the gas, as opposed to EWS, where only a fraction of the energy is outcoupled into the surrounding medium. In any fiber chemical sensor it is important to have sufficient sensitivity and selectivity to unequivocally measure the optical properties of the material to be studied. For example, in an application to monitor and detect methane (CH_4) gas, it is crucial that the sensor's sensitivity is below the explosion limit of CH_4. Furthermore, the sensor must be able to distinguish between the absorption in CH_4 and that for other gases whose absorption spectra may lie close to those of CH_4; i.e., the sensor must have high selectivity.

8.1.1 Extrinsic fiber in/out systems

Extrinsic fiber optic chemical sensors are very simple in construction. One IR fiber is used to transmit source energy to the sample, and then another fiber can be used to go from the sample to a remote detector. A sketch of a generic extrinsic sensor arrangement is shown in Fig. 8.1(a). Most often the fibers are configured for transmission measurements, but sometimes they are used in a reflective mode. In the reflective mode the same fiber is used to transmit and receive information.

An excellent example of this type of sensor used to measure small gas concentrations is the aforementioned CH_4 sensor. This fiber sensor was one of the earliest applications of chalcogenide fibers, and it illustrates well the essential features of an extrinsic fiber sensor. In developing their CH_4 sensor, Saito et al.[5] were addressing the need for a nonelectrical remote optical sensor that could detect and also measure explosive gases in a hostile environment. They were motivated by the fact that sensor systems using silica fibers and LED sources operating between 1.3 and 1.7 μm required long-pass gas cells with path lengths of at least 0.5 m in order to detect gas concentrations at the explosion limit. These gas cells were large, required too much space, and were subject to alignment errors. In short, the absorption was too weak in the 1- to 2-μm region to permit smaller gas samples to be used. To overcome these obstacles, Saito et al. constructed a fiber sensor system comprised of transmitting and receiving As_2S_3 fibers, an IR HeNe laser operating at 3.39-μm, and an LN_2-cooled InSb detector. Their system is shown in Fig. 8.5(a). The gas cells had lengths of 20, 50, and 100 mm. There are several key features to their system. First, their choice of an IR fiber is As_2S_3. Referring to Fig. 5.10, it can be seen that As_2S_3 fibers have a rich absorption spectrum, which must be taken into account in designing the sensor. In particular, they chose As_2S_3 because it has a reasonable transmission window near 3.4 μm where there is strong absorption by C−H bonds. The 1- and 1.5-m long As_2S_3 fibers, with core sizes ranging from 500 to 1000 μm, had a transmission between 63% and 67%, including Fresnel end-face losses at 3.39 μm. The 3.39-μm laser source that they chose is close to the peak absorption of the gas. Further, note from Fig. 8.5(a) that they increased the sensitivity of the system by modulating the source energy. Using different chopper configurations, they were able to provide a stable reference and to null out fluctuations in the source energy.

The results of Saito et al. absorption measurements are given in Fig. 8.5(b). The concentration of the CH_4 gas, c, can, in principle, be simply obtained from Beer's

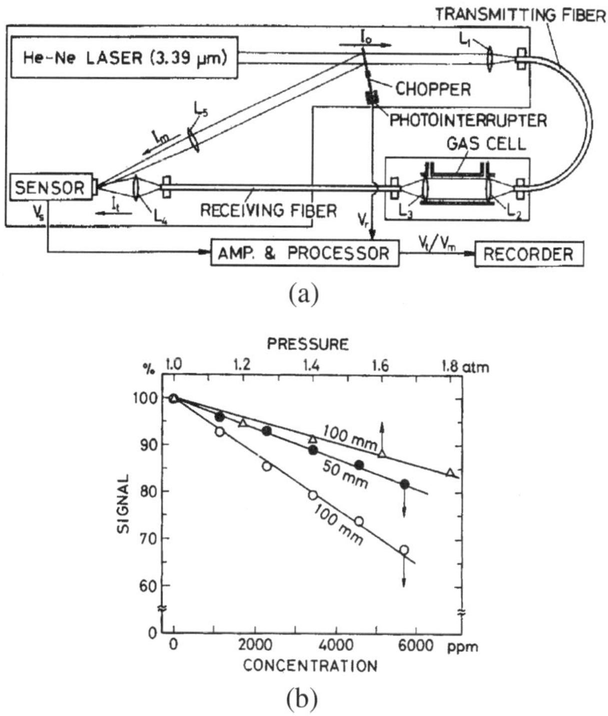

Figure 8.5 Remote sensing of hydrocarbon gases: (a) setup, and (b) dependence of signal intensity on methane gas pressure and concentration.[5] [Reprinted with permission.]

law for gases,

$$\frac{I}{I_o} = e^{-\alpha cpl} \quad \text{or} \quad c = -\frac{1}{\alpha pl} \ln \frac{I}{I_o}, \tag{8.3}$$

where p is the total pressure in the cell and l is the length of the cell. The data in Fig. 8.5(b) gives the results for both concentration and pressure variations as a function of the normalized transmission through the cell. The concentration of CH_4 is varied by mixing it with air. From this data the authors conclude that the the detection limit for methane is ~700 ppm and ~300 ppm for the 50-mm- and 100-mm-length cells, respectively. This translates to ~0.6% of the lower explosion limit for the 100-mm cell. They also found that their sensor was also a good means to measure pressure. Specifically, they calculated that the minimum detectable pressure change was $\sim 6 \times 10^{-4}$ atm for 100% CH_4. In this work the authors made only a breadboard fiber sensor, for longer fibers would be needed for most real-world applications. Nevertheless, their system serves to illustrate the fundamental aspects of an IR fiber extrinsic sensor.

Artjushenko[6] and his colleagues in Russia have developed a more current version of a fiber optic gas sensor. Their system consists of transmitting and receiving IR fiber links to remote multipass gas cell (White cell) fibers coupled to cooled sources and detectors. Sensitivity is gained by using discrete tunable laser diode sources from 1.8 to 3.6 μm and 4 to 16 μm along with HgCdTe detectors, all in one LN_2 dewar. The input and output fiber ends are placed in the dewar next to the source and detectors. They require a broad spectral response from their system; therefore, they use Ag-halide, chalcogenide, and silica fibers to cover the spectral range from 1 to 20 μm. Fibers of varying length were used to couple to remote gas cells that were 30 cm long. Some of the cells had a multipass configuration with an effective path length of 30 m.

The work of Artjushenko et al. clearly indicates that there is a distinct advantage to using a tunable laser source for greater sensitivity. Some recent work of McAleavey et al.[7] using a fiber laser as a source has shown that it is possible to measure methane concentrations as low as 50 ppm per meter of gas cell. These authors have developed a Tm^{3+}-doped fluoride glass fiber laser that has a tuning range of $\Delta\lambda = 130$ nm and a very small bandwidth, which they have used with a conventional multipass gas cell for the detection of CH_4 gas. The unique feature of their fiber laser is that it operates near a peak absorption in CH_4 at 2.3 μm. Furthermore, the 1.3-m-long fiber laser is operated in the grazing incidence configuration so that they are able to obtain a very narrow laser emission of 207 MHz, or significantly narrower than the methane linewidth. In the actual experiment, the laser was tuned to 2373 nm and the sensitivity for measurement of the concentration of CH_4 was found to be 50 ppm. This corresponded to a change of 1 ppm in the 50-m-long gas cell. These authors have used the same fiber laser as a fluorescent source for the measurement of water. For this application the laser operated at 1.47 μm.

An interesting variation on the gas sensors described above is the more recent work of Lu et al.[8] They set out to develop a fiber optic sensor to measure the gas concentrations in a chemical vapor deposition (CVD) chamber. Specifically, they sought to provide a system for the accurate, real-time reaction control of CVD processes used in semiconductor device manufacturing. To do this accurately they needed to cover the IR spectrum from about 2 to 20 μm. Since no single solid-core IR fiber exists to cover this broad range efficiently, they bundled 6-ZBLAN fluoride glass and 2-Ag-halide fibers into a 1-m-long cable. This cable is used to connect the CVD chamber to a remote FTIR spectrometer. Their experimental setup is shown in Fig. 8.6(a). Note the use of off-axis parabolic reflectors M1–M6 in the setup. As a test of the system they chose to measure the reaction between tetraethoxysilane (TEOS) with ozone, which is used to produce thin SiO_2 films. In Fig. 8.6(b) the spectral data of the reactants is shown.

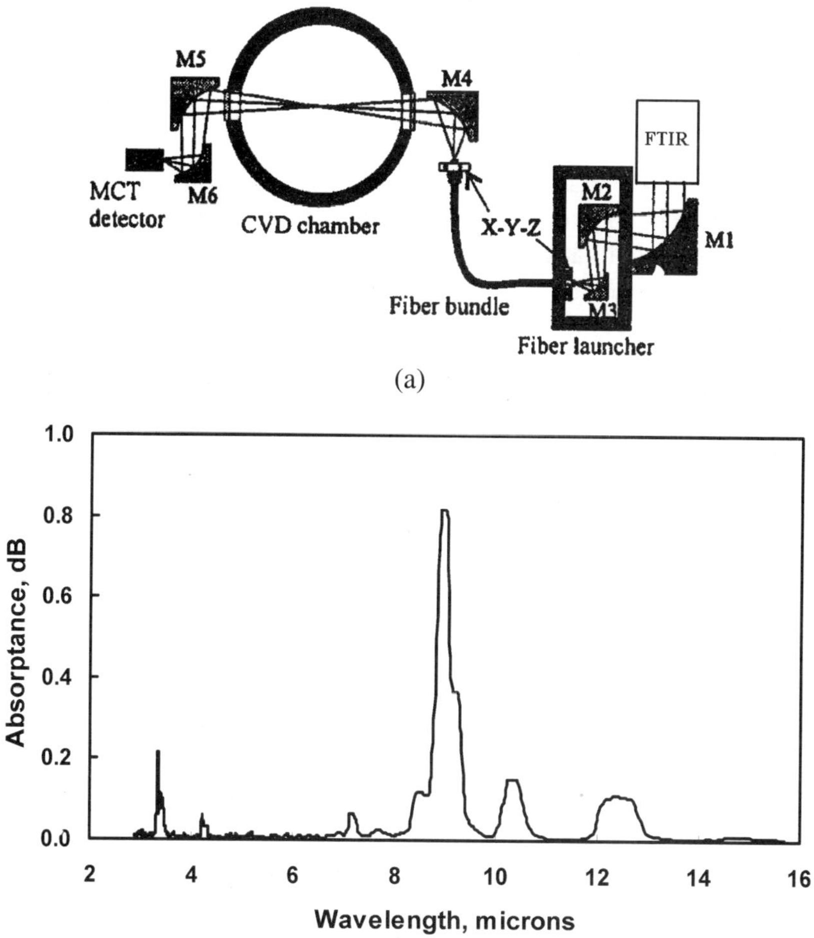

(a)

(b)

Figure 8.6 (a) Experimental arrangement for measuring gas concentration in CVD chamber using an IR fiber bundle composed of both HMFG and Ag-halide fibers, and (b) absorption spectrum for TEOS measured using the IR fiber bundle shown in (a).[8] [Reprinted with permission.]

The simple extrinsic sensors described so far are based on the transmission of light from a source (transmitting) fiber through the sample and then back to the detector through a receiving fiber. It is also possible to do similar measurements based on reflection rather than transmission principles. The basic idea is to use a transmitting and receiving fiber but to position the fibers so that the light is reflected off a sample/mirror combination as shown in Fig. 8.7, rather than being transmitted through the sample to the receiving fiber. Tugendhaft et al.[9] have devised such a fiber-reflection sensor for the analysis of surface films on reflecting substrates. They have used two fiber configurations coupling light to their FTIR spectrometer. One fiber system employed two chalcogenide fibers as shown in Fig. 8.7, and the other used only one fiber with a coupler to decouple the light to a detector. The AsSeTe fibers had a core diameter of 750 μm, and the gap height from the reflectance surface was about 500 μm. Using this setup, the authors were able to easily measure the absorption spectrum of a polymer coating on the surface of an aluminum can and also the spectrum from the diffuse scattering off of a sheet of drafting paper. They found that the two-fiber system was more sensitive than the single-fiber system.

Bunimovich et al.[10] made a similar noncontact, reflectance sensor employing Ag-halide fibers. Their system again involved positioning two fibers at nonnormal incidence with respect to a reflecting surface. The setup is shown in Fig. 8.8. As in the system of Tugendhaft et al., the source and detector is a commercial FTIR spectrometer. Two $AgCl_{0.5}Br_{0.5}$ PC fibers, each 20 cm in length, were used in their setup. The fibers were positioned so that the optic axes were at right angles to each other, as shown in Fig. 8.8. They found that the optimum parameters for this configuration were a fiber spacing of $d = 2$ mm and a height of $h = 1$ mm above the reflecting surface. To demonstrate the sensitivity of their system they measured the absorption spectrum of a 40-μm-thick sample of ethanol placed on the reflecting surface. The methods proposed by both Tugendhaft et al. and Bunimovich et al. illustrate the basic principles of how a simple chemical sensor can be configured

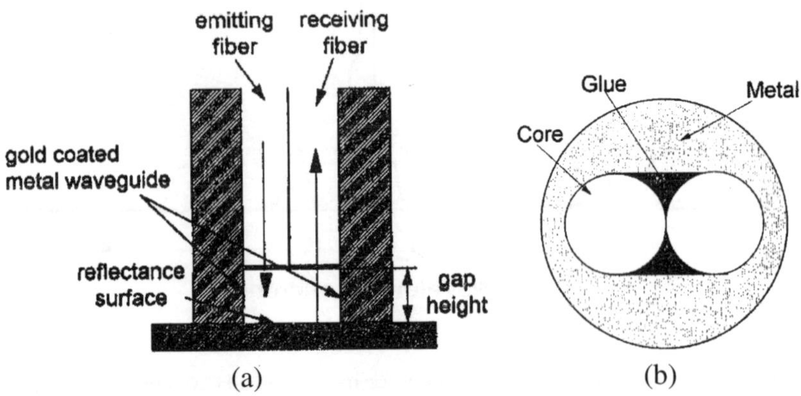

Figure 8.7 Simple fiber optic reflectance sensor made using 750/1000 μm, AsSeTe glass fibers.[9] [Reprinted with permission.]

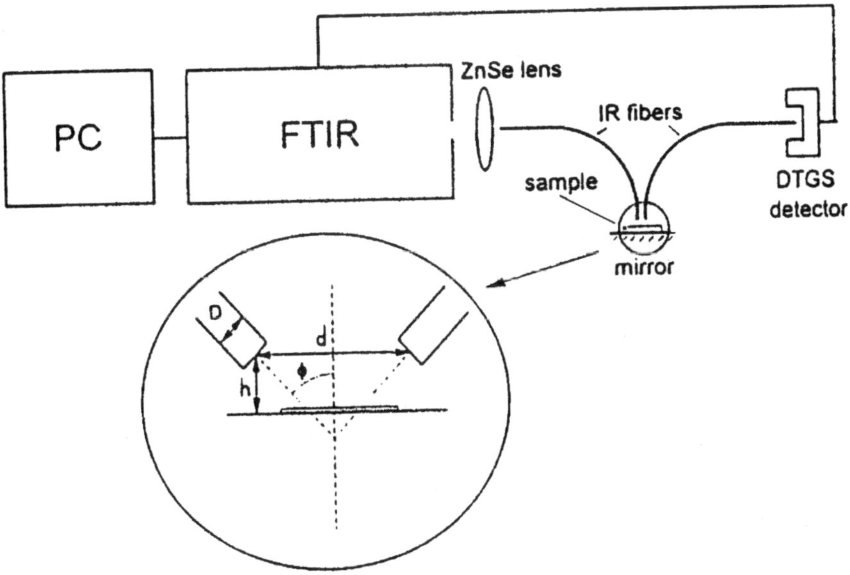

Figure 8.8 Experimental arrangement for a noncontact IR reflectance spectroscopy using Ag-halide fibers.[10] [Reprinted with permission.]

using IR fibers to measure the absorption spectrum in the sensitive fingerprint region of the spectrum.

8.1.2 Hollow waveguides for gas sensing

In the applications discussed in Sec. 8.1.1, the fibers used to relay information from source to detector were all solid-core IR fibers. It should be mentioned that it is possible to substitute hollow waveguides in place of the solid-core fibers for the extrinsic fiber sensor applications given above. This has not generally been done, because many of the fiber chemical sensors involve EWS sensing (covered in Sec. 8.1.3), and for this technique the surface of the core needs to be coated with the analyte. Hollow waveguides cannot be easily configured to act as EWS fiber sensors and, therefore, they are not so commonly employed for liquid and solid chemical detection systems. There is, however, an excellent opportunity for hollow waveguides in gas sensing. The advantage of a hollow guide in gas sensing is that the core can be filled with the gas analyte so that all of the light passes through the medium. This is in contrast to EWS, in which only part of the light is outcoupled into the medium. In a sense, the hollow guide acts like a multipass White cell. Used this way, hollow waveguides afford the benefit over White cells of a smaller gas volume, a faster response time, and reduced cost.

There are two ways to configure a hollow guide as a gas sensor. The simplest means and that used most often is to have gas enter one end of the guide and exit the other end as shown in Fig. 8.9(a). Note in Fig. 8.9(a) that the guide may be coiled for a longer path length and greater sensitivity while keeping the overall

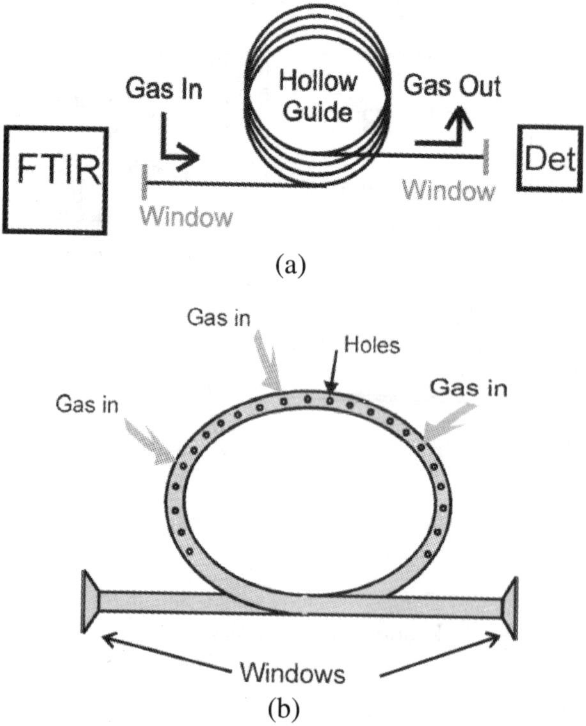

Figure 8.9 Gas sensing using hollow waveguides filled with a nonreactive gas. (a) Gas flow is via inlet and outlet fittings on either end of the sealed waveguide, and (b) the "holey" hollow waveguide structure in which gas flows into and out of holes drilled, with a CO_2 laser, through the walls of the guide.[11]

profile small. The other method is to make holes in the wall of the waveguide so that gas can diffuse into the guide at will. This "holey" hollow guide is shown in Fig. 8.9(b). The advantage of the holey design is that the guide can be used in a real-time *in situ* manner to sense and/or monitor gas concentration or related properties. Harrington and his coworkers made a holey guide from a hollow glass waveguide with Ag/AgI coatings. Numerous holes were drilled into the glass annulus of the coated fiber using a CO_2 laser. Most of the holes were in a line, but sometimes they were placed in a helical pattern. What these researchers found is that the loss of the guides was not affected by the holes unless too many holes were drilled into the glass. The reason for this is that the electric field is largely confined to the core and little light is lost through the holes. This is a very convenient configuration for monitoring, for example, benign gases such as CO_2 in a closed room.

Most of the limited data on hollow-guide gas sensing to date has involved guides that are straight or with little bending. The real advantage for these fibers, however, lies in using long lengths coiled to preserve space. Coiling the fiber introduces a bending loss, which can be rather large if the bending radius, R, is small, and the number of coils is large. Bending losses for hollow guides were discussed in Sec. 7.3.4.3, and Eq. (7.10) summarizes the bending loss for coils bent to a constant bending angle, θ:

$$\alpha_{total} = \alpha_{\infty} L + \frac{m}{R} \cdot R\theta = \alpha_{\infty} L + m\theta. \tag{7.10}$$

From Eq. (7.10) it may be seen that the loss depends on θ and the slope of the bending loss curve, m. Written this way α_{total} is independent of the radius of curvature. Generally, m varies between 0.13 and 0.60 dB. Equation (7.10) may be written in terms of the total loss and the number of turns, N, of the coil as

$$Loss_{tot} = \alpha_{\infty}(a)L + 2\pi m N. \tag{8.4}$$

Clearly, the design of any coiled system needs to take into account the last (bending) term in Eq. (8.4), as this may contribute significantly to the overall system loss.[11]

One of the earliest examples using a hollow guide as a gas-sensing absorption cell is the work of Saito et al.[12] Their setup is similar to that shown in Fig. 8.9(a), except that the source was a tunable CO_2 laser. They chose to measure the gas concentration of NH_3 because this gas is highly absorbed at 10.719 μm and non-absorbing at 10.697 μm. The ratio of these two CO_2 laser wavelengths was then used to determine the minimum detectable limits of ammonia gas. Their waveguide was a 1500-μm-bore, straight 1-m-long Ge-coated Ni tube made by Miyagi's group and discussed in Sec. 7.3.2. Using this setup they were able to measure NH_3 concentrations as low as 5 ppm.

A potentially important application of a hollow guide gas cell is in the area of environmental sensing. Micheels et al.[13] developed a portable gas cell/FTIR spectrometer for rapid field screening of volatile organic compounds (VOCs). Their instrument, shown in Fig. 8.10, employs headspace sampling in small sample vials to measure the VOC content of soil or water samples. The instrument employs either a straight or coiled hollow-waveguide gas cell to achieve an absorption pathlength of 1 to 2 m with a very low cell volume (in the range of 2 to 5 cc). This is in contrast to White cells with pathlengths in the 1- to 20-m range and gas volumes of 1 to 6 l. The small gas volumes used by Micheels et al. permit high detection sensitivity along with a rapid response time. Two types of hollow guides were used. One is a flexible polymer tube with Ag/AgI coatings, and the other is Pyrex or silica glass tubing. When the bore size of the silica tubing was equal to or less than 1300 μm, the tubing was sufficiently flexible to be bent into a semicircular shape, as shown in Fig. 8.10. For the 2000-μm-bore tubing, it was necessary to form a permanent 0.2-m-radius bend in the tubing first, and then coat the inside with Ag/AgI films. However, the glass blowing required to permanently bend the tubing led to some irregularities in the bore and bending radius that in turn degraded the optical properties of these permanent coils. Each guide had the ends sealed with ZnSe windows.

The system of Micheels et al. shown in Fig. 8.10 was calibrated using a set of 11 certified VOC gas mixtures with concentrations in the range of 20 to

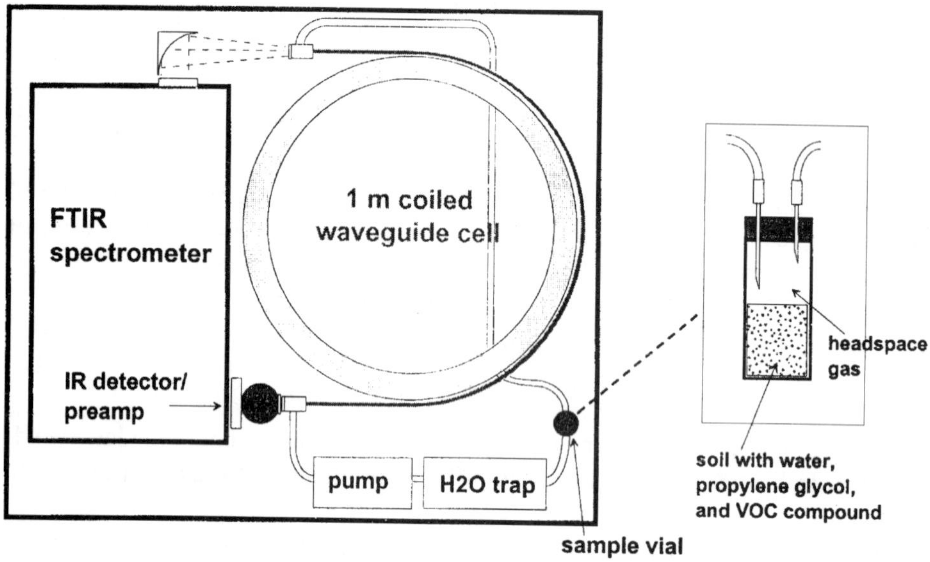

Figure 8.10 HGW used in a portable VOC gas monitor. The hollow waveguides were 1-m-long bent 180° with bore sizes ranging from 1000 to 1300 μm. In some cases, hollow plastic waveguides with 2000-μm bores were used for greater throughput.[13]

200 ppm. After calibration, 5 VOCs—benzene, toluene, M-xylene, ethylbenzene, and trichloroethylene—were separated from three different soil samples and the gas phase spectra were measured from 4000 to 700 cm^{-1}. From 17% to 70% of the VOC compounds were volatized with a gas phase detection sensitivity of 3 to 13 ppm. They concluded that a 2- to 3-m-long coiled waveguide cell coupled to an FTIR spectrometer should be capable of VOC detection sensitivities below 1 ppm.

A similar experimental arrangement was used by Kozodoy et al.[14] to measure the concentration of CO_2 gas mixed with N_2. They used Ag/AgI HGWs with bore sizes ranging from 530 to 1000 μm and lengths of only 20 cm in the final setup. Nevertheless, they were able to monitor the strong 4.26-μm (2350 cm^{-1}) CO_2 band with sensitivities as low as 200 ppm.

One method to increase the sensitivity of the hollow-guide gas cells, other than increasing the pathlength, is to use a more intense source. Hvozdara et al.[15,16] used a quantum cascade laser (QCL) tuned around 10 μm to investigate the absorption of atmospheric ethane. Their 1130-μm-bore HGW was 43.5 cm in length, and the signal from the QCL was transmitted through the HGW to an FTIR spectrometer for analysis. They measured the absorption of ethane mixed with helium using various laser lines, but the best data was taken at the strongest absorption line at 999.09 cm^{-1} or 10.009 μm. At this laser wavelength, the minimum detectable level of absorption was estimated to be 250 ppm. They concluded, however, that if they had been able to tune their laser to a stronger absorption line at 949.2 cm^{-1}, then they would have had a minimum detectable limit of about 20 ppm. Worrell and Gallen[17] also used a CO_2 laser and HGWs to measure trace amounts of volatile organics at the sub-ppm levels. Using a similar HGW gas cell and the CO_2 laser, they found that the minimum detectable limits for trichloroethylene and 2-furaldehyde were both less than 10 ppm.

A more sophisticated HGW gas cell has been recently described by Fetzer et al.[18,19] They used a distributed feedback diode laser along with up to a 9-m length of coiled, 1000-μm-bore Ag/AgI HGW for their sensor system. Their apparatus was similar to that shown in Fig. 8.10. The difference is that they filled their hollow guides with CO_2 and NH_3 through holes drilled in the sides of the guide similar to that shown in Fig. 8.9(b). The absorbance for these two gases was measured near 1.5 μm and they found that the minimum detectable absorbance was 3.5×10^{-5}. In a very recent study they used a quantum cascade laser operating near 5.4 μm to measure nitric oxide (NO) from exhaled breath. The small gas volume and sensitivity of their system using a 9-m-long HGW enabled them to achieve a minimum detectable limit of 58.8 ppb of NO in N_2 with a response time of 0.48 s.[19] A similar arrangement has also been used by Blake et al. to measure NO gas using a coiled HGW and a diode laser.[20]

8.1.3 Intrinsic evanescent wave sensors

Evanescent wave sensor (EWS) fiber sensors have a long history both for IR and conventional silica-based fibers. The basic principle of an EWS sensor was discussed in Sec. 8.0. The simple idea is to outcouple some portion of the light energy from the core into a medium surrounding the fiber as shown in Fig. 8.1(b). The evanescent wave energy coupled into the medium can be absorbed or cause the material to fluoresce. In this way, the light transmitted through the fiber is altered. In some sense, the fiber can be thought of as "flypaper" in that the material to be analyzed sticks to the core's surface, where it can be interrogated by light traveling down the core. This means that for most applications of interest, the fiber must have the cladding removed so that the core can be exposed to the analyte. Furthermore, it is clear that this technique will work most easily with liquid analytes, as they provide the greatest surface coverage and, therefore, the highest sensitivity. In practice, solid materials may be studied by adding solid particulate matter to a liquid binder. In some cases, solids alone and even gas samples have been measured using EWS methods.

Harrick[1] gave a very straightforward method for determining the amount of evanescent wave energy that couples out of a high-index core material into a surrounding material of lower index. The evanescent wave energy decays exponentially into the medium with a penetration depth given by Eq. (8.1). As mentioned in Sec. 8.0, the closer the light travels to the critical angle, the greater is the penetration depth. Therefore, it is often desirable to increase the launch angle for greater sensitivity. A simple example will serve to illustrate how much light can be absorbed in a relatively short distance. Assume a single ray is totally internally reflected in a slab or ATR plate[1] with a thickness of 500 μm incident at an angle of 30° with respect to the optic axis. Using Eq. (8.2) with $\theta = (90° - 30°) = 60°$, the number of bounces per length, N/L, is 1.15/mm. Suppose that the ATR plate has a length of 100 mm (typical for many EWS IR fiber sensors) and that there is an absorption of 1% per bounce. The light will then bounce 115 times over the 100 mm pathlength, and the approximate total amount of light transmitted after traversing 100 mm is

$$T = (1 - A)^N = (1 - 0.01)^{115} = 0.315; \tag{8.5}$$

that is, only about 30% of the light remains after interacting with the absorbing medium.

This is a very simplistic approach to EWS fiber sensors, yet it illustrates how powerful this technique can be. EWS sensors do not have to be very long because light interacts strongly with the surface material, especially when small-core fibers are used; i.e., the number of bounces is large. A more realistic analysis based on fiber geometry and not on trapezoidal geometry (ATR plates) considered by Harrick is needed for EWS fiber sensors. There have been a number of good reviews of EWS fiber sensors, but one of the best for IR fibers is that given recently by Messica et al.[21] Their comprehensive analysis is based on a 3D geometrical optics model, which includes the effects of different launching conditions, multiple Fresnel reflections from the fiber ends, and both bulk and evanescent wave absorption. Messica et al. obtain an expression for the power transmitted by a fiber in terms of a complicated integral that involves the product of a general transmission function and an intensity distribution function. They have solved the integral for the specific case of the absorption of SF_6 gas using an Ag-halide fiber with a core index of 2.15. Figures 8.11(a) and (b) give the results of their calculations for a gas sensor that was comprised of a gas cell containing SF_6, a CO_2 laser source, and an FTIR spectrometer. This data is not shown for any details on SF_6 absorption *per se*; rather, it serves to illustrate some of the key properties of EWS fiber sensors in general. From Fig. 8.11(a) it can be seen that the absorbance increases as the length of the gas cell increases. The dependence of absorbance on length is quite linear, as would be expected from an exponential decay of transmission with length. In Fig. 8.11(b), the absorbance is seen to increase as the fiber radius decreases, as expected since smaller-core fibers have more bounces per length than large-core fibers. The authors also show that the sensitivity of the sensor can be increased, i.e., have greater absorption, by tilting the fiber with respect to the input beam or decreasing the f/# of the input coupling lens. A smaller f/# gives a large input cone, which in turn leads to more coupling into larger ray angles or higher-order modes.

While the use of solid-core IR fibers for gas sensing via EWS is somewhat unusual, Messica et al.[21,22] were able to get good sensitivity using their 29-cm-long AgClBr PC fiber and a gas cell to measure the partial pressure of the atmospheric contaminant SF_6. Specifically, their PC fiber had a 900-μm diameter, and a lead salt diode laser tuned near the strong absorption peak at 10.562 μm was used as the source. To increase the absorption of light by the gas, they tilted the input laser beam so that it made an angle of 20° with respect to the fiber axis. Their results indicate a minimum detection limit of 1% SF_6. Another way of increasing the

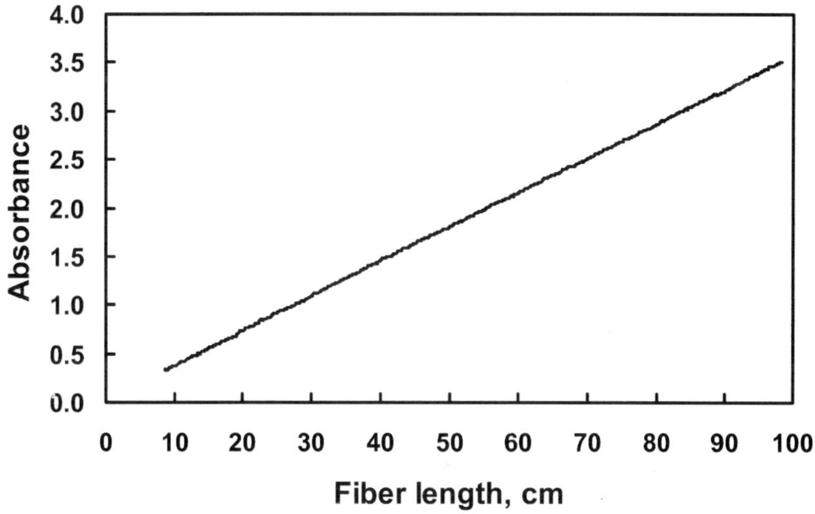

(a)

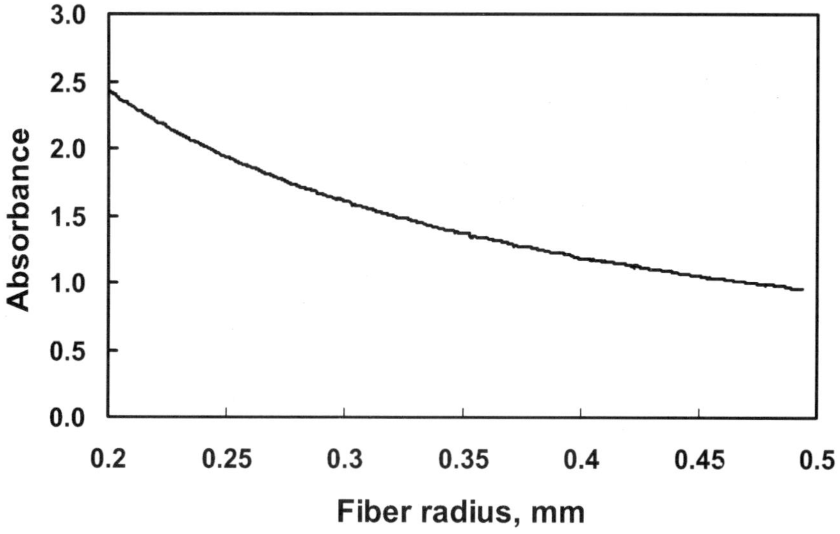

(b)

Figure 8.11 EWS fiber sensors employing Ag-halide fibers: (a) increasing absorbance as the length of the sensing element increases, and (b) decrease in absorbance as the fiber radius increases due to fewer bounces, i.e., less interaction with the absorbing species surrounding the fiber.[21] [Reprinted with permission.]

sensitivity of their fiber sensor is to coat the fiber with a polymer that absorbs the analyte. This has been done with liquid analytes as noted below.

The most common application of IR fiber EWS sensors is the monitoring of absorbing liquid samples that surround the core-only portion of the fiber.[23] A simple fiber absorption cell was developed by Simphony et al.[24] for the detection of various liquid samples. Their cell, shown in Fig. 8.12(a), is a 900-μm Ag-halide fiber placed on a microscope slide. The input angle of the FTIR spectrometer light could be changed, as well as the length of the absorbing medium. The entire absorption cell was placed in the sample chamber of an FTIR spectrometer. For calibration, Simphony et al. used water. Their spectral results are shown in Fig. 8.12(b) for fiber immersed in water with lengths of water from 1 to 65 mm. Obviously, the greater the length of water, the greater the absorbance. The authors also increased the absorbance by increasing the tilt angle from 10° to 34°.

An effective method to increase the sensitivity of liquid detection is to coat the fiber with a polymer, which absorbs selectively or nonselectively the liquid analyte. Walsh et al.[25] and Krska et al.[26] coated Ag-halide fibers with various absorbing polymers for the detection of trace amounts of chlorinated hydrocarbons (CH) found in water. Walsh and his coworkers used polyisobutylene (PIB), low-density polyethylene, silicone rubber, as well as other polymers to coat and fabricate a 9-cm-long fiber cell. This unclad PC fiber was coupled to two, 1.5-m-

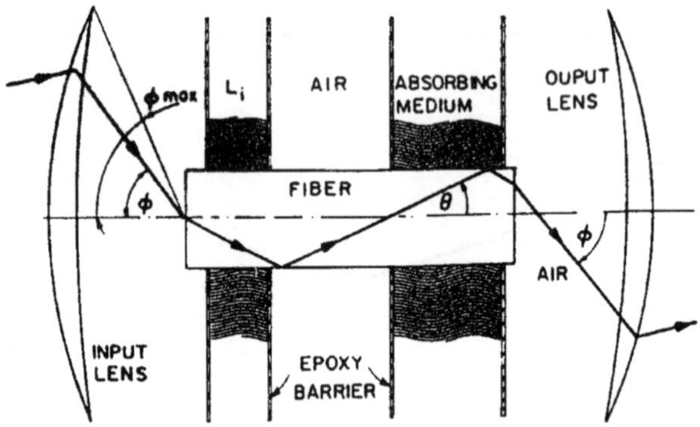

(a)

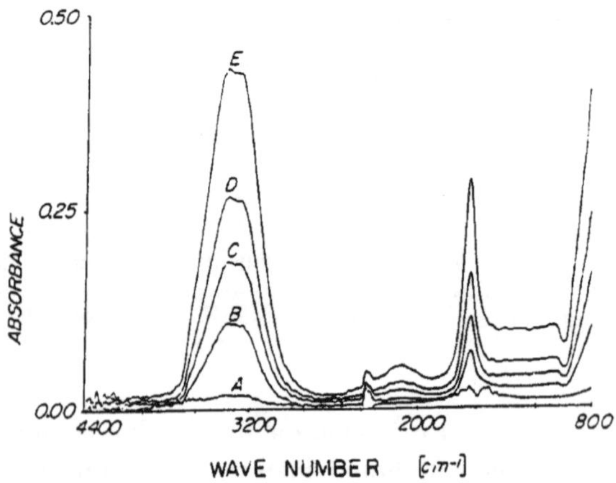

(b)

Figure 8.12 (a) Simple liquid cell surrounding core-only Ag-halide fiber for EWS, and (b) absorption of water for fiber immersion lengths of A) 1 mm, B) 10.5 mm, C) 20.5 mm, D) 33.5 mm, and E) 65.5 mm.[24] [Reprinted with permission.]

long, core/clad PC fibers—one end going to an FTIR spectrometer and the other to a HgCdTe detector. The coated fiber cell was surrounded by another liquid cell in which a mix of trichloroethylene (TCE) and water flowed. The PIB coating proved to be a good choice of polymer for enriching the concentration of TCE. For the study of pesticides, they found that better enrichment was obtained using polyvinylchloride (PVC). The minimum detectable limit of TCE was found to be 3 ppm.

Walsh et al. found that by beveling the input end of the fiber they were able to greatly increase the sensitivity. A beveled end, as shown in Fig. 8.13(a), increases the ray angle in the fiber and, thus, the energy coupled into the liquid. As an example of this effect, Fig. 8.13(b) shows their absorption spectra for acetone as a function of the bevel angle as defined in Fig. 8.13(a). The absorption increases

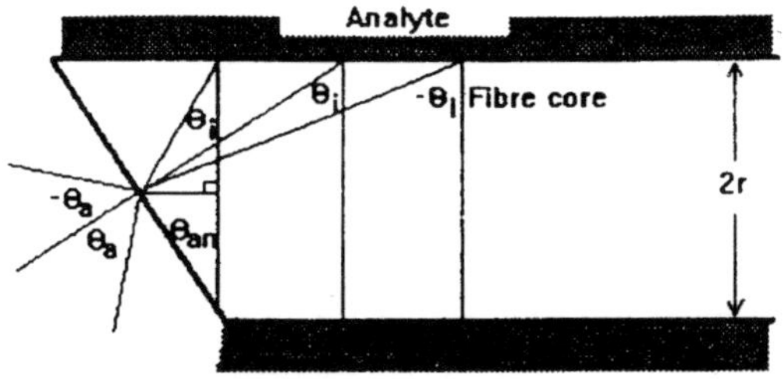

(a)

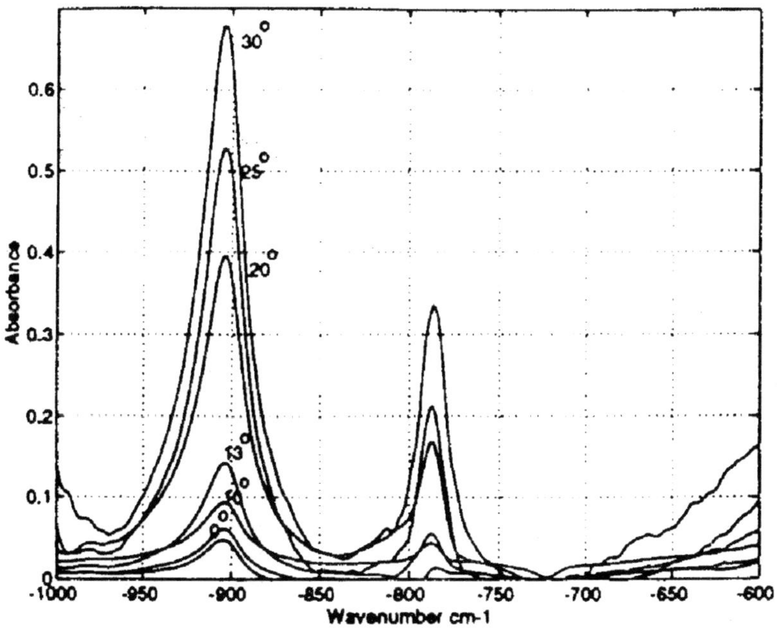

(b)

Figure 8.13 Effect of beveling the front face of a fiber used for EWS. (a) Geometry of bevel angle on front face, and (b) 1000-μm-core Ag-halide fiber used to detect acetone. Note increased absorption with increasing bevel angle.[25]

by about a factor of 10 for a 30° bevel compared to a flat-end fiber. No polymer coating is used in this case, although a polymer coating would presumably also increase sensitivity. They also increased sensitivity by coiling a 45-cm-long fiber into 3-cm-diameter loops. As in the case of an end bevel, light in a bent fiber will travel at a greater ray angle.

Krska et al. also coated an Ag-halide fiber with low-density polyethylene as means to make a reversible enrichment cell for the measurement of chlorinated hydrocarbons (CHs). They found that polyethylene worked well to exclude water and, therefore, enrich the CHs and yet have minimal absorption in the region of CH absorption. The cell was 10-cm long and an FTIR spectrometer was used to scan the peak absorption of chlorobenzene at 13.51 μm. They found that they had to expose their CH/water sample to the fiber cell for about 3 minutes to allow for sufficient diffusion into the 7-μm-thick polyethylene film. The minimum detectable limit of absorption was 10 mg/l.

Chalcogenide fibers are a very common fiber for use in EWS. There are several reasons for this. One is that these materials are quite insoluble in most liquids, especially solvents and VOCs. Another is that they have high refractive indices. As indicated by Eq. (8.1), a high refractive index means that there is less penetration into the surrounding liquid. A smaller penetration depth is often important because one does not want too much of the energy absorbed by the analyte; otherwise there is too little signal at the detector. An early illustration of the use of chalcogenide glass fibers was the work of Heo et al.[27] Heo and his coworkers at Rutgers University used 380-μm core-only GeSeTe fibers in a liquid cell to detect separate mixtures of acetone, ethyl alcohol, and sulfuric acid in water as well as CCl_2F_2 gas. The cell length was 15 cm and spectra were recorded from 5 to 11 μm. Sensitivity was limited to approximately 5 volume % of analyte in water. A particularly interesting feature of their study was the ability to immerse their fiber cell in concentrated sulfuric acid without damage to the fiber.

Chalcogenide fiber EWSs have been used to detect numerous other solvents and organic fluids. In particular, Sanghera et al.[3,28,29] have made a reflectance probe using chalcogenide fibers for the detection of ground water contamination. Their probe was placed in a cone penetrometer so that it could be forced into the ground at depths up to 20 m. A schematic of their cone penetration system is shown in Fig. 8.14.[3] Detection limits were 130 ppm of marine diesel fuel in sea sand. Another example is the use of fiber optic dipstick probe for monitoring the degradation of motor oil. The authors suggest that the active monitoring of engine oil will allow for oil changes only when the oil needs changing, as opposed to changes at regular intervals. Sanghera et al.[3] also describe coating fibers with a polymer to enhance the detection of certain organic compounds. They etched the cladding layer of a chalcogenide fiber and then coated the core with polydimethylsiloxane and low-density polyethylene. These polymers prevent the absorption of water while enriching the concentration of benzene and CHs in water mixtures, respectively.

Lucas' group at the University of Rennes has been very active in developing long-wavelength chalcogenide and chalcohalide glass fibers for EWS. In particular, they have devised some variations on the usual liquid cells to increase the

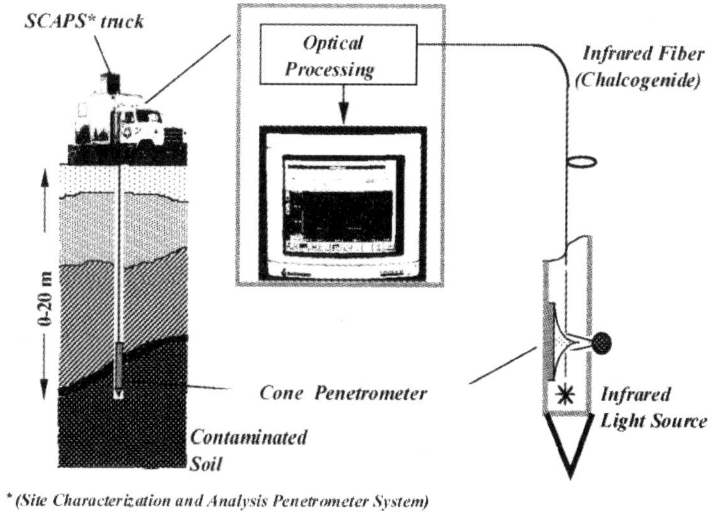

Figure 8.14 The cone penetrometer system using chalcogenide fiber for detection of contaminants and water in soil. [Reprinted from Ref. 3, Copyright 2001, with permission from Elsevier.]

sensitivity of their sensors. The method that Hocde et al.[30,31] have used is to draw a biconic taper in a short section of unclad fiber. Katz et al.[32] have also investigated the benefits of a fiber taper. In Fig. 8.15(a) the tapered fiber section is shown along with the complete fiber sensor in Fig. 8.15(b).[30] The advantage of a taper over a straight section is that the bounce angle of the light increases as the light travels from the large to small portion of the taper. This leads to a greater number of bounces and a greater penetration depth and, therefore, greater absorption of the light, especially in the small center section of the taper. The taper is placed at the bottom of a U-shaped fiber like that shown in Fig. 8.15(b). Hocde et al. formed their taper in TeAsGe fibers either during the fiber-drawing process or by a post-fabrication etching process. The tapers were between 1 and 6 cm in length. As a test of their tapered fiber system, Hocde et al. measured the spectra of ethanol for fiber tapers made from TeAsGe fibers with core diameters ranging from 300 down to 76 μm. Figure 8.16 gives their results for ethanol absorption measured using tapers with these different core diameters in the 9- to 12-μm band.[31] Clearly the sensitivity increases as the core size decreases. A similar TeAsSe fiber device was recently used by Michel et al. to detect pollutants in groundwater.[33]

Lucas' group has also carried out similar studies using chalcohalide fibers.[34] Their TeX fibers are composed of TeSeAsI, and both straight and tapered core fibers have been used. One of the novel applications that they pursued using TeX fibers was the study of fermentation processes. Zhang et al.[35] monitored the decrease in glucose absorption at 9.27 and 9.68 μm and the concomitant increase in ethanol concentration at 9.56 μm during the alcoholic fermentation of grape juice into wine. The same authors also studied the process of lactic fermentation. Their results are shown in Fig. 8.17 for the conversion of lactose into glucose and finally

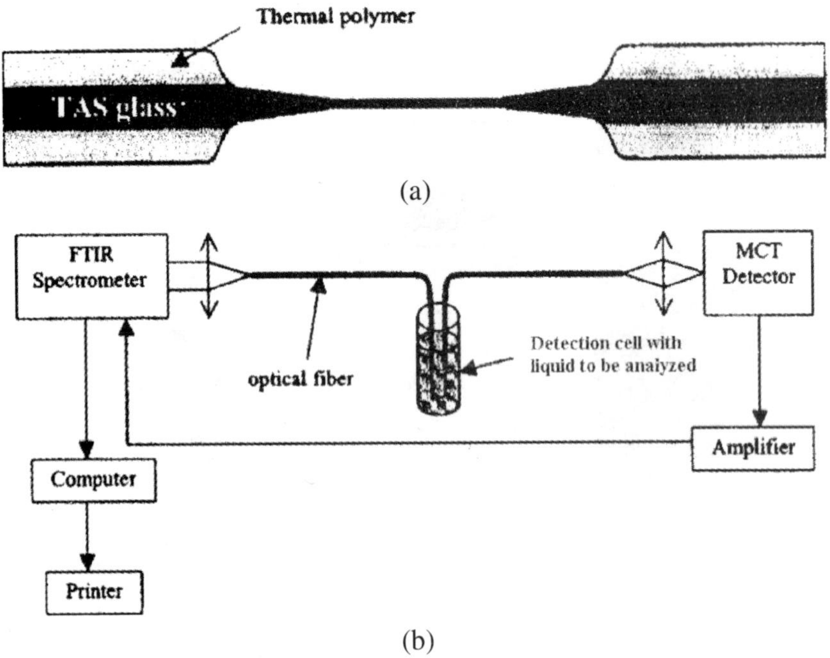

Figure 8.15 (a) Tapered chalcogenide glass fiber for EWS used in (b) in a liquid cell. The fiber is bent into a U shape to further increase coupling of the evanescent wave energy into the liquid. [Reprinted from Ref. 30, Copyright 2001, with permission from Elsevier.]

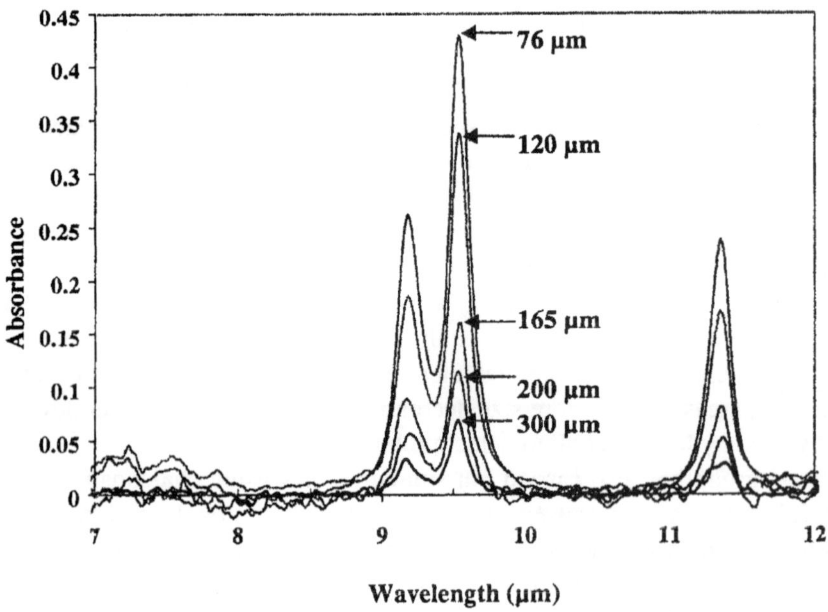

Figure 8.16 Absorption of ethanol for different fiber tapers. The smallest diameter taper (76 μm) gives the strongest absorbance as would be expected based on the increasingly acute angle of incidence of rays propagating through such a small taper. [Reprinted from Ref. 31, Copyright 2000, with permission from Elsevier.]

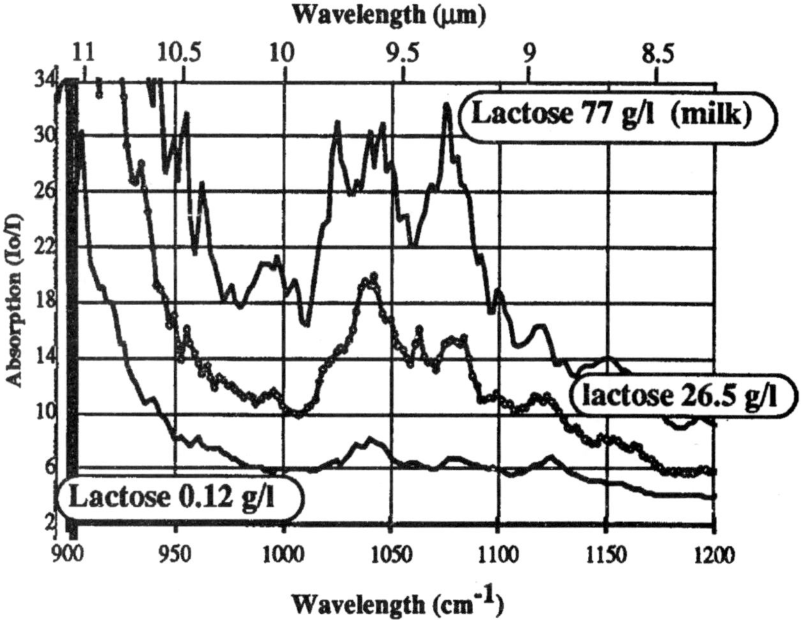

Figure 8.17 TeX glass fibers used as EWS to monitor the process of lactic fermentation. By analyzing the intensity of the absorption peaks at 1041 and 1074 cm^{-1} it is possible to monitor the real-time decrease in lactose as a function of time.[35]

into lactic acid. For example, this process is representative of the fermentation of milk into yogurt in which the end product, yogurt, is lactose free.

Chalcogenide fibers for EWS have been used to form a probe to be used much like an optrode to measure the IR spectra of many materials. Generally, the fiber is U shaped. The greatest sensitivity is the bottom of the U, which may be tapered or straight. Making a U-shaped probe is not easy, as the fiber must be bent at a very sharp angle at the bottom so that the probe has a small profile. Only the bottom or U-shaped portion of the fiber is exposed. In this way the probe can be placed in contact with a variety of samples types. Spectra-Tech (Thermo Nicolet) makes U-shaped, ATR needle probes as well as reflectance probes, based on principles illustrated in Fig. 8.7. In addition to liquid sensing these probes are also useful for the study of biological tissue samples.[36]

There has been much more limited use of fluoride glass fibers for EWS. Saggese et al.[37] did employ ZBLAN fibers in a study of various organic solvents mixed with water or other solvents. They were able to obtain modest sensitivities using their fluoride fiber cells but nothing better than that obtained with chalcogenide fibers. There are several reasons why fluoride fibers are not so well suited for EWS. The primary reason is that they are more soluble in water and other solvents compared to chalcogenide or even Ag halide fibers. One would clearly not want the surface of the fiber to be attacked by the analyte, as would certainly be the case for most HMFG materials. Additionally, fluoride fibers do not transmit far enough into the more sensitive fingerprint region of the IR spectrum. Most of the

IR fiber EWS examples given so far are for sample absorptions greater than about 4 μm, and this region is not transmitted by fluoride fibers.

8.2 Temperature measurements

Noncontact techniques for the measurement of temperature have been used for many years. Pyrometers and radiometers are standard instruments for line-of-sight measurements of, for example, high-temperature furnaces to low-temperature biological tissue. Sensitive IR cameras employing high-resolution uncooled microbolometers are today used for thermal imaging in applications ranging from defense systems to medicine. Fiber delivery systems have been coupled to a variety of radiometric systems as a convenient means of measuring non–line-of-sight temperatures. In a sense, the fiber bundles merely serve, for the most part, to transmit blackbody radiation from the source to the detector. As mentioned in Sec. 8.0, IR fibers are an excellent candidate for this role as they transmit blackbody radiation over a wide spectral band. In particular, IR fibers transmit the long-wavelength radiation of low-temperature objects such as the human body temperature. This may be seen in the blackbody curves in Fig. 8.3. Note that the peak radiation from a perfect blackbody at room temperature (300 K) is at 10 μm and, therefore, well suited for transmission by the Ag-halide PC fibers or hollow waveguides.

IR fiber radiometers have been proposed and studied for many years. The basic concept of such a radiometer is really quite simple. The block diagram in Fig. 8.18, taken from the early work by Shimizu and Kachi[38] in 1983, illustrates the key elements of a simple IR fiber thermometer. The IR fiber is near but not in contact with the hot object. Light is transmitted to a detector, usually a lock-in amplifier, for signal processing. While the system in Fig. 8.18 appears very straightforward, measurement of the actual, true temperature is not so easy. This is due to the difficulty in obtaining an accurate reference temperature. That is, the object is unlikely to be an ideal blackbody with unit emissivity. Instead, the object will be a gray body with an emissivity less than 1.

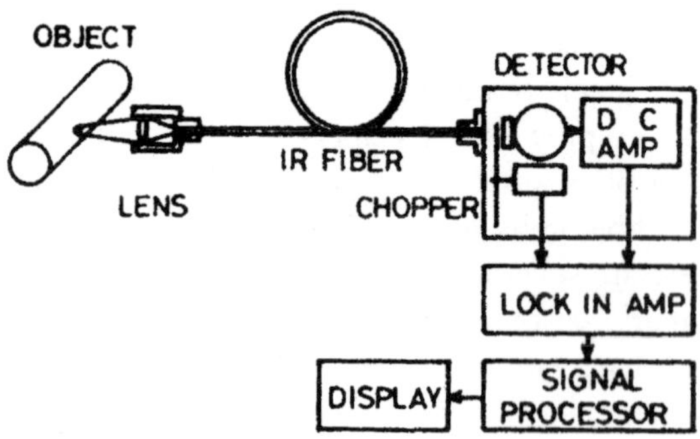

Figure 8.18 Simple low-temperature radiometer used to measure blackbody radiation.

An object at temperature T radiates according to Planck's law given by

$$Q(\lambda, T) = \varepsilon \frac{c_1}{\lambda^4} \frac{1}{\exp(c_2/\lambda T) - 1}, \tag{8.6}$$

where $Q(\lambda, T)$ is the spectral radiant photon emittance, i.e., the number of photons emitted from the body per unit wavelength, per unit area, per unit time into a hemisphere (units of photons s^{-1} cm^{-2} μm^{-1}); c_1 and c_2 are constants; and ε is the emissivity. The emissivity of a body is dependent on the wavelength, temperature, surface quality, and shape. A further interpretation of Eq. (8.6) is given by Eyal et al.,[39] who point out that to detect the smallest temperature changes it is important to look at the radiant energy within a small wavelength region. This region is found by solving $d/d\lambda(dQ/dT) = 0$. For example, at 300 K this corresponds to approximately 10 μm, which is the same as noted above from the graphical data in Fig. 8.3.

There are many variables and constraints in the design of a sensitive fiber optic thermometer. In most applications it is important not only to measure the absolute temperature but also to have a small minimum resolvable temperature difference (MRΔT). For example, in tissue studies the desirable MRΔT is usually about 0.1°C. To achieve this temperature sensitivity requires careful choices of detectors, fibers, and methods for data processing. Katzir and his group at Tel Aviv University have done extensive analysis on the best approaches to measure temperatures near room temperature.[39–41] Their analysis is rather complex, and only a summary of the critical factors will be given here. The fiber that they chose is unclad Ag-halide fiber. This fiber has a large effective NA and is ideal for collecting nonuniform (Lambertian) emission from the target. Eyal et al.[39] calculate the total power emitted from the hot body and collected by the fiber. The overall setup and the geometry used are shown in Fig. 8.19. Note that they use no collecting optics between the target and fiber. The radiation out of the fiber is efficiently coupled to the detector using two spherical mirrors arranged somewhat like an integrating sphere. A special chopper was designed to not only provide the sample and reference signals, but also to stabilize the radiometer against changes in temperature. The output from a LN_2-cooled HgCdTe detector was fed into a lock-in amplifier for signal processing. As may be seen in Fig. 8.19, Eyal et al. employed a CO_2 laser as a pulsed heat source. They irradiated their target, usually biologic tissue, with a train of CO_2 laser pulses synchronized with the chopper so that only thermal radiation reached the detector. Using this system, they were able to heat tissue (a sausage slice) to about 85°C with a MRΔT of about 0.3°C.

One common problem associated with the radiometers that rely on the measurement of a radiometric signal over only one wavelength range, like that described by Eyal et al., is that it is difficult to get an accurate measure of the true temperature. The reason for this is that the single radiometric signal is a function of the target's emissivity as well as certain geometric factors such as the distance of the target from the detector. There are several methods to overcome this difficulty. One method proposed by Eyal and Katzir[42] is to measure the radiometric signal

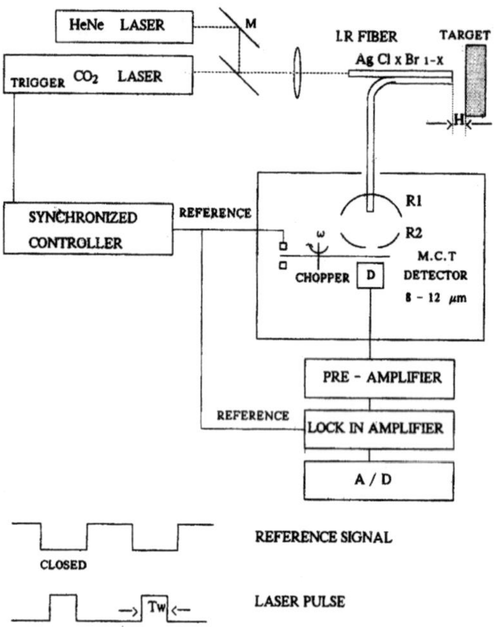

Figure 8.19 A radiometer employing Ag-halide fibers, a sensitive HgCdTe detector, and phase-sensitive detection.[48] [Reprinted with permission.]

in two separate wavelength bands. In two-bandpass radiometry, signals from the two wavelength bands can be ratioed to essentially nullify the effects of geometry and emissivity. To demonstrate the two-bandpass concept, Eyal and Katzir[42] and Dankner et al.[43] used an unclad 900-μm-core Ag-halide fiber, 1-m-long, coupled to a two-color LN_2-cooled detector. This detector was composed of a sandwich of InSb/HgCdTe for detection in the 1- to 5-μm (InSb) and 5.5- to 12.5-μm (HgCdTe) bands. After calibration against a black body, they used their system to measure the temperature of a thin photoresist layer on a silicon wafer during dry etching. It is important to monitor the temperature of the ~1-μm-thick photoresist layers to prevent crosslinking, which can make the layer hard to remove. In this application the emissivity of the photoresist layer changes during heating, and this makes even the two-bandpass approach difficult because the change in emissivity is different in the two wavelength regions. Nevertheless, they were able to obtain a ratio of the signal from the two detectors that was independent of emissivity. They concluded that their two-color radiometer was able to reliably measure temperatures in the critical crosslink temperature range of 105°C without calibration against a thermocouple each time some parameter was changed.

Another method used by Eyal et al.[41] to improve the performance of near-room-temperature radiometry is to irradiate the sample with a laser pulse as shown in Fig. 8.19, but to determine the temperature from the time dependence of the emitted infrared signal. They call this technique pulsed photothermal radiometry. Its advantage over ordinary radiometry is that it enables temperature measurements of short-lived phenomena, such as phase transitions in superconductors, and of moving targets. The target temperature is determined from the shape of the photothermal signal decay curve and not from the signal amplitude. The basic idea is to measure the decay time of the excitation laser pulse in the target. Eyal et al. found a linear increase in sample decay time with increasing temperature. In their prototype system, two PC fibers were used. One delivered the heating laser pulse, and the other received and transmitted the photothermal signal to the detector.

There has been considerable interest in measuring the temperature of human tissue using IR fiber-based thermometers.[44] As mentioned above, IR fibers are a natural choice for these applications because of their high transparency in the 10-μm region. One interesting and potentially important example of this is the measurement and control of tissue temperature during exposure to therapeutic laser energy. Simhon et al.[45] have devised a two-fiber system for controlling the temperature during laser anastomosis or tissue welding. One Ag-halide fiber is used to deliver CO_2 laser energy to tissue that has been coated with biological solder, and the other fiber is used to measure and control the temperature of the tissue. The basic idea is to control the heating of the tissue to a temperature of about 70°C. What Simhon et al. found was that using feedback from the radiometer, they were able control the CO_2 laser radiation so that the temperature was held to 69 ± 7°C. This was somewhat higher than desired, but their device has proven to be an effective method to control as well as measure temperature.

Solid-core IR fibers other than the Ag-halide PC fibers for radiometric applications have been less well studied. KRS-5 fibers were used by Artjushenko et al.[46] in a setup virtually identical to that shown in Fig. 8.19. As with the Ag-halide fibers, the objective was to heat a sample with a CO_2 laser and monitor the radiation with the KRS-5 fiber. Fluoride glass fibers were studied by Tran et al.,[47] but these fibers are not as sensitive to low temperatures as the PC fibers because they cut off at about 4 μm. For some unknown reason there is even less work using chalcogenide fibers in these thermal applications.

Hollow waveguides have also been used to transmit thermal radiation in radiometry. Small et al.[48] used a 2-m-long 700-μm-bore HGW with Ag/AgI coatings to make a two-bandpass radiometer. They used two separate thermoelectrically cooled HgCdTeZn detectors for the collection of the 2- to 6-μm and 2- to 12-μm wavelength bands. The output portion of their device is shown in Fig. 8.20(a). As in the work of Eyal et al., they have the two signals from the wavelength bands to nullify the effects of changing emissivity and sample geometry. In Fig. 8.20(b) the output from each detector is given from near room temperature up to 450°C. The solid curves in Fig. 8.20(b) are an exponential fit to the detector output as a function of the temperature of a blackbody source. Their fit to the data accurately reflects the dependence on temperature predicted by the Planck function in both the low- (the Wien limit) and high-temperature limits. Alaluf et al.[49] used a single hollow plastic waveguide with Ag/AgI coatings to measure temperatures near 37°C for biomedical applications. The bore size of their waveguides varied between 1.0 and 2.4 mm. For the largest bore size and a time constant of 1 s, they measured a MDΔT of 0.15°C.

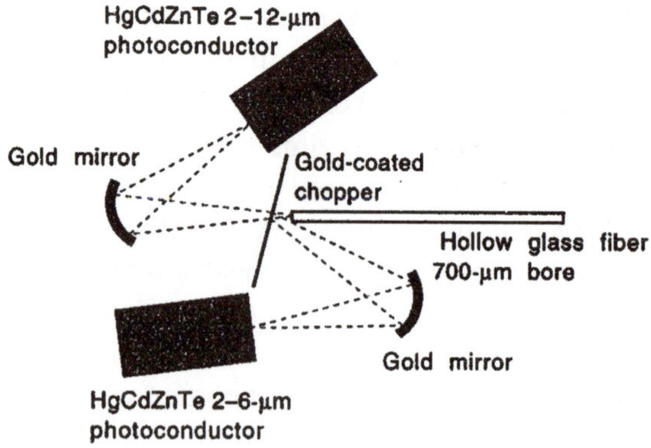

(a)

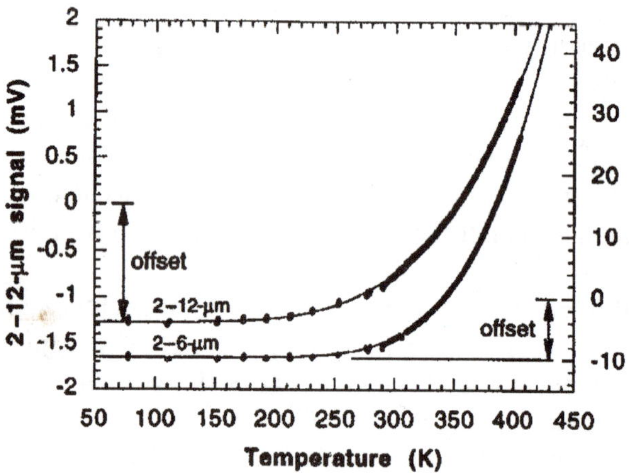

(b)

Figure 8.20 (a) Configuration of two-color radiometer employing an HGW and two HgCdTe detectors, and (b) calibration of blackbody signals from each detector calibrated in terms of temperature.[48] [Reprinted with permission.]

Not all temperature measurements have been made in the low temperature ($\lesssim$ 150°C) range where IR fibers have distinct advantages over their oxide fiber counterparts. Latvakoski et al.[50] have used HGWs up to 10 m in length to measure the temperature of the combustion sections of turbine engines. The objective of their work was to measure the turbine blade temperature from about 425 to 1425°C in gas turbine aircraft engines. This is a high-temperature range and it would seem that silica fibers would function well in this high-temperature range where the blackbody radiation (see Fig. 8.3) is shifted to shorter wavelengths. Sapphire fibers or rods would also work well as a waveguides here, and in fact they have been used for this purpose. The problem, however, is that modern engine blades are coated with thermal barrier coatings (TBCs) so that they may operate at high temperatures. The TBC-coated engine blades have lower emissivity than the uncoated blades and, therefore, radiate poorly at short wavelengths. Instead, it has been found that it is better to measure the radiation between 10 and 13 μm where the emissivity is approximately 1. Latvakoski et al. used 1000-μm-bore HGWs to transmit the thermal energy from the blades back to a cooled HgCdTe detector for phase-sensitive detection and signal processing. The input end of the HGW was inserted into an engine probe, but it was protected from the hot engine gases by a window at the end of the probe. Their measurements indicated a statistical error of 8°C for a 6-m-long waveguide.

8.3 Coherent bundles for thermal imaging

Coherent bundles made from oxide glass fiber optics have been used for many years. In fact, one of the earliest uses of fiber optics was in imaging. As discussed in Sec. 8.0, there are two primary methods of making an image bundle. One is the wrapping method, in which fibers are spooled on a drum to form ribbons and then the ribbons stacked into a bundle. The other is the leached-bundle technology, in which fiber canes are drawn and redrawn into a final high-count fiber bundle. Both methods are used to make visible fiber bundles for endoscopes and borescopes. The fiber count for these oxide fiber bundles can exceed 1 million fibers, and the fiber core size can approach 3 μm. It would seem natural that the same methods could be merely adopted for IR fibers. However, this has not been the case, primarily because IR fibers are delicate, cladding layers are too thick, there is no easy analog to the leached bundle technology for IR fibers, and the cost of IR fiber optics is high.

Most of the applications of IR image guides are in thermal imaging. For example, a flexible bundle could be used to transmit information from a remote, non–line-of-sight location to a central detector for signal processing. Today there are numerous IR cameras with room temperature microbolometers that have pixel elements of 25 μm or less. In some applications it would be much more helpful to have a flexible bundle coupled to these cameras as a means of increasing their versatility and efficiency. To do this, however, the IR fibers used in the bundle must have core sizes near the pixel size for maximum resolution. So far this has presented a major challenge for IR fiber researchers.

Not surprisingly, most work to date on IR fiber coherent bundles has been with the chalcogenide glass fibers. This is because these fibers transmit well into the IR spectrum and, thus, they are capable of transmitting images of low-temperature objects. Klocek et al.[51] considered some of the basic properties of an image bundle, including the modulation transfer function (MTF), crosstalk, resolution, tapered bundles, and reformatters. These authors made bundles using 25-, 75-, and 300-μm-diameter GeSbSe unclad fibers and a wrapping method. Their early bundles had a low fiber count and the resolution was not too good.

Saito et al.[52] used a somewhat different method to make a bundle of AsS fibers with core sizes varying from 200 to 1000 μm. They drew a monofilament fiber coated with Teflon FEP, which was then cut into 10- to 30-cm long pieces; 200 to 1000 of these FEP/AsS fibers were further bundled in an FEP tube for redrawing.

The final bundle was about 2 mm in diameter. The FEP coating/cladding does not absorb beyond about 6 μm, so their bundle was suitable only for the 2- to 6-μm range. They were able to obtain a low-resolution image of a hot iron through a bundle composed of 200 90-μm-core fibers 1.3 m in length. The minimum loss for the bundle was 0.6 dB/m.

A more extended study of chalcogenide fiber bundles was carried out by Nishii et al.[53,54] Like Saito et al., they used a FEP-coated AsS fiber to form their bundles. One bundle was prepared in a manner identical to that used by Saito, i.e., using a redrawing process. The other bundle was assembled from ribbons made using individual, 1-m-long monofilament. The ribbons were stacked into a 2.8 × 3.4-mm array. The cross section of the latter bundle is shown in Fig. 8.21.[54] This 1-m-long bundle is made up of 1386 fibers with 55-μm cores and 75-μm cladding (55/75). Nishii et al. coupled this bundle to an IR camera with an InSb detector, and they used this system to image an operating integrated circuit heated to 60°C. They found that they could measure the temperature of the circuit to an accuracy of 0.5°C once their system was calibrated against a blackbody. They also concluded that the best-quality bundles were made using the wrapping as opposed to the redrawing method. In particular, the alignment of the fibers was better in the wrapping/ribbon approach. The losses for the AsS/FEP fibers are like those shown in Fig. 5.10. The FEP coating begins to absorb strongly beyond about 4 μm. Nishii et al.[53] have also made bundles with a fiber count as high as 8400.

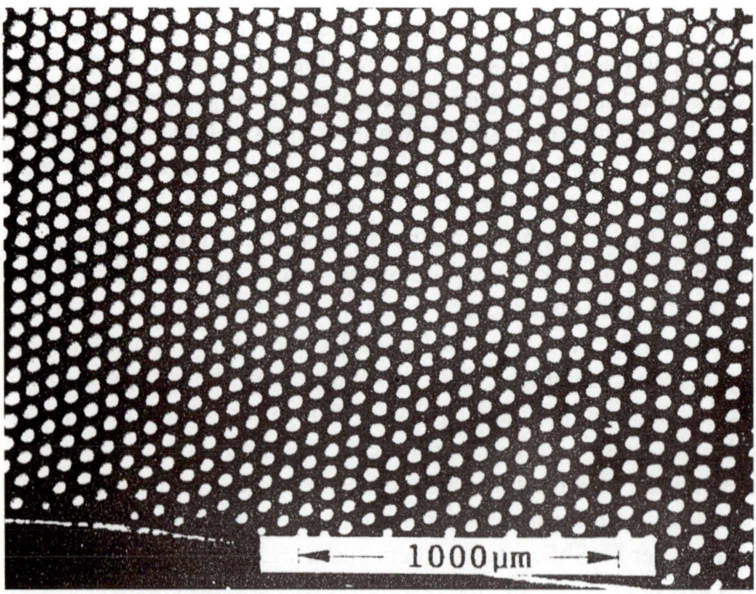

Figure 8.21 Cross-section of image bundle made using 1386 As_2S_3 fibers with an FEP cladding. The fibers had a 55/75-μm core/clad structure and the bundle was 100 cm long.[54] [Reproduced by permission of Taylor & Francis, Inc., http://www.routledge-ny.com.]

Saito et al.[55] employed the chalcogenide glass bundles developed by Nishii for a novel application in spectroscopy. They used a 5×8 mm^2 bundle of 8400 AsS/FEP fibers to image gas flow. Their experimental setup consisted of a gas cell illuminated by either the 3.4-μm line of a HeNe laser or a glower source and a PtSi CCD sensor array that detected radiation in the 1- to 5-μm band. The single AsS source fiber was arranged so that light passed through the gas cell and reflected back through the gas cell to the coherent bundle for imaging. The 3.4-μm wavelength was chosen as the source wavelength because it is highly absorbed by the CH_4 and CO_2 gases used in their tests. The sampling rate used in their measurements was $1/60$ s, so that they were able to study the real-time flow of gas in the cell. It should be noted that the use of a 3.4-μm HeNe laser for the detection of methane gas is along the lines of the early work of Saito and Takizawa[5] as described in Sec. 8.1.1. In that work only a single chalcogenide fiber was used as the receiving fiber as opposed to a fiber bundle used in this study. The advantage of the image bundle approach is that the researchers could image a volume of gas rather than merely measure the concentration of the gas. As in the early work of Saito and Takizawa, Saito et al. also measured small concentrations of both CH_4 and CO_2 gases near the explosion limit for CH_4.

More recently, Yang et al.[56] also made coherent bundles by forming wrapped ribbons of AsS fibers. In their case they used optically clad fiber instead of the FEP polymer. Their core material was $As_{38}S_{62}$ and the cladding was $As_{35}S_{65}$. Fiber bundles with 1000 (32×34) and 4000 (64×64) fibers were made using 70/80 and 60/70 fibers, respectively. They were also able to transmit a thermal image using their 30-cm-long bundle coupled to an IR camera.

Substantial efforts to fabricate high-quality, long-length chalcogenide fiber bundles have been made by Hilton and his coworkers at Amorphous Materials.[57] They have used both AsS and AsSeTe unclad fibers to make bundles transmitting in the 1- to 6-μm and 2- to 12-μm regions, respectively. To make high-quality bundles, they used the method of wrapping fibers on drums to make ribbons. The longest bundles were 10 m; shorter 1- and 3-m lengths were also made. The 10-m circumference drum used by Hilton is shown in Fig. 8.22(a). The drum in Fig. 8.22(a) rotated continuously at a rate of 0.5 to 10 m/min. Once one ribbon was formed, an automatic space was left before the next ribbon was wound. Up to 40 ribbons and a total of 30 km of fiber were fabricated in one run. The coherent bundles were formed using 1000 to 2500 fibers; fiber core sizes were from 75 to 100 μm, and bundle dimensions were as large as 7×7 mm. As an example, the 10-m AsS bundle had an active area of about 40%, with a total transmission of about 25% and a resolution of 10 line pairs/mm (see Sec 8.0). The ends of the bundle were AR coated to improve the transmission. Figure 8.22(b) shows the image of a line chart transmitted through a 1-m-long bundle connected to an IR camera.

An alternative approach to using glass fiber to make a coherent bundle is the use of PC fibers. Rave et al.[58,59] have used clad AgClBr fibers to make bundles with 100 to 9000 fibers. They have developed a unique multiple-extrusion approach to make their guides. Rave et al. started with a 700- to 900-μm-diameter core/clad

(a) (b)

Figure 8.22 (a) 10-m drum for winding fiber bundles and (b) best 1-m long bundle made with 100-μm-core chalcogenide glass fibers. Resolution is approximately 10 line-pairs/mm.[57]

monofiber. Then this fiber was cut into shorter sections, which were then inserted into a pure AgCl tube. This tube, with the ordered monofibers, was then extruded into a multiple-core fiber. Finally, this multiple-core fiber was cut into small pieces and these pieces were again inserted into a AgCl tube for a final extrusion. One difficulty encountered by Rave et al. was that multiple extrusions with decreasing core size led to a degradation of the fiber's transmission. Specifically, they found that the initial loss of the monofiber increased in the 9- to 11-μm region more than a hundredfold, i.e., from 0.2 dB/m (starting monofiber) to 0.7 dB/cm (a single fiber in a 100-fiber bundle). Furthermore, they observed 30% crosstalk even though the fiber was clad. Nevertheless, they were able to obtain a thermal image of two hot wires using a small, 100-fiber bundle with 40-μm-diameter fibers. Clearly, more work needs to be done to reduce die friction so that the overall integrity the fiber's geometry and especially the core/clad region is preserved.

A novel approach to making a coherent bundle was taken by Gopal and Harrington.[60] They used hollow waveguides to make rigid and flexible bundles for broadband IR imaging. The advantage of using hollow guides is that both the wrapping and leached-bundle technologies are, in principle, available to form the coherent waveguide structure. That is, one can begin with thin-wall silica tubing and form a ribbon by wrapping this tubing into ribbons. Then the ribbons can be stacked into coherent arrays for coating. Alternatively, a flexible bundle of holes can be made using a leached-bundle technique. That is, it is possible to use hollow glass tubing coated with an acid-leachable glass to form the multifiber canes. These canes can then be drawn, cut and stacked, and redrawn, etc., into the final bundle. In a sense, the difficulty with this approach is not in forming the hollow core bundle, but rather the fundamental issue of the high loss that can result for the small-bore cores needed for IR imaging. In fact, it is much easier to form a bundle of oxide glass tubes than it is to form chalcogenide or silver halide fiber bundles. This is due in part to the high strength of oxide glass fibers and in part to the advanced state of oxide fiber image guide technology. The real obstacle to the hollow guide approach is the potentially high loss for bore sizes in the 25- to 50-μm range. For example, the traditional Ag/AgI-coated HGW has a calculated loss for a 50-μm bore of 240 dB/m at 10.6 μm. This is intolerably high for any reasonable length of fiber. As pointed out by Gopal and Harrington, if a three-layer coating of Ag/AgI/Ge/AgI were deposited in the same bore size, the loss would be 60 dB/m. This is still quite high. Therefore, these authors conclude that multilayer-dielectric coatings are essential to having a reasonable loss for a long flexible bundle of small-bore hollow guides.

Because the loss for a 50-μm-bore waveguide is so high, Gopal and Harrington decided to coat rigid coherent bundles. Rigid bundles would be useful in transferring images, as is done with fiber optic faceplates or in reformatting images. To this end, tubing composed of an array of glass tubes was obtained from collimated holes in a variety of bore sizes. This collimated hole tubing can be drawn into rather long lengths, but it is not flexible. Gopal and Harrington used as the starting material short, 4- to 8-cm long collimated hole tubes composed of a matrix of small-bore tubing with bore sizes of 50 and 150 μm. The glass matrix, shown in cross section in Fig. 8.23, is then coated with the standard Ag/AgI coatings. The 50-μm and 150-μm bore collimated hole guides had 900 and 500 total capillary tubes, respectively. As may be seen from Fig. 8.23, the alignment of the holes is excellent. Also in Fig. 8.23, the spectra of the two samples are shown. The interference peaks are well defined, indicating that the coatings are quite uniform in each of the holes. The authors were not able to use these collimated hole guides to transmit an image, as the overall losses even for these short lengths were still rather high.

An alternative to using a multifiber array to transmit an image is the use of a single-fiber or hollow waveguide that is scanned over the target. Dekel et al.[61] used both a silver halide fiber and a HGW as the scanning fiber. The basic idea

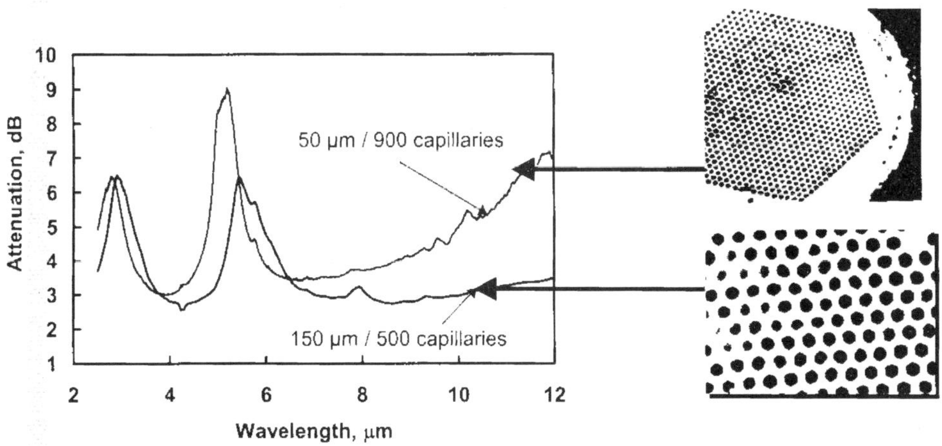

Figure 8.23 Spectral response of rigid bundle of Ag/AgI-coated hollow waveguides.[63] [Reprinted with permission.]

is to scan the output end of the fiber across the thermal image plane to obtain a two-dimensional image. Small magnetic metal plates were placed on the sides of the end of the fiber, and then the fiber end plus plates was placed in the oscillating magnetic field. The modulated magnetic field was used to rapidly move the fiber end in the x or y directions by as much as 10 mm at velocities up to 5 mm/s. The target was scanned in a 24×24 array with a dwell time of 300 ms per pixel. The data was analyzed using phase-sensitive detection methods. Dekel et al. determined a MDΔT of 0.5°C using their Ag halide fibers to image a 120°C target.

8.4 Miscellaneous sensor applications

While most of the sensor applications using IR fibers fit in the three categories above, there are some other applications worth noting, as they may lead to some significant uses of these fibers in the future. Generally, these other uses may be loosely described as new IR components or instruments that extend the range of existing instrumentation into the infrared region.

8.4.1 Spectroscopy

Zhao et al.[62] proposed the use of fluoride glass fiber in an all-fiber-optic Fourier transform spectrometer (FTS). The basic idea of this unusual application is to use SM fluoride glass fibers as the two arms of a Michelson interferometer to make an FTS. In a traditional FTS, the input light beam is split into two beams. One beam reflects off of a stationary mirror and the other off of a moving mirror. The moving mirror provides the optical path difference (OPD) necessary to form the interferogram once the two beams are recombined. Zhao et al. proposed that the moving mirror be replaced by an SM fiber so that their FTS would have no moving parts. Furthermore, since the fiber arms of their FTS were made of fluoride glass, they would transmit in the 2- to 4-μm region. The main problem that they faced was to provide the optical path difference (OPD) now that there was to be no moving mirror. They solved this problem by wrapping fiber from one leg of the spectrometer around a piezoelectric cylinder, which could be modulated to stretch the fiber and thus provide the requisite OPD. The SM fluoride glass fiber chosen for their experiments had a cutoff wavelength of 1.95 μm. A fiber coupler with a 46/54 split was used to divide the energy into the two fiber arms. Their initial experiments showed that they could indeed obtain an interferogram but that nonlinearities in the piezoelectric cylinder and OPD dispersion created difficulties that needed to be overcome before their FTS functioned as well as conventional FTS instruments.

Phase-modulated fiber optic sensors employing SM silica fibers are quite common. However, almost nothing has been done to exploit IR fibers for use in phase-modulated sensors. The principle reason for this is that there are few SM IR fibers, and most phase-modulated fiber sensors require SM fibers. However, Wang et al.[63] showed that they could use a sapphire fiber as part of an intrinsic Fabry-Pérot interferometer for temperature sensing. These researchers formed their Fabry-Pérot interferometer using conventional SM silica fiber, a coupler, and a 1.3-μm laser source. The transmission/reflection arm of the interferometer, shown in Fig. 8.24, was terminated with a short length of MM sapphire fiber. The advantage of using a sapphire fiber-sensing element over silica fiber was that sapphire fiber could withstand temperatures in excess of the 1000°C limit for silica fiber. The sapphire fiber formed the interference cavity. Optical interference occurred as a result of reflections from the fusion splice between the SM silica/MM sapphire fiber and the light reflected from the output (free) end of the sapphire fiber. To enhance the amount of reflected light, the free end of the sapphire fiber was coated with chromium. That is, there is a phase difference or shift, $\Delta\phi$, due to the path-length difference that occurs as a result of the reflections from the two ends of the sapphire fiber optical cavity. This phase difference is given by

$$\Delta\phi \cong \frac{4\pi}{\lambda_0}(n\Delta L + L\Delta n), \tag{8.1}$$

where λ_0 is the source wavelength in vacuum, and L is the length of sapphire fiber. As Eq. (8.7) shows, there is a phase shift that results from either a change in length or refractive index. Sapphire fiber is sensitive to changes in both temperature and strain, so both terms in Eq. (8.7) can contribute to the observed interference. To test their Fabry-Pérot interferometer, Wang et al. used a blowtorch to heat the sapphire fiber tip from 310°C to 976°C. The MRΔT was 0.2°C over this range.

8.4.2 Microscopy

Scanning near-field optical microscopy (SNOM) or scanning near-field infrared microscopy (SNIM) is a technique that has been used to image the surface morphology of semiconductor devices, biological samples, and organic films. Normally, a tapered SM silica fiber is used as the fiber probe and this taper is drawn to a very fine point. Schaafsma et al.[64] have extended the basic technique of SNOM

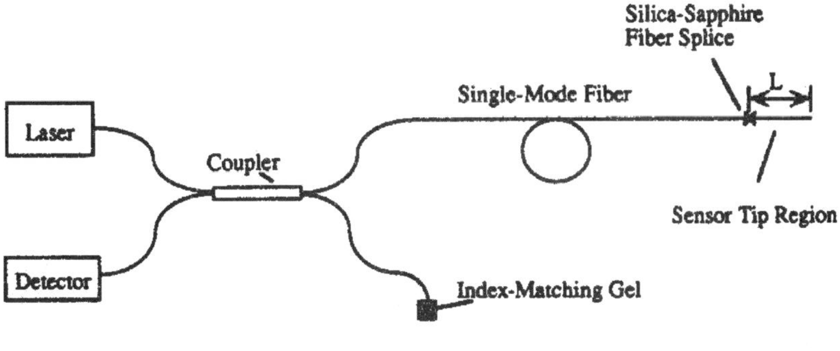

Figure 8.24 Sapphire/silica Fabry-Pérot interferometric fiber sensor for use in temperature sensing. The sapphire tip was placed in a furnace and the sensor was calibrated for temperatures from 310°C to 976°C with a temperature resolution of 0.2°C.[63]

to the infrared by using a tapered SM chalcogenide glass fiber. They formed a microtip in an SM AsS fiber with a tip diameter estimated to be between 100 and 200 nm. To test their tips they used a free electron laser (FEL) as a source of 3.5-μm radiation. The sample that they studied was polycrystalline diamond film. They obtained a spatial resolution of 80 to 100 nm. Their conservative estimate of the lateral resolution was $\lambda/15$, which is excellent for optical microscopy.

Hong et al.[65] also used chalcogenide fibers that have been tapered by pulling and/or etching to a radius less than the wavelength of light used. The source laser was the FEL set to operate at 6.05 μm. They made a series of measurements as the tip moved in toward the underwater samples of single motile fibroblasts. Under the best conditions they were able to obtain a spatial resolution of $\lambda/3$. They noted that the spatial resolution was highest at the most strongly absorbing wavelengths (water absorbs near 6 μm).

Quartel and Dainty[66] used a CO_2 laser and a tapered tip of Ag-halide fiber for their SNIM measurements of troughs in silicon. They were clearly able to distinguish features in the silicon that were well below the 10.6-μm wavelength used to illuminate the samples. Finally, Sanghera et al.[67] used tapered chalcogenide fibers to resolve polycrystalline and pancreatic cells to a 100-nm resolution at a wavelength of 6.1 μm.

References

1. N. J. Harrick, *Internal Reflection Spectroscopy*, Wiley Interscience, New York (1967).
2. M. Saito and K. Kikuchi, "Infrared optical fiber sensors," *Optical Review*, Vol. 4, pp. 527–538 (1997).
3. J. Sanghera, L. B. Shaw, and I. Aggarwal, "Applications of chalcogenide glass optical fibers," *C. R. Chimie*, Vol. 5, pp. 873–883 (2002).
4. E. Rave, D. Shemesh, and A. Katzir, "Thermal imaging through ordered bundles of infrared-transmitting silver-halide fibers," *Appl. Phys. Lett.*, Vol. 76, pp. 1795–1797 (2000).
5. M. Saito, M. Takizawa, K. Ikegawa, and H. Takami, "Optical remote sensing system for hydrocarbon gases using infrared fibers," *J. Appl. Phys.*, Vol. 63, pp. 269–272 (1988).
6. E. V. Stepanov, A. I. Kouznetsov, P. V. Zyrianov, V. G. Plotnichenko, Y. G. Selivanov, and V. Artjushenko, "Multicomponent fiber-optical gas sensor based on MIR tunable diode lasers," *Infrared Phys. & Tech.*, Vol. 37, pp. 149–153 (1996).
7. F. J. McAleavey, J. O'Gorman, J. F. Donegon, J. Hegarty, and G. Maze, "Extremely high sensitivity gas detection at 2.3 microns using a grazing incidence Tm^{3+} fibre laser cavity," *Sensors and Actuators B*, Vol. 87, pp. 107–112 (2001).
8. P. Lu, X. Bao, T. Widden, and S. Y. Lee, "Application of a mid-infrared fiber bundle in remote measurement of gas concentrations in a chemical vapor deposition chamber," *Appl. Opt.*, Vol. 39, pp. 1112–1117 (2000).

9. I. Tugendhaft, A. Bornstein, Y. Weissman, and A. Hardy, "Reflection intensity optical fiber sensors for the mid-infrared," *Appl. Opt.*, Vol. 36, pp. 1297–1302 (1997).
10. D. Bunimovich, E. Belotserkousky, and A. Katzir, "Noncontact infrared reflection spectroscopy using optical fibers," *Rev. Sci. Instrum.*, Vol. 65, pp. 2731–2732 (1994).
11. D. J. Haan and J. A. Harrington, "Hollow waveguides for gas sensing and near-IR applications," Proc. SPIE, Vol. 3596, pp. 43–49 (1999).
12. Y. Saito, T. Kanaya, A. Nomura, and T. Kano, "Experimental trial of a hollow-core waveguide used as an absorption cell for concentration measurement of NH_3 gas with a CO_2 laser," *Opt. Lett.*, Vol. 18, pp. 2150–2152 (1993).
13. R. Micheels, K. Richardson, D. J. Haan, and J. A. Harrington, "FTIR-based instrument employing a coiled hollow waveguide cell for rapid field analysis of volatile organic compounds," Proc. SPIE, Vol. 3540, pp. 66–74 (1999).
14. R. L. Kozodoy, R. Micheels, and J. A. Harrington, "Small-bore hollow waveguide infrared absorption cells for gas sensing," *Appl. Spectrosc.*, Vol. 50, pp. 415–417 (1996).
15. L. Hvozdara, N. Pennington, M. Kraft, M. Karlowatz, and B. Mizaikoff, "Quantum cascade lasers for mid-infrared spectroscopy," *Vibrational Spectro.*, Vol. 30, pp. 53–58 (2002).
16. L. Hvozdara, S. Gianordoli, G. Strasser, W. Schrenk, K. Unterrainer, E. Gornik, C. S. S. S. Murthy, M. Kraft, V. Pustogow, B. Mizaikoff, A. Inberg, and N. Croitoru, "Spectroscopy in the gas phase with GaAs/AlGaAs quantum-cascade lasers," *Appl. Opt.*, Vol. 39, pp. 6926–6930 (2000).
17. C. A. Worrell and N. A. Gallen, "Trace-level detection of gases and vapours with mid-infrared hollow waveguides," *J. Phys. D.*, Vol. 30, pp. 1984–1995 (1997).
18. G. Fetzer, A. Pittner, W. Ryder, and D. A. Brown, "Tunable diode laser absorption spectroscopy in coiled hollow optical waveguides," *Appl. Opt.*, Vol. 41, pp. 3613–3621 (2002).
19. G. Fetzer, A. Pittner, and P. Silkoff, "Mid-infrared laser absorption spectroscopy in coiled hollow optical waveguides," Proc. SPIE, Vol. 4957, pp. 124–133 (2003).
20. T. A. Blake, J. F. Kelll, T. L. Stewart, J. S. Hartman, S. W. Sharpe, and R. L. Sams, "Absorption spectroscopy in hollow-glass waveguides using infrared diode lasers," Proc. SPIE, Vol. 4817, pp. 216–232 (2002).
21. A. Messica, A. Greenstein, and A. Katzir, "Theory of fiber-optic, evanescent-wave spectroscopy and sensors," *Appl. Opt.*, Vol. 35, pp. 2274–2284 (1996).
22. A. Messica, A. Greenstein, A. Katzir, U. Schiessl, and M. Tacke, "Fiber-optic evanescent wave sensor for gas detection," *Opt. Lett.*, Vol. 19, pp. 1167–1169 (1994).
23. H. M. Heise, L. Kupper, and L. N. Butvina, "Novel infrared optical probes for process monitoring and analysis based on next-generation silver halide fibers," *Anal. Bioanal. Chem.*, Vol. 375, pp. 1116–1123 (2003).

24. S. Simhony, I. Schnitzer, A. Katzir, and E. M. Kosower, "Evanescent wave infrared spectroscopy of liquids using silver halide optical fibers," *J. Appl. Phys.*, Vol. 64, pp. 3732–3734 (1988).
25. J. E. Walsh, B. D. MacCraith, M. Meaney, J. G. Vos, F. Regan, A. Lancia, and V. Artjushenko, "Midinfrared fiber sensor for the in situ detection of chlorinated hydrocarbons," Proc. SPIE, Vol. 2508, pp. 233–244 (1995).
26. R. Krska, E. Rosenberg, K. Taga, R. Kellner, A. Messica, and A. Katzir, "Polymer coated silver halide infrared fibers as sensing devices for chlorinated hydrocarbons in water," *Appl. Phys. Lett.*, Vol. 61, pp. 1778–1780 (1992).
27. J. Heo, M. Rodrigues, S. J. Saggese, and G. H. Sigel, "Remote fiber-optic chemical sensing using evanescent-wave interactions in chalcogenide glass fibers," *Appl. Opt.*, Vol. 30, pp. 3944–3951 (1991).
28. J. Sanghera and I. Aggarwal, "Active and passive chalcogenide glass optical fibers for IR applications: a review," *J. Non-Cryst. Solids*, Vol. 256&257, pp. 6–16 (1999).
29. G. Nau, K. J. Bucholtz, S. T. Ewing, J. S. Vohra, J. Sanghera, and I. Aggarwal, "Fiber optic reflectance sensor for the cone penetrometer," Proc. SPIE, Vol. 2504, pp. 291–296 (1995).
30. S. Hocde, C. Boussard-Pledel, G. Fonteneau, and J. Lucas, "Chalcogens based glasses for IR fiber chemical sensors," *Solid State Sciences*, Vol. 3, pp. 279–284 (2001).
31. S. Hocde, C. Boussard-Pledel, G. Fonteneau, D. Lecoq, H. L. Ma, and J. Lucas, "Recent developments in chemical sensing using infrared glass fibers," *J. Non-Cryst. Solids*, Vol. 274, pp. 17–22 (2000).
32. M. Katz, A. Bornstein, I. Schnitzer, and A. Katzir, "Evanescent wave spectroscopy using chalcogenide glass fiber: theoretical analysis and experiments," Proc. SPIE, Vol. 1591, pp. 236–245 (1992).
33. K. Michel, B. Bureau, C. Pouvreau, J. C. Sangleboeuf, C. Boussard-Pledel, T. Jouan, J. L. Adam, K. Staubmann, H. Steinner, T. Baumann, A. Katzir, J. Bayona, and W. Konz, "Development of a chalcogenide glass fiber device for in situ pollutant detection," *J. Non-Cryst. Solids*, Vol. 326&327, pp. 434–438 (2003).
34. K. LeFoulgoc, L. LeNeindre, S. Hocde, F. Smetktala, X. H. Zhang, and J. Lucas, "Preparation, optimization and infrared applications of TeX glass optical fibers," Proc. SPIE, Vol. 3416, pp. 66–73 (1998).
35. G. Zhang, H. L. Ma, C. Blanchetiere, K. LeFoulgoc, J. Lucas, J. Heuze, P. Collardelle, P. Froissard, D. Picque, and G. Corrieu, "Tellurium halide IR fibers for remote spectroscopy," Proc. SPIE, Vol. 2131, pp. 90–99 (1994).
36. J. Keirsse, C. Boussard-Pledel, O. Loreal, O. Sire, B. Bureau, B. Turlin, P. Leroyer, and J. Lucas, "Chalcogenide glass fibers used as biosensors," *J. Non-Cryst. Solids*, Vol. 326&327, pp. 430–433 (2003).
37. S. J. Saggese, M. R. Shahriari, and G. Sigel, Jr., "Fluoride fibers for remote chemical sensing," Proc. SPIE, Vol. 929, pp. 106–114 (1988).

38. M. Shimizu and S. Kachi, "Low-temperature radiometer using infrared fiber," *Jap. J. Appl. Phys.*, Vol. 22, pp. 275–276 (1983).
39. O. Eyal, A. Zur, O. Shenfeld, M. Gilo, and A. Katzir, "Infrared radiometry using silver halide fibers and a cooled photonic detector," *Opt. Eng.*, Vol. 33, pp. 502–509 (1994).
40. O. Eyal, V. Scharf, and A. Katzir, "Temperature measurements using pulsed photothermal radiometry and silver halide infrared optical fibers," *Appl. Phys. Lett.*, Vol. 70, pp. 1509–1511 (1997).
41. O. Eyal, V. Scharf, and A. Katzir, "Fiber-optic pulsed photothermal radiometry for fast surface-temperature measurements," *Appl. Opt.*, Vol. 37, pp. 5945–5950 (1998).
42. O. Eyal and A. Katzir, "Temperature measurements utilizing two-bandpass fiber optic radiometry," *Opt. Eng.*, Vol. 34, pp. 470–473 (1995).
43. Y. Dankner, O. Eyal, and A. Katzir, "Two-bandpass fiber-optic radiiometry for monitoring the temperature of photoresist during dry processing," *Appl. Phys. Lett.*, Vol. 68, pp. 2583–2585 (1996).
44. S. Sade and A. Katzir, "Fiber optic infrared thermometer for accurate measurement of the core temperature of the human body," *Opt. Eng.*, Vol. 40, pp. 1141–1143 (2001).
45. D. Simhon, A. Ravid, M. Halpern, I. Cilesiz, T. Brosh, N. Kariv, A. Leviav, and A. Katzir, "Laser soldering of rat skin, using fiberoptic temperature controlled system," *Lasers in Surgery and Medicine*, Vol. 29, pp. 265–273 (2001).
46. V. G. Artjushenko, V. V. Voitsekhovsky, V. J. Masychev, J. V. Zubov, and V. K. Sysoev, "Fibre optic device for simultaneous laser power transmission and temperature measuring of irradiated object," *Elect. Lett.*, Vol. 20, pp. 983–984 (1984).
47. D. Tran, K. Levin, and R. Mossadegh, "IR fiber temperature sensing system," Proc. SPIE, Vol. 843, pp. 148–154 (1987).
48. W. Small, P. M. Celliers, L. B. DaSilva, D. L. Matthews, and B. Soltz, "Two-color mid-infrared thermometer with a hollow glass optical fiber," *Appl. Opt.*, Vol. 37, pp. 6677–6683 (1998).
49. M. Alaluf, J. Dror, A. Katzir, and N. Croitoru, "Infrared radiometry measurements using plastic hollow waveguides based on thin films," *J. Phys. D (Appl. Phys.)*, Vol. 26, pp. 1036–1040 (1993).
50. H. Latvakoski, J. Markham, J. A. Harrington, and D. J. Haan, "Sensors for performance monitoring of advanced gas turbines," Proc. SPIE, Vol. 3535, pp. 115–123 (1999).
51. P. Klocek, M. Roth, and R. D. Rock, "Chalcogenide glass optical fibers and image bundles: properties and applications," *Opt. Eng.*, Vol. 26, pp. 88–95 (1987).
52. M. Saito, M. Takizawa, S. Sakuragi, and F. Tanei, "Infrared image guide with bundled As-S glass fibers," *Appl. Opt.*, Vol. 24, pp. 2304–2308 (1985).
53. J. Nishii, T. Yamashita, T. Yamagishi, C. Tanaka, and H. Sone, "Coherent infrared fiber image bundle," *Appl. Phys. Lett.*, Vol. 59, pp. 2639–2641 (1991).

54. J. Nishii, T. Yamashita, T. Tamagishi, C. Tanaka, and H. Sone, "As_2S_3 fibre for infrared image bundle," *Int. J. Optoelectronics*, Vol. 7, pp. 209–216 (1992).
55. M. Saito, K. Kikuchi, C. Tanaka, H. Sone, S. Morimoto, T. Yamashita, and J. Nishii, "Spectroscopic gas-flow imaging through an infrared optical fiber bundle," *Rev. Sci. Instrum.*, Vol. 70, pp. 4308–4312 (1999).
56. K. Yang, P. Wu, G. Wei, and B. Yu, "IR image guide with As-S glass fibers," *J. Non-Cryst. Solids*, Vol. 184, pp. 40–43 (1995).
57. A. R. Hilton, "Infrared imaging bundle development at Amorphous Materials," Proc. SPIE, Vol. 3894, pp. 60–66 (1999).
58. E. Rave and A. Katzir, "Ordered bundles of infrared transmitting silver halide fibers: attenuation, resolution and crosstalk in long and flexible bundles," *Opt. Eng.*, Vol. 41, pp. 1467–1468 (2002).
59. E. Rave, L. Nagli, and A. Katzir, "Ordered bundles of infrared-transmitting AgClBr fibers: optical characterization of individual fibers," *Opt. Lett.*, Vol. 25, pp. 1237–1239 (2000).
60. V. Gopal and J. A. Harrington, "Coherent IR bundles made using hollow glass waveguides," Proc. SPIE, Vol. 4204, pp. 216–223 (2001).
61. B. Dekel, A. Inberg, and N. Croitoru, "Fiber optic thermal imaging system based on hollow glass waveguides or silver halide fibers as scanning elements," *Opt. Eng.*, Vol. 39, pp. 941–946 (2000).
62. P. Zhao, J. Mariotti, P. Lena, V. C. du Foresto, and G. Maze, "Performance analyses of an infrared single-mode all-fiber-optic Fourier-transform spectrometer," *Appl. Opt.*, Vol. 34, pp. 4200–4209 (1995).
63. A. Wang, S. Gollapudi, K. A. Murphy, R. G. May, and R. O. Claus, "Sapphire-fiber-based intrinsic Fabry-Perot interferometer," *Opt. Lett.*, Vol. 17, pp. 1021–1023 (1992).
64. D. T. Schaafsma, R. Mossadegh, J. Sanghera, I. Aggarwal, M. Luce, R. Generosi, P. Perfetti, A. Cricenti, J. M. Gilligan, and N. H. Tolk, "Fabrication of single-mode chalcogenide fiber probes for scanning near-field infrared optical microscopy," *Opt. Eng.*, Vol. 38, pp. 1381–1385 (1999).
65. M. K. Hong, A. G. Jeung, N. V. Dokholyan, T. I. Smith, H. A. Schwettman, P. Huie, and S. Erramilli, "Imaging single living cells with a scanning near-field infrared microscope based on a free electron laser," *Nucl. Instr. and Methods in Physics Research B*, Vol. 144, pp. 246–255 (1998).
66. J. C. Quartel and J. C. Dainty, "An infrared photon scanning tunnelling microscope for investigations of near-field imaging," *J. Phys. A*, Vol. 1, pp. 517–522 (1999).
67. J. Sanghera, L. B. Shaw, L. Busse, V. Nguyen, P. Pureza, B. Cole, B. B. Harbison, and I. Aggarwal, "Development and infrared applications of chalcogenide glass optical fibers," *Fiber and Integrated Optics*, Vol. 19, pp. 251–274 (2000).

Chapter 9

Laser Power Delivery Through IR Fibers

9.0 Introduction

The use of fiber optics to transmit laser energy has important applications in areas as diverse as laser surgery, printing and marking, and industrial welding. For example, industrial welding, cutting, and heat treating require delivering high laser power to remote locations. In these and other applications fiber optics has some clear advantages over bulky and expensive articulated arms, i.e., moving tube-and-mirror systems. Fiber delivery systems are small, much more flexible than articulated arms, and less expensive. Large-core, MM silica fiber optics have been used to deliver several kilowatts of Nd:YAG laser power over tens of meters. In surgical laser systems, surgeons prefer delivery via fiber optics because the distal (output) end of a fiber gives the surgeon the tactile feel of a surgical instrument, and fibers can be inserted directly into the body through endoscopes for least-invasive procedures. From the earliest beginnings of IR fiber technology, it has been the dream of many to reliably transmit IR laser power for applications analogous to those employing silica fibers. As will be seen, the goal of a reliable IR laser power fiber optic has not been fully realized, but great progress has been made, largely through the use of hollow waveguides.

Back in the mid-1970s when IR fibers were first being seriously studied, it was felt by many that one of the most important and exciting uses for non-glass IR fibers was in the transmission of CO_2 laser power. At that time the CO_2 laser was already an important medical and industrial laser, yet the only beam delivery systems were articulated arms. It was generally felt that the new IR fibers being developed would find an immediate home as power fibers for 10.6 μm, just as silica fibers were being used to transmit Nd:YAG laser power. When the first PC, KRS-5 fiber was developed at HRL, it was felt that this fiber would revolutionize CO_2 laser surgery. Unfortunately, some serious problems developed within only a few years after the first extrusion of this fiber. The major problem was that the melting point of the thallium as well as the silver halide fibers is quite low compared to silica fiber (see Table 6.1). During transmission of laser power, the KRS-5 fiber heats up due to absorption. Heating at the output end is greatest because of the high reflection (~17%) at the end face, which leads to large scattering losses. In time, the thermal energy at the output end softens the crystalline material and the end surface rounds to a lens. At this point, the laser power refocuses back into the fiber and the fiber burns up like a fuse. The author was involved in the early work on these fibers at HRL and saw many fibers ignite and burn from the output back to the input end. While unusual to watch, it is nevertheless dangerous, as the smoke and vapor from the burning TlBrI fiber is toxic. The other problem with KRS-5 fiber is that thallium is a toxic material and its use in surgical applications, particular those *in vivo*, are essentially prohibited.

A more exciting early development in CO_2 laser delivery was the announcement by Garmire et al.[1] of the transmission of more than 1000 W of power through a rectangular hollow waveguide. This was an enormous power, exceeding the highest power transmitted by a solid-core IR fiber by a factor of 10. Garmire showed how these simple, aluminum-plate waveguides, described in Sec. 7.1.1, could be coupled to an industrial CO_2 laser and used in machining-type applications. The rectangular shape and overall size of these guides unfortunately effectively eliminated their use in medical applications because a surgical fiber needs to be small enough to be inserted into the body through an endoscope. Nevertheless, Garmire and her coworkers at University of Southern California demonstrated the great potential for hollow waveguides in delivering significant amounts of laser power.

9.1 Solid-core fibers

The early work using KRS-5 fibers was disappointing in that these PC fibers were so susceptible to laser damage. This fact coupled with transmission-aging effects has eliminated them as a good power fiber candidate. However, other solid-core fibers such as the fluoride and chalcogenide fibers and the Ag halide fibers are today still an important fiber for low average power transmission. Their clear advantage over the hollow waveguides is that there is no loss on bending—an important issue for some applications requiring little variation in output power during fiber flexure.

9.1.1 Fluoride glass fibers

One of the first groups to recognize the importance of HMFG fibers in the transmission of Er:YAG laser energy was Bonner et al.,[2] working at NRL and NIH. Tran at NRL drew MM fluoride glass fibers, and Bonner and coworkers at NIH evaluated their use in conjunction with a 2.94-μm Er:YAG laser for medical applications. At the time of their work the Er:YAG laser was thought to have great potential as a surgical laser owing to the very high absorption of this wavelength in tissue ($\alpha_{tissue} \cong 10^4$ cm^{-1}). That is, the 2.94-μm wavelength is highly ablative and therefore a good wavelength for controlled cutting of both hard and soft tissue. Bonner et al. compared the ability of an Er:YAG laser/ZBLAN fiber to ablate human aorta to both KrF and XeCl excimer laser ablation of the same tissue. They found that the depth of ablation was about the same for the 2.94-μm radiation as it was for the 248-nm KrF laser. They also measured threshold Er:YAG laser energies in excess of 300 mJ for a 200-μm-core ZBLAN fiber.

This early work was followed by more detailed studies on the laser-induced damage (LID) thresholds of the fluoride fibers themselves. Two groups, Whitehurst et al.[3] in England and Wuthrich et al.[4] in Switzerland, did extensive testing on ZBLAN fibers with different core sizes using pulsed Er:YAG lasers. Whitehurst et al. used ZBLAN fibers from three different sources and with core sizes ranging from 100 to 300 μm. They prepared the ends of the 2-m-long fibers by removing 2 cm of the polymer coating and then carefully cleaving the fiber. For their power testing, they used a 10-Hz repetition rate, Er:YAG laser with about 200-μs pulses. Wuthrich et al. used a very similar setup. Their fiber core sizes ranged from 11 to 350 μm, and their Er:YAG laser had a fixed rep rate of 4 Hz. As in the work of Whitehurst et al., Wuthrich et al. removed the polymer coating and cleaved the fiber ends. The results of the energy measurements from both studies are shown Fig. 9.1. This combined data should be taken as a range of energy that can be transmitted by a given core-size fiber. The black bar indicates the maximum transmitted energy without fiber failure for the work of Wuthrich et al., whereas the dotted bar indicates the maximum energy without LID obtained by Whitehurst et al. The fiber core sizes for each group differ somewhat, so the data in Fig. 9.1 is collected and displayed for each small range of core sizes used in the experiments. It is clear from Fig. 9.1 that Whitehurst et al. obtained significantly higher LID thresholds than Wuthrich et al. It is not clear just why this is the case, but there are many factors contributing to laser damage in fiber optics. For one, the Er:YAG lasers are usually very multimode and the beam shape and intensity can change with increasing laser power. In addition, end preparation is critical. It is often necessary to make multiple cleaves to get a good fiber end finish that can withstand high laser energy. Finally, the fiber ends can be contaminated with a small amount of water, which can absorb and lead to end face damage. Such damage is discussed in greater detail in the work of Wuthrich et al.

The study performed by Wuthrich et al. involved measuring the LID of the various core-size fibers multiple times. They noted the number of laser pulses before

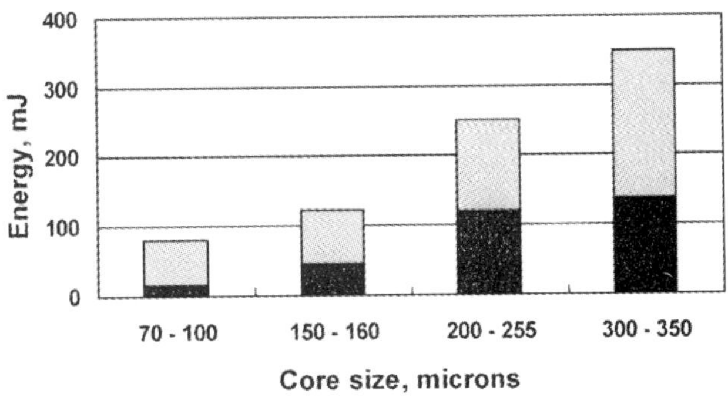

Figure 9.1 Maximum energy threshold of Er:YAG laser transmission by ZBLAN fibers for different core sizes. Data taken from Wurthrich, et al. (black) and Whitehurst, et al. (dotted gray).[4] [Reprinted with permission.]

the fiber was damaged and then obtained a range of energies that can be transmitted by the fibers. From these measurements they concluded that the limit for safe transmission was 8.0×10^4 W/cm^2. The maximum power density transmitted was 2.7×10^6 W/cm^2, which corresponds to 136 mJ. This translates into a maximum average power of 0.544 W (= 136 mJ $\times$ 4 Hz). They also noted that either the input or the output end could be damaged. It is not uncommon to have the output end fail before the input end once a proper end finish is applied to the fiber. The reason for this is that the electric fields are higher in the denser medium (ZBLAN fiber). Therefore, the output end can be more readily damaged than the input end.

The average power-handling capability of ZBLAN fiber is relatively low. Other fluoride glass fibers, however, have significantly higher laser power thresholds. In particular, the fluoroaluminate fibers (see Sec. 4.2.2.2) developed by Itoh et al.[5,6] at Hoya Corporation in Japan are capable of delivering up to 8.7 W of average power. The reason for the significantly higher damage threshold for the AlF_3-based glasses is the much higher T_g for the fluoroaluminate compared to the fluorozirconate glass. From Table 4.1 it can be seen that the T_g for the AlF_3-based glasses is 50% higher than for ZBLAN. The simple fact of the softening point being greater can account for most of the higher LID threshold. In addition, of course, the fiber loss is important, but both of these fibers have about the same loss as may be seen from the data in Figs. 4.15 and 4.17. For the energy measurements, Itoh et al. used a multimode Er:YAG laser operating at 1 Hz. Figure 9.2 gives the results of their energy-out/energy-in experiments using a 500/550-μm, 2.4-m-long fluoroaluminate fiber. From the data in Fig. 9.2 it can be seen that the maximum average output power is approximately 8.7 W. This may not seem like a large power in comparison to, say, a CO_2 laser, but it is quite high and suitable for almost all surgical applications envisioned for this laser.

Recently, Anastasopoulou et al.[7] described their loss measurements using a Q-switched Er:YAG laser and a variety of IR fibers including ZBLAN. Their work

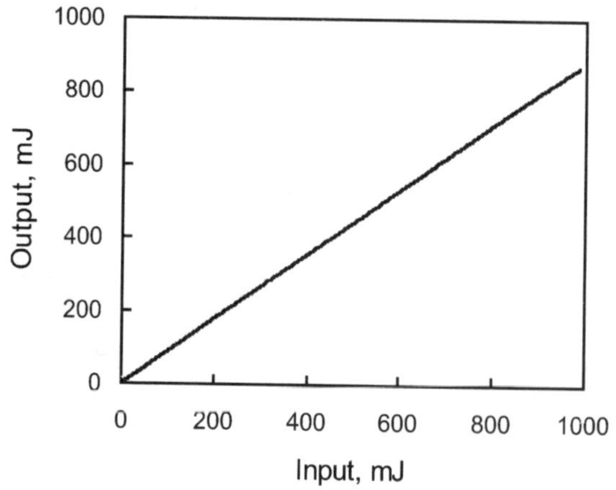

Figure 9.2 Energy transmission of an of an Er:YAG laser by a 500-μm-core fluoroaluminate glass fiber. Maximum average power is about 8.7 W for this fiber.

emphasized the attenuation of the fibers rather than a determination of the LID threshold. Nevertheless, their experiments were some of the first using the short pulses of a Q-switched 2.94-μm laser. Specifically, their Q-switched laser had a fixed energy of 6 mJ, a rep rate of 5 Hz, and a pulse width of 280 ns (or roughly 10^3 times shorter than the typical Er:YAG laser). At this energy level there was no damage to the ZBLAN fibers. Furthermore, the authors noted that there was also no significant difference in the loss for the fluoride glass fiber when measured with either a Q-switched or a free-running Er:YAG laser.

9.1.2 Chalcogenide glass fibers

In the early development of chalcogenide glass fibers it was thought that they would be good candidates for delivering the power of most IR lasers, including the important CO_2 laser. For the most part this will be seen not to have been the case. There are several reasons for this. One is that the loss for chalcogenide fibers is not as low as it is for the fluoride glass fibers. This means that the fiber will heat up and this will lead to melting of the ends and destruction of the fiber. Another reason is the high refractive index of the fibers, which results in multiple end reflections. In turn, this can lead to high scattering loss and excessive heating at the output end. But one of the most important limiting factors is the temperature dependence of the refractive index. The dn/dT for almost all of the chalcogenide glasses is rather high. This means that as the fiber's temperature increases the index will change, and this will cause thermal lensing and melting at the fiber end face.

Most of the early laser-power transmission testing was done using cw CO (5.25 μm) and cw CO_2 (10.6 μm) lasers. As_2S_3 fibers were used extensively in measurements with a CO laser. From the data in Fig. 5.10 it can be seen that the losses for As_2S_3 fibers are about 0.5 dB/m at 5.2 μm. Arai et al.[8] studied the CO laser-power handling capability of an FEP clad, As_2S_3 fiber with a 400-μm-core fiber up to 1.3 m in length. The maximum power output was found to be 16.9 W, or 13.5 kW/cm^2. They attributed this power threshold to melting of the output end of the fiber. At about the same time, Sato et al.[9] were able to transmit 62 W of CO laser power through a 700-μm-core AsS fiber.

A major breakthrough in CO laser-power delivery was made by Sato et al.[10] in 1993. Again using FEP/AsS and also FEP/GeAsS fiber, they were able to deliver 226 W of power at a wavelength of 5.4 μm! Their fiber had a core diameter of 1000 μm and a length of 1 m. The loss for both fibers was about 0.4 dB/m at 5.4 μm. The power-out/power-in curve for both of these fibers is given in Fig. 9.3. The power densities for the AsS and GeAsS fibers shown in Fig. 9.3 were 29 kW/cm^2 and 23 kW/cm^2, respectively. Neither fiber had an AR coating applied to the end face. There are several reasons for the high power thresholds obtained by Sato et al. One is the fact that both fiber ends are cooled with room-temperature N_2 gas during power delivery. The other more fundamental reason is that the dn/dT for AsS is less than 1×10^{-5} deg^{-1}. As may be seen from the dn/dT data in Table 5.4, this is in contrast to dn/dT for other chalcogenide glasses such as GeAsSeTe, which is 14×10^{-5} deg^{-1}. The smaller dn/dT leads to less self-focusing and a higher damage threshold. Sato et al. also applied a single-layer PbF_2 AR coating to the same 700-μm-core fibers. While the damage threshold was only about 100 W, the transmission of the fiber improved with the reduction in the Fresnel reflections. Presumably the lower power threshold was due to the absorption by the AR coating.

Nishii et al.[11] have given an excellent summary of the power-handling capability of chalcogenide fibers. Their extensive study included both the AsS-type glass for the shorter-wavelength CO laser and the Se- and Te-containing glasses for the longer-wavelength CO_2 laser. A summary of their power-testing results is given in Table 9.1. The best fiber for the CO_2 laser wavelengths was found to be the GeSeTe glass fiber, which has the loss spectrum shown in Fig. 5.12. The lowest

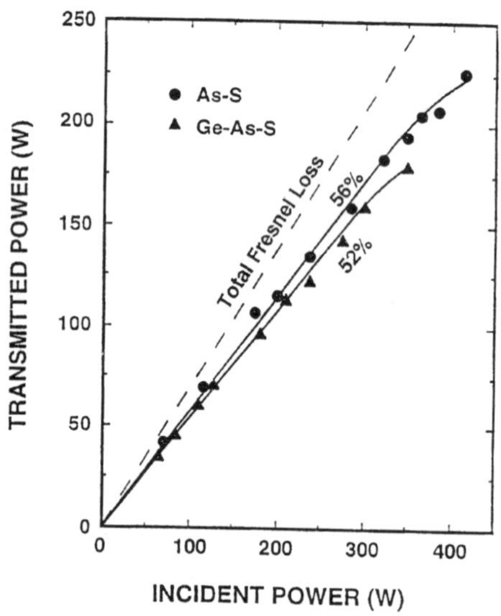

Figure 9.3 Power transmission of cw CO laser radiation through 1000-μm-core chalcogenide glass fibers. Note high maximum power threshold of 226 W for the As_2S_3 fiber. Percent transmission for each fiber is also shown next to each curve.[10]

Table 9.1 Maximum input/output powers of CO and CO_2 lasers for several different chalcogenide glass fibers under different end conditions.

Core/clad	Core diameter, μm	AR coating and/or water	Loss, dB/m 5.4 μm/10.6 μm	CO laser power, P_{in}/P_{out}, W	CO_2 laser power, P_{in}/P_{out}, W
GeSeTe/ GeAsSeTe	450	AR water	1.0/2.1	—	19.4/10.7
GeAsSe/ GeAsSe	450	water	0.5/5.5	—	19.4/3.3
AsSSe/AsS	780	none	0.6/—	169/85	—
GeAsSe/ GeAsSe	700	none	0.5/5.5	13/8.5	—

loss at 10.6 μm for this fiber is about 2 dB/m. While this is certainly rather high for an IR fiber, it is one of the lowest losses reported for a chalcogenide at this wavelength. Note from Table 9.1 that the maximum CO_2 laser power out of the fiber is about 10 W. This power level was achieved with PbF_2-AR coated, water-cooled fiber ends. Nishii et al. attribute this low power threshold to thermal focusing resulting from the large dn/dT for this material. Water cooling is not too practical, but even with gas cooling the upper limit is only about 9 W. The power thresholds given in Table 9.1 for the AsS fibers at 5.4 μm are not as high as in the work of Sato et al.,[10] but again they do show the dramatic difference in the power delivery between the AsS and other chalcogenide glass fibers. In fact, one can conclude from this data that a key property needed for a chalcogenide fiber to operate at high-CO_2 laser powers (i.e., greater than 100 W) is a dn/dT that is nearly zero. To achieve this goal the glass community would need to engineer a chalcogenide glass that has low loss at 10.6 μm and also a very small dn/dT.

Similar studies for both sulfide and telluride glass fibers were more recently reported by Sanghera et al.[12] These authors measured a maximum power output for 200/330-μm sulfide fiber of about 3.3 W (~125 kW/cm^2) at 5.4 μm and an output power for 162/275-μm telluride fiber of about 0.6 W (~54 kW/cm^2) at 10.6 μm. Again these cw laser powers are not high, but the fiber core diameters are rather small, and, therefore, potentially useful for some low-power military applications. Specifically, Sanghera et al. suggest that this power could be useful in IR countermeasures in which the fiber would deliver power from the IR laser to a remotely located jam head mounted on the surface of an aircraft.

Most of the work done on the chalcogenide fibers has been done at 5 and 10 μm. Arai et al.[13] extended this work somewhat by using AsS fibers to simultaneously deliver Nd:YAG and CO laser power. This would be a useful combination of lasers for medical applications because the CO laser is highly ablative, while the Nd:YAG operating at 1.06 μm is good at coagulating bleeding. The AsS/FEP fiber used by Arai et al. had a core diameter of 400 μm, with a loss of approximately 8 dB/m at both 5.4 and 1.06 μm. The maximum cw laser power transmitted when both lasers were used simultaneously was 14 W (power ratio of 1:1). Curiously, the loss of the fiber during exposure to the high laser power increased to 2.6 dB/m.

The bulk of laser-power delivery studies using chalcogenide glass fibers involve cw lasers. There has, however, been some limited work using pulse lasers. Sanghera et al.[12] used 525/620-μm sulfide fibers to transmit about 18 mJ of 2.94-μm energy from an Er:YAG laser. Their Er:YAG laser energy delivery data is shown in Fig. 9.4(a). They have also used their sulfide fibers to transmit pulsed energies from a free electron laser (FEL). The FEL has an unusual two-pulse output structure. This two-pulse profile is composed of macropulses with 3- to 15-μsec pulsewidths and short micropulses with about 1-psec pulsewidth. Additionally, the FEL has a very broad range of tunability, covering most wavelengths of common IR lasers. In Fig. 9.4(b), the transmission of FEL energy is shown for a sulfide fiber. These measurements were carried out with the FEL set to operate at 2.94 μm, or the same wavelength as the Er:YAG laser. Similar studies using GeAsSeTe glass-

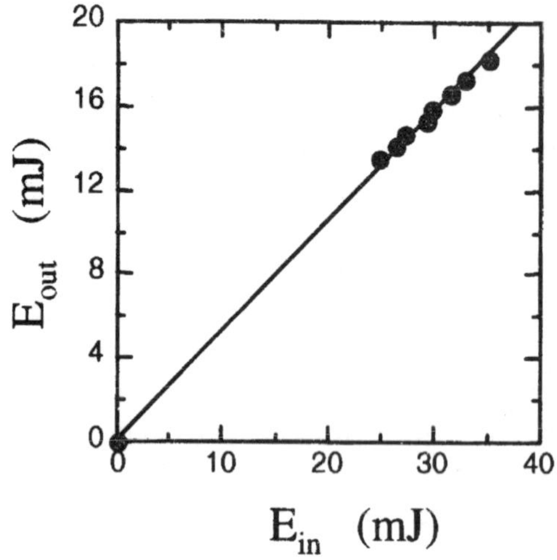

(a)

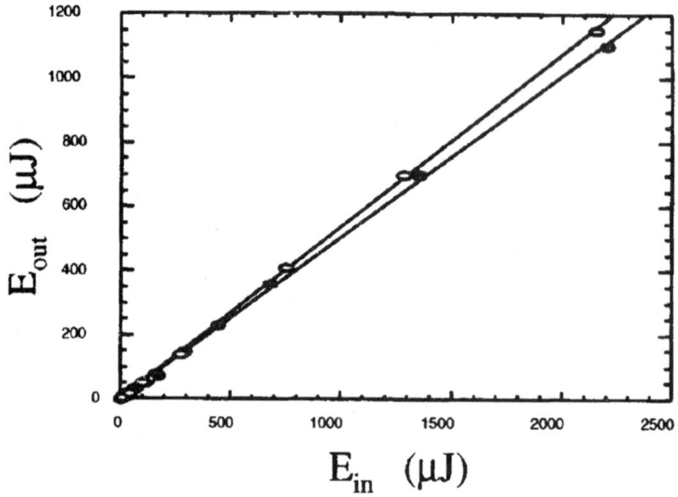

(b)

Figure 9.4 (a) Er:YAG laser energy delivery by 525/620-μm chalcogenide glass fiber. The maximum power density is 271 kW/cm^2. (b) Free electron laser pulse energy at the same wavelength of 2.94 μm as the Er:YAG laser through a sulfide glass fiber.[12] [Reproduced by permission of Taylor & Francis, Inc., http://www.routledge-ny.com.]

clad fibers were conducted by Awazu et al.[14] using an FEL operating between 5 and 7 μm.

The chalcohalide fibers discussed in Sec. 5.6.1 have also been used to transmit low-power CO_2 laser radiation. Smektala et al.[15] used a TeX glass fiber composed of $Te_2Se_3As_{4.5}I_{0.5}$ with a 600-μm core, 1 m in length, coupled to a tunable CO_2 laser for operation in the 9.3-μm band. This wavelength region was chosen because these fibers have a lower loss at 9.3 μm (0.7 dB/m) compared to 10.6 μm (>2 dB/m). They were able to transmit 2.6 W of 9.3-μm power through this fiber.

9.1.3 Germanate glass fibers

Germanate glass fibers are a relatively new class of IR-transmitting fibers which have proven to be very useful for the delivery of high-power Er:YAG laser radiation.[16] These glasses have not been discussed up to now, as their primary application so far seems to be in laser-power delivery, and, therefore, it is most appropriate to introduce them here. Heavy-metal germanate glass fibers are composed largely of heavy-metal oxides, with the major component being GeO_2. The heavy-metal oxides shift the IR absorption edge to longer wavelengths. That is, the germanate-oxide glasses have a cutoff somewhat greater than 3 μm, compared to the normal 2-μm cutoff for silica-based oxide glasses. One particularly attractive feature of these germanate glasses is that they have a higher glass transition temperature compared to either chalcogenide or HMFGs and, therefore, higher laser-damage thresholds. But the loss for the HMFG fibers is lower. As an example of their composition, a typical GeO_2-based glass fiber is composed of GeO_2 (30–76%)—RO (15–43%)—XO (3–20%), where R represents an alkaline-earth metal and X represents an element of Group IIIA.[17] The oxide-only germanate glasses have glass transition temperatures as high as 680°C, excellent durability, and a relatively high refractive index of 1.84. In Fig. 9.5, loss data from Infrared Fiber Systems, Silver Spring, MD, is given for a typical germanate glass fiber. These fibers have an exceptionally high damage threshold at 3 μm. Specifically, more than 20 W (2 J at 10 Hz) of Er:YAG laser power has been transmitted by these fibers.

9.1.4 Single-crystal fibers

Single-crystal sapphire fibers are potentially an excellent choice for Er:YAG laser-energy delivery. Their extremely high melting point (>2000°C), high bulk crystal LID threshold, and relatively low refractive index ($n = 1.72$ at 2.94 μm) are key

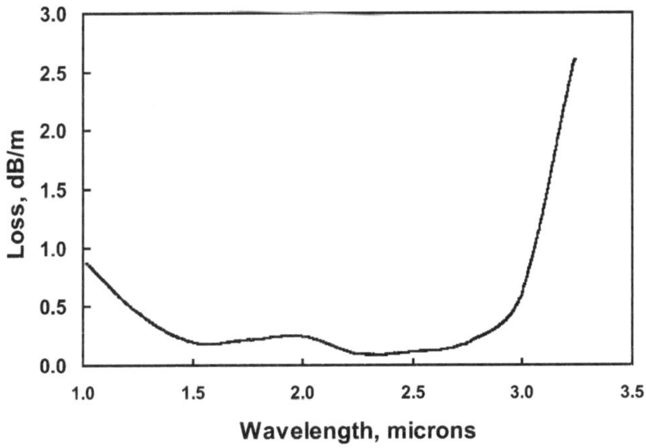

Figure 9.5 Low-loss germanate glass fibers made by Infrared Fiber Systems. These fibers have a particularly high Er:YAG laser damage threshold owing in part to their higher T_g than fluoride glass fibers.

to their ability to transmit high average power Er:YAG laser radiation. Most of the studies to date involve the use of multimode Er:YAG lasers with pulsewidths ranging from 100 to 200 μsec. The multimode output of these lasers is not the most desirable from the viewpoint of either minimizing laser damage or maximum transmission. In fact, it has been found that many Er:YAG lasers have hot spots, i.e., areas of high peak energy, in their spatial profile; and these hot spots often initiate the laser damage occurring at the fiber surfaces.

The earliest work on Er:YAG laser transmission through SC sapphire fiber is that of Jundt et al.[18] at Stanford University, using LHPG sapphire fibers with core diameters of 110 μm. The lowest loss for their fibers at 2.94 μm was 0.88 dB/m. The Er:YAG laser used in their measurements had a pulsewidth of 110 μsec and a repetition rate of 3 Hz. The maximum incident energy that could be launched without end-face damage was 130 mJ, corresponding to an average power of 0.39 W. Assuming uniform intensity over the input fiber end, this means that the damage threshold is greater than 1.2 kJ/cm^2.

The results by Jundt et al. were largely supported in more recent studies by Chang et al.[19] at the University of South Florida. Chang and his group also used LHPG fiber with diameters ranging from 100 to 300 μm for their power testing. The losses for their 100-μm-diameter fiber are given in Fig. 6.18. From this data it can be seen that the loss at 2.94 μm is approximately 0.6 dB/m. Chang et al. measured the energy transmitted from an Er:YAG laser through four different fiber diameters. Their data, along with that of Stanford, Saphikon, and Rutgers, is shown in Fig. 9.6. Their highest energy density was 1.3 kJ/cm^2 for the 100-μm-core fiber. To improve the energy throughput yet preserve the flexibility of their SC sapphire fibers, Chang et al.[19] grew a new, novel SC fiber with two flared ends. The purpose of flaring the fiber ends was to permit greater energy input and output. Specifically, they flared each end of a 100-μm-diameter fiber to 200 μm.

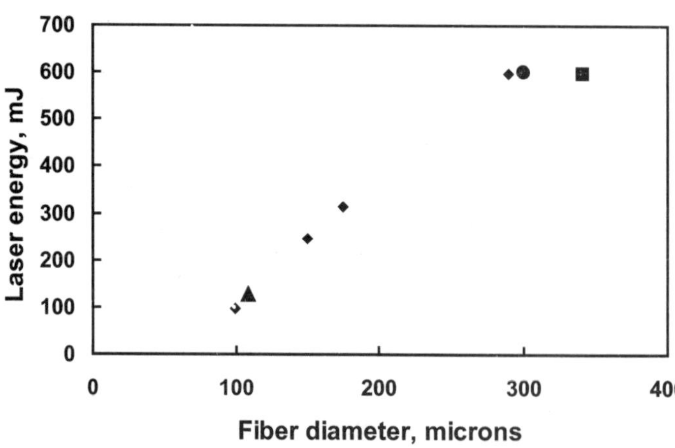

Figure 9.6 Laser damage threshold energies at the output end of SC sapphire fibers at the 2.94-μm Er:YAG laser wavelength. Data from (◆) USF, (▲) Stanford, (●) Saphikon, and (■) Rutgers.

Since the laser damage scales approximately as the area, a flared end will have a higher laser damage threshold. The small 100-μm-diameter core fiber is quite flexible.

LHPG fiber grown at Rutgers University has also undergone extensive power testing. The losses for the fiber grown by Nubling and Harrington[20] are given for the 3-μm region in Fig. 6.19. The lowest loss is approximately 0.2 dB/m at 2.94 μm. The fiber used for the power testing had losses of ~0.4 dB/m when a nearly Gaussian beam from their apertured Er:YAG laser was used as the input. Their Er:YAG laser had a pulsewidth of ~200 μsec and a repetition rate of 10 Hz. The results of their power testing are given in Fig. 9.7 for a 300-μm core, 1.08-m-long sapphire fiber.[21] Note that the data is in terms of average power in watts rather than energy, in mJ. Nubling and Harrington found that the loss measured for their fiber depended greatly on the modal properties of the laser. For example, the loss of 0.4 dB/m at 2.94 μm for a nearly Gaussian input beam increased to 1 dB/m when the full, unapertured multimode laser beam was used for the loss measurement. The situation becomes even worse as the laser power is increased. From 1 dB/m measured at 0.5 W, the loss increases monotonically to 2.5 dB/m at the highest input power of 10 W (1000 mJ).[21] What these investigators observe is a loss that is due to the increasing multimode character of the beam profile as the laser energy is increased. That is, the higher-order modes generated at the highest powers lead to increased absorption. Ideally, one would like a high-power (>10 W) Er:YAG laser that retains the modal quality of the low-power (~1 W) lasers. In practice it is not easy to find this well-controlled beam quality laser at high average powers. Nevertheless, the energy of 1000 mJ (10 W) at the input end of the fiber corresponds to an energy density of ~1.4 kJ/cm^2, which is in line with the results of Jundt et al. and Chang et al.

In addition, Nubling and Harrington irradiated this fiber (see Fig. 9.7) with 10 W input for 20 min (more than 10,000 pulses) with no degradation in transmis-

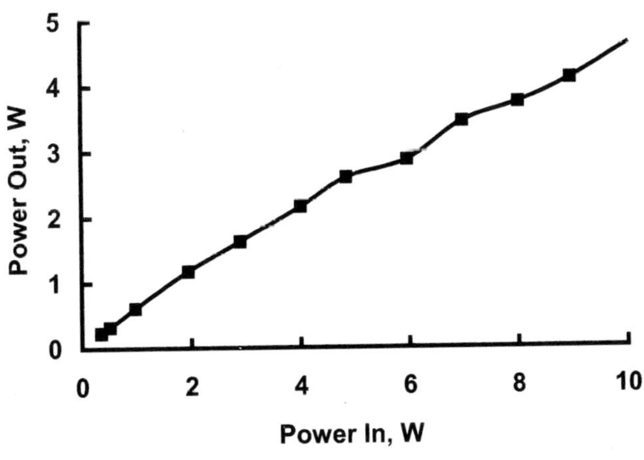

Figure 9.7 Transmission of Er:YAG laser power through a straight, 300-μm-diameter, 1.08-m-long sapphire fiber.[21] [Reprinted with permission.]

sion. They found that there was no damage to the output end of the fiber, but the input end showed some small pits due to laser damage. This laser damage is shown in the microphotographs in Figs. 9.8(a) and 9.8(b). Fig. 9.8(a) is a 300× photo showing that the damage pits are scattered randomly across the surface of the fiber end face. One of the pits is magnified 3000× in Fig. 9.8(b). Nubling and Harrington attribute these damage sites to hot spots in the laser beam. Oddly enough, damage to the input end did not lead to catastrophic failure of the fiber. Presumably, however, further exposure to the high energies of the laser beam would lead to a reduction in fiber transmission and ultimately to fiber failure.

In addition to the LHPG-grown fibers, EFG-grown SC sapphire fibers have also been carefully investigated as a potential Er:YAG laser delivery system for medical and dental applications. The loss for EFG fibers are higher than for the LHPG fibers, as may be seen in the data in Fig. 6.20. Nevertheless, they are capable of delivering similar laser energies as the lower-loss LHPG fibers. Clarke et al.[22] measured 300-μm-diameter EFG fibers grown by Saphikon using an average input power of up to 8.4 W. The maximum power delivered was about 6 W for a 35-cm-long fiber. This corresponds to an input energy density of 1.2 kJ/cm^2, again in line with the results of other investigators. They also found, as shown in other studies, that the damage threshold depends significantly on the polishing conditions of the end faces. They also found a somewhat curious effect during their damage measurements. Namely, there was a laser conditioning effect in which the LID threshold increased after the fiber was exposed to high-power laser irradiation. Specifically, the fiber end face was mechanically polished and sufficient laser energy was coupled into the fiber to cause front end-face damage. The fiber was then cleaved and repolished, and the damage threshold was observed to increase by at least 10%. Similar effects have also been observed in silica fibers.

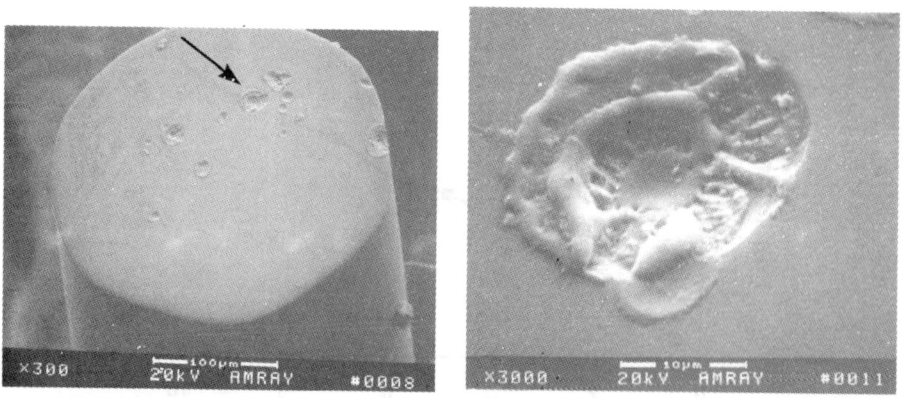

(a) (b)

Figure 9.8 Input-face surface damage to sapphire fiber from 10-W, 10-Hz Er:YAG laser at (a) 300× and (b) 3000× magnifications. The arrow in (a) points to the damage site shown in (b).[21] [Reprinted with permission.]

9.1.5 Polycrystalline fibers

The first extruded polycrystalline (PC) fibers were made from KRS-5 which, as may be seen from Fig. 6.10, transmits well at CO_2 laser wavelengths. In their classic paper detailing the first extrusion of KRS-5 fiber, Pinnow et al.[23] pointed out the potential of this PC fiber to transmit 10.6-μm laser power for surgical applications even though they reported only 2 W of transmitted power. Once they published their results, other investigators began to study not only KRS-5 but other crystalline fiber materials for the purpose of delivering CO_2 laser energy for a myriad of surgical applications. This was a particularly active quest because laser surgery was developing rapidly in the 1980s, especially using the Nd:YAG laser and readily available MM silica fiber delivery systems. It was clear that a flexible, small-size fiber optic that could replace the awkward and expensive articulated arm delivery systems employed on all surgical CO_2 lasers would be a very desirable medical instrument. Unfortunately, the initial euphoria over the use of KRS-5 fibers soon subsided once some of the limitations of these and other PC fibers were realized. In general, the problems for high laser power delivery included the low melting point of the thallium and silver halides, scattering at the output end of these fibers, and the toxicity of the thallium halides that limits their use in medical applications. What most researchers found when they tried to increase the output power of the CO_2 laser was that the output end of the PC fiber would heat up as a result of the high reflection from the output end face and scattering from small voids and surface imperfections. The heating of the end resulted in slight melting of the flat end to form a convex lens on the end face. This lensed end in turn refocused the light, melting the fiber from the output back toward the input end. It is quite exciting to observe this fiber fuse effect but ominous when one considers that the vaporized thallium halides are toxic.

Several years after the initial extrusion of KRS-5 fiber at HRL, Sakuragi et al.[24] working at Horiba in Japan greatly increased the power throughput of these fibers. Horiba researchers used a 1000-μm-core-only KRS-5 fiber to deliver 97 W of CO_2 laser power. The power density at the input end for this measurement was 36 kW/cm^2, and the fiber loss was 0.4 dB/m at 10.6 μm for fiber lengths less than 1 m. Sakuragi et al. also observed that the end finish was critical for high power damage thresholds, and that small damage spots ranging in size from 10 to 100 μm appeared on both fiber ends after exposure to the laser beam for about 7 s.

The highest powers for KRS-5 fibers, however, were obtained by Ikedo et al.[25] They performed a very thorough study of the effects of transmission, grain size, impurity size, and end finish on the power threshold for 500-μm-core KRS-5 fibers. Using their best 1.5-m-long fiber (loss of 0.128 dB/m or 93%/1.5 m at 10.6 μm) and a 200-W CO_2 laser, they were able to transmit 138 W of power for periods of up to 10 min. This corresponded to a power density of 66 kW/cm^2. This fiber also had a three-layer AR coating of As_2Se_3-KCl-As_2Se_3 on each fiber end face. Although the AR coating only reduced the Fresnel reflectance of the uncoated surface from 14%/surface to about 11%/surface for the AR-coated surfaces, the coatings did raise the damage threshold slightly. One reason for the less than expected improvement with the AR coatings was the damage to the As_2S_3 layer in the coating stack. Ikedo et al. also investigated the lifetime of their fibers under exposure to continuous CO_2 laser radiation. Not unexpectedly, they found that the lifetime was strongly dependent on the input power of the laser. Their lifetime data is shown in Fig. 9.9 for the 500-μm-diameter fibers with AR-coated ends. The lifetime is defined by Ikedo et al. to be the time until the fiber transmission decreases by 10% from its initial value. From their data it can be seen that the lifetime of 10 h for a 100-W incident beam can be increased dramatically to 1000 h when the input power is decreased to 40 W. A fiber optic cable containing KRS-5 fibers for both CO and CO_2 laser-power delivery has been described by Alejnikov et al.[26]

PC silver halide fibers have also undergone extensive power testing, again mostly for applications in laser surgery. Takahashi et al.[27,28] made a thorough study of mixed AgX fibers. They were able to extrude fiber with losses as low as 0.07 dB/m at 10.6 μm in core sizes ranging from 500 to 1500 μm. The results for their power testing are given in Fig. 9.10 for three different fiber core sizes. As expected, the data indicates that the power thresholds increase with increasing core area. Takahashi et al.[27] also did some lifetime testing of their fiber and found that they could transmit 28 W of CO_2 laser power for 230 h.

Katzir and his group at Tel Aviv University investigated both cw and pulsed CO_2 laser damage thresholds in their mixed AgX fibers.[29] Their results for cw radiation indicated that the fibers could transmit about 30 W (~40 kW/cm^2) of power without damage. When a pulsed TEA CO_2 laser is used as the input

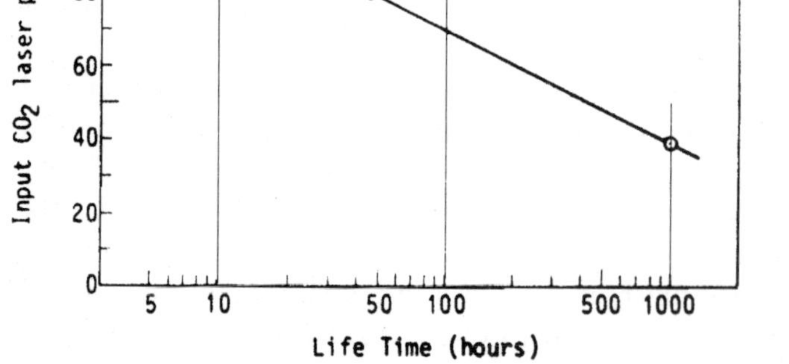

Figure 9.9 Lifetime of a 500-μm core, 1.5-m long KRS-5 fiber exposed to a cw CO_2 laser. The data represents the time that a fiber will transmit a maximum input power. For example, the lifetime is 1000 h for 40-W incident or 10 h for a 100-W incident beam. Note that the ends of the fiber are AR coated to minimize reflection losses.[25]

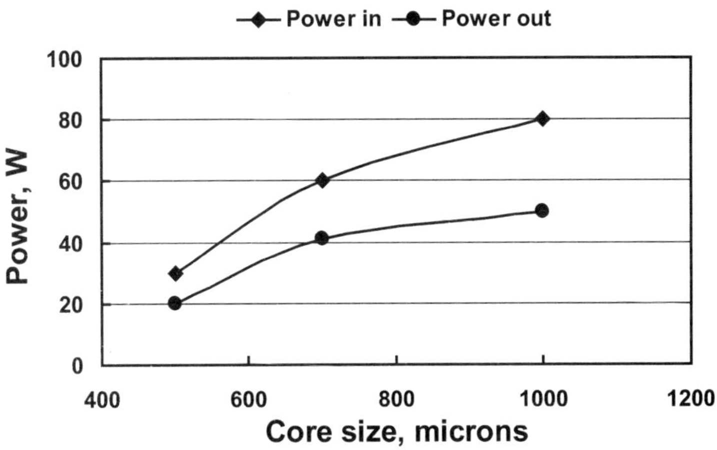

Figure 9.10 CO_2 laser power delivery though different bore size Ag halide fibers. The power density is greatest for the 500-μm core size fibers, being 42.4 kW/cm^2 at the input end and 10.2 kW/cm^2 at the output end.

source, laser-induced breakdown (LIB) occurs with only 6.8 mJ of energy. For their 60-ns-pulsewidth laser, this means that the LIB for the fiber tested was about 2×10^8 W/cm^2. This energy of 6.8 mJ/pulse is rather low when compared with the pulse energies for the fluoride glass fibers and the Er:YAG laser. However, Moser et al.[29] did observe that the LIB for energies from a long-pulse (0.1 to 100 ms) CO_2 laser were substantially higher than for the TEA laser. Specifically, they found that the LIB for their long-pulse laser was 0.3 J.

Finally, Alimpiev et al.[30] made similar studies on both clad and unclad AgX fibers using both cw and pulsed CO_2 lasers. Their results for cw radiation were from 7 to 10 kW/cm^2 for core-only AgBrCl PC fibers, and 14 kW/cm^2 for the core/clad fiber. These researchers also found a strong dependence on TEA CO_2 laser pulsewidth on the LIB for the fibers. For 10-μs pulses, LIB occurred with input energies ranging from 25 to 50 mJ, but the LIB increased to about 100 mJ for 40-μs input pulses.

In summary, PC fiber power delivery has been somewhat of a disappointment, as more was expected for these fibers when they were first extruded in 1978 at HRL. During the succeeding 10 years, many attempts were made to improve the optical quality of these fibers and to increase the cw laser power thresholds. The 1986 work of Takahashi et al. stands today as probably the highest power delivered by these fibers. So why are they not used in power delivery applications today? There are many fundamental and practical reasons for their lack of acceptance in laser surgery and related applications. First, PC fibers have low melting points and weak bending strength. These two characteristics have led to fibers which can easily be damaged in any application in which the fiber output end may be contaminated or for which the fiber undergoes repeated cyclic flexure. Continual bending can cause grain boundary separation, which leads to scattering and, ultimately, laser damage. However, one of the most serious problems is a practical one. That is, the fibers are exposed to contamination that results from the vaporization or ablation of material near the output end of the fiber. Even when the fibers are protected by a window, for example, AR-coated ZnSe, the output end can still become hot enough to melt slightly and self destruct. Additionally, these fibers all have high refractive indices of 2.0 or greater and, therefore, the ends are highly reflective. The reflected power, together with any contamination of the output end, can easily destroy the fiber. In light of all of these potential problems, it is still possible to use the fibers for low-power CO_2 laser transmission. As a rule of thumb, a safe operating level for cw power delivery is an input power of about 15 W or $\sim$10 kW/cm^2 for fiber core sizes of 800 μm or larger.

9.2 Hollow waveguides

An obvious approach to transmitting high average laser powers through fiber optics is to eliminate the solid core in favor of a hollow core. The intrinsic advantages of hollow waveguides are transmission through air or some other gas and minimal contact with optical material. In short, hollow-core fibers are in principle an ideal fiber for power delivery, as they minimize the interaction of damaging laser radiation with optical materials. Furthermore, a hollow-core fiber should be able to withstand short-pulse laser radiation as well as significant levels of cw power. In the latter case, Nubling and Harrington[31] have calculated that if the waveguides are sufficiently cooled, then it should be possible to transmit many kilowatts of CO_2 laser power through the guides.

9.2.1 Rectangular waveguides

The work that initiated the interest in using hollow waveguides to transmit power was described in the classic review paper by Garmire et al.[1] in 1980. In that paper, Garmire and her group at USC reported the transmission of >200 W of CO_2 laser power through the rectangular waveguides made using two aluminum metal strips held apart by shim stock spacers, as described in Sec. 7.3.1. The first rectangular waveguides had a very simple construction, as they were made from commercial-grade aluminum strips held together by clamps. In the next few years, Garmire refined these rectangular guides somewhat by using polished aluminum strips, and with these guides and a more powerful laser she reported output powers of 840 W with an input power of 960 W for guides with a 0.5 $\times$ 10 mm cross section. Her work showed for the first time that one could indeed reliably transmit nearly kilowatt laser power, and these results helped launch the study of high power laser delivery in hollow guides.

The early work of Garmire was followed by the work of Kubo and his group at Kinki University in Japan. Kubo et al.[32] fabricated 0.5 $\times$ 8 mm cross-section guides using aluminum strips held apart with plastic spacers. Their structure is shown in Fig. 7.2(a). The entire waveguide was then overcoated with plastic for protection and to give the appearance of a finished product. Most of their testing was done using a TEA CO_2 laser with an 80-ns pulsewidth and an energy of 1.8 J/pulse. In one set of tests they were able to transmit 36 mJ (450 kW peak power) with a 40 mJ (500 kW peak power) input energy. The waveguides were 1 m long. Kubo also used the same waveguides very successfully to transmit KrF (248 nm) excimer laser energies.

9.2.2 Circular waveguides

Rectangular cross-section waveguides, as mentioned in Sec. 7.3.1, have the disadvantage of not being flexible in all directions and, furthermore, they have rather large dimensions for some applications. For example, they are essentially unsuitable for least-invasive surgical procedures as they are just too big, but they could work fine as delivery systems for industrial lasers. Instead, circular cross-section waveguides are the clear choice, as they have smaller dimensions and they can be quite flexible.

Hongo et al.[33,34] at Hitachi Cable in Japan did a very thorough study of the early Miyagi-type hollow metallic waveguides (see Sec. 7.3.2) for high-power CO_2 laser delivery. Their ultimate application was industrial laser welding and cutting; therefore, they configured their waveguides to be able to work on 3-kW industrial laser systems. The waveguides studied were similar to those shown in Fig. 7.4, except that the inner coating was Ge over Ag rather than ZnSe over Ag. Since their objective was high power, they used large-bore-size guides with diameters of 1700 and 1500 μm. The waveguides, which were essentially Ni tubes with inner reflective Ge/Ag coatings, were 2 m in length. To protect the guides from overheating at high laser powers, they surrounded the guides with a water jacket that had gold-plated input ends to further reduce the chance of laser damage at the input end. Finally, they used an Ar-gas purge through the inner bore to provide additional cooling. A diagram of the hollow-guide delivery system is shown in Fig. 9.11. The source laser was a slightly multimode 3-kW CO_2 laser.

The results of their work are impressive. As shown in Fig. 9.12, they were able to transmit 2.6 kW (114 kW/cm^2) through a 1700-μm-bore waveguide and 2.1 kW (118 kW/cm^2) through a 1500-μm guide for a period of 30 s. Hongo et al.[34] observed that the loss for the guides is somewhat higher (see Fig. 9.12) than that measured if a TEM_{00}-input laser beam was used as the source laser. The higher loss is a result of the multimode output of their 3-kW laser. The full angle beam divergence of both waveguides was between 60 to 70 mrad. The investigators also made an analysis of bending losses and the modal properties of the output beam when the guides were both straight and bent 90°. Specifically, they showed that the output of the waveguide is multimode. This presents a problem for many cutting and welding applications, as most of these applications require a good beam profile, such as a Gaussian or top-hat profile for proper materials processing. For example, the kerf needs to be sharp in many cutting applications. This is usually accomplished using a laser with small M^2 value. However, if that same laser beam is transmitted through a hollow guide in which mode mixing occurs, then the kerf

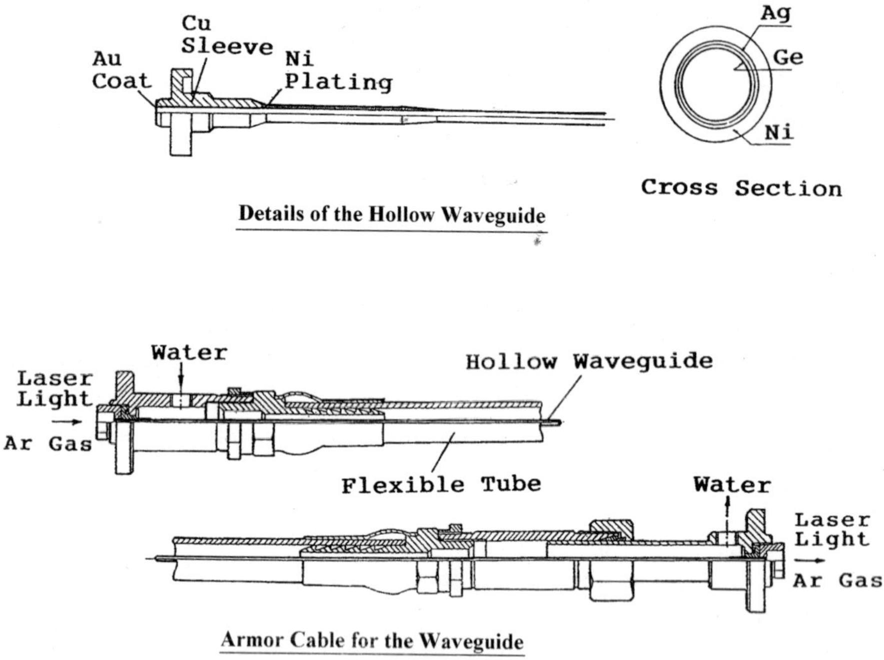

Figure 9.11 Structure of hollow waveguide and armor cabling with a water jacket used for high-power CO_2 laser delivery.[34] [Reprinted with permission.]

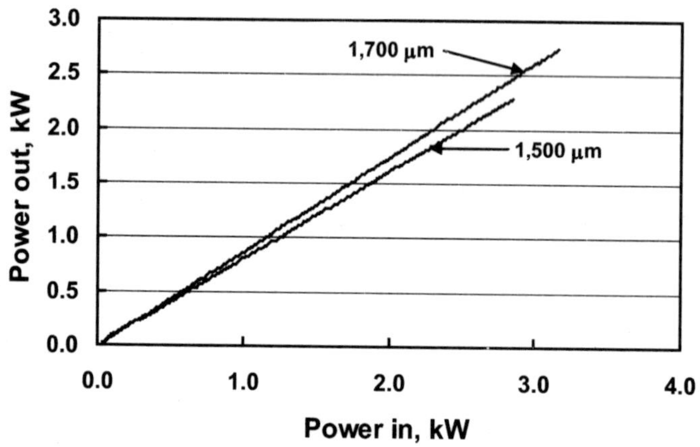

Figure 9.12 Highest recorded power transmission by a hollow waveguide. The 1700-μm bore guide, packaged in the water-cooling jacket shown in Fig. 9.11, was 2 m long and delivered a maximum power of 2.6 kW.[34] [Reprinted with permission.]

will not be sharp and clean. In fact, it is the poor modal properties of the output beam rather than the laser power that can be delivered that has so far limited the acceptance of hollow waveguides in industrial laser applications. As an example of how the guides might be used in a welding application, Hongo et al. used their waveguide to deliver CO_2 laser power to weld two 1-mm-thick steel plates at a welding speed of 2 m/s. For this test the input laser power was 2.6 kW.

The work of Hongo and Garmire shows that it is possible to transmit very high laser power through hollow guides that are larger than 1 mm and for guides whose structure is essentially a metal tube with or without coatings. Nubling and Harrington[35] studied the ability of the HGWs (see Sec. 7.3.4), and the hollow sapphire, $n < 1$ waveguides (see Sec. 7.3.5), to transmit high-CO_2 laser power. As in the work of Hongo, Nubling and Harrington fitted their hollow guides with a water jacket for cooling. These authors calculated the heat transfer in a water-jacketed hollow guide and found that the temperature of the guides decreases exponentially from the input to the output end of the guide. For example, if a cooling flow of 5 l/min at 18°C is used, then the input end of the fiber would not rise above 200°C for either the HGWs or the sapphire guides for input powers up to 5 kW. To prevent damage to the input end of the guides, these investigators used an inverted, gold-coated cone over the input end. This turned out to perform very well, as any excess laser power was directed away from the glass or sapphire tube.

Nubling and Harrington used three CO_2 lasers for the high-power testing: 1.5 kW, 2.0 kW, and 3.0 kW. The output of the two highest-power lasers was a mix of $TEM_{00} + TEM_{01}$ (D-mode with an estimated $M^2 \sim 2$). The D-mode is a popular mode for industrial applications, but it does not couple as well into hollow guides. Furthermore, the 2- and 3-kW laser beam diameters are a function of laser output power; that is, the diameter of the beam increases with increasing power. This effectively lowers the f/# as the power increases, which in turn leads to higher-order mode loss in the guides. The output of the 1.5-kW laser was nearly single mode, with an M^2 between 1.2 and 1.5.

Hollow sapphire waveguides with bore sizes of 1070 and 790 μm, each with a length from 1.2 to 1.5 m, were tested using the three CO_2 lasers. The losses for the 1070-μm-bore guides were 0.83 and 0.99 dB/m when measured with the 1.5- and 3.0-kW lasers, respectively. The results for both the large- and small-bore straight sapphire guides are shown in Fig. 9.13. There are several important points to take from the data. First, the losses depend on which laser was used for the measurement. The higher-order modes present in the output of the higher-power lasers lead to higher waveguide loss. In fact, the loss for the 1070-μm bore guide measured using a TEM_{00} input beam from a low-power probe laser is 0.45 dB/m or below the value obtained with the 1.5-kW laser. Second, the maximum transmitted power for the waveguides was 1.8 kW (200 kW/cm^2) and 825 W (168 kW/cm^2) for the 1070 and 790-μm bore guides, respectively. The power density for these guides is higher than that obtained by Hongo et al., yet the researchers felt that if they had a better 3-kW laser output beam quality and different optics, then even higher power levels could be achieved. Finally, Fig. 9.13 also shows the multimode output from the two waveguides. The output is somewhat better for the smaller-bore guide as expected, but it is still multimode. As mentioned above for the circular metallic waveguides, this multimode output is deleterious for many industrial welding and cutting operations.

The hollow glass waveguides were only tested using the 1.5-kW laser. As may be seen in the data for the HGWs in Fig. 9.14, the highest power delivered through a straight 1.5-m-long 700-μm-bore waveguide was 1.01 kW (262 kW/cm^2). This waveguide had a loss of 0.46 dB/m. The data in Fig. 9.14 also shows the loss increase when the guide is bent into a single loop with a 15-cm radius. The loss of the bent fiber increased to 0.89 dB/m. As for the hollow sapphire guides, the

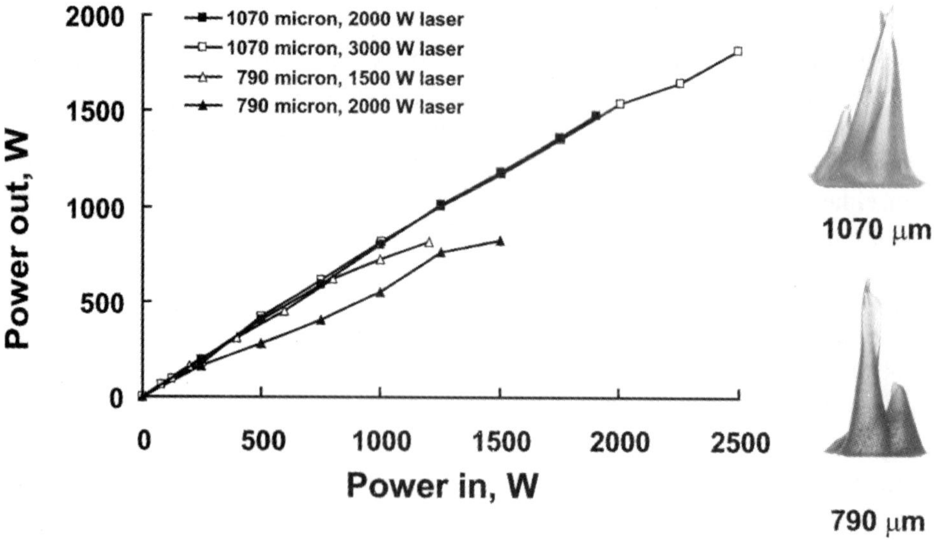

Figure 9.13 CO_2 laser power delivery by hollow sapphire waveguides using different high-power laser sources.[35] [Reprinted with permission.]

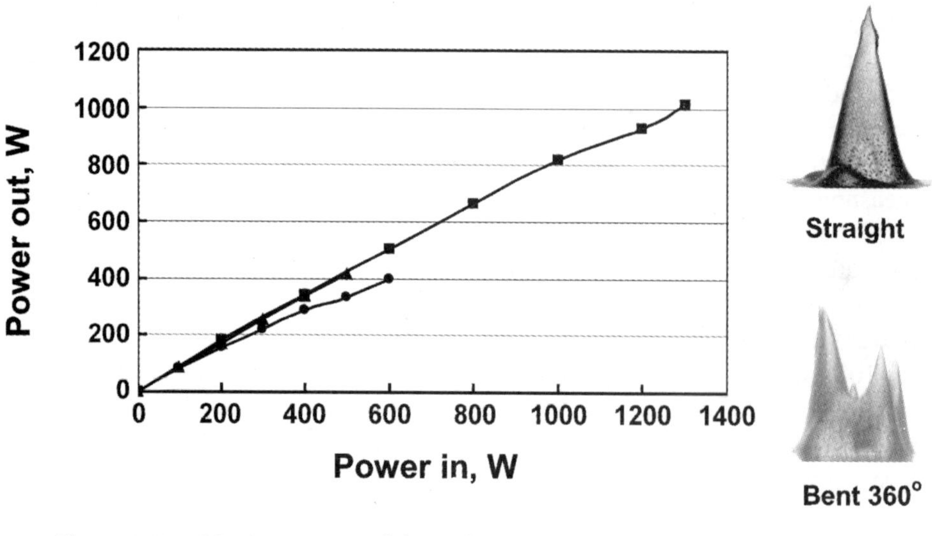

Figure 9.14 CO_2 laser power delivery by 700-μm bore HGWs, 1.54-m long in a water-cooled jacket. Maximum power delivered by straight guide was 1010 W; (■) straight, (▲) 180° bend, (○) 360° bend.[35] [Reprinted with permission.]

HGW output was multimode. From Fig. 9.14, however, it can be seen that the mode remains nearly single mode when the guide is straight, but the spatial profile becomes decidedly more multimode on bending. It should be noted that the power density for this HGW is the highest recorded to date for any hollow waveguide.

The laser powers described above are clearly extremely high. These results are mainly presented to indicate how advantageous hollow guides can be in the delivery of laser power that is typical of what may be used in heavy industrial applications. Many other power applications, however, require much less power. For these low power needs, the data in Fig. 9.15 should serve as a useful reference. This data is for AgI/Ag-coated HGWs for three small-bore sizes measured using an amplitude-stabilized TEM_{00} CO_2 laser source with a maximum input power of about 85 W. None of the waveguides had any cooling such as a water jacket or a gas purge. The guides were straight, 1 m in length, and the output was essentially single mode. As expected, the small-bore guides delivered less power, but the modal output for the smallest bore (250 μm) was excellent.[36] Furthermore, these power levels could be sustained for reasonably long time periods of approximately one hour. The issue of mode mixing and modal purity was discussed in Sec. 7.3.4.1. As mentioned in this section, the best approaches to preserve and propagate the lowest-order mode are to use a thicker glass wall or a small-bore waveguide.

HGWs have been used very successfully to transmit high peak powers from TEA CO_2 lasers. Dai and Harrington[37] made a series of measurements on 700-μm-bore HGWs using a 100-ns-pulsewidth TEA laser with a maximum peak power of 0.7 MW and a maximum average power of about 30 W. They found that they were able to reliably deliver 50–100 mJ/pulse energy through the HGWs when no purge gas was used, but that the LID threshold increased substantially when a 3-l/min N_2 purge flowed in the bore of the guide. Specifically, with the gas purge they were able to deliver 300–350 mJ/pulse energies or an average power of 24 W without

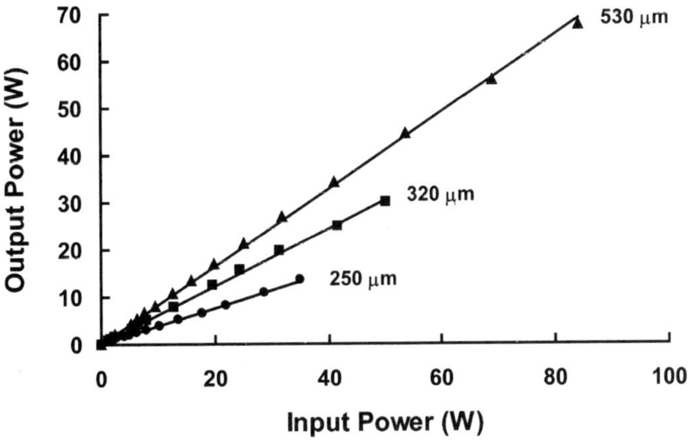

Figure 9.15 CO_2 laser power delivery by uncooled HGWs for low-power applications.[38] [Reproduced by permission of Taylor & Francis, Inc., http://www.routledge-ny.com.]

damage. The gas purge served to prevent the air breakdown that occurred at lower energies in the unpurged guides.

Hollow plastic waveguides are not capable of delivering such high CO_2 laser power as are their glass counterparts. Naturally, they are more vulnerable to laser damage, because of the low-melting polymer used as the supporting tube. Power testing by both Croitoru[39] and his group at Tel Aviv University and by Harrington[40] and his group at Rutgers University, revealed that most of the plastic waveguides would not withstand more than about 25 W for any length of time. Harrington and his group did test some of Croitoru's larger-bore guides and found that they could deliver up to 65 W through a 1850-μm-bore plastic guide, but at this power level the input end of the guide would soon melt. Also the rougher inner surface of the plastic waveguides led to even more mode mixing than the HGWs. It is felt that the best applications for the hollow plastic guides are in thermal or chemical sensing.

Hollow waveguides are also quite useful in delivering other IR laser wavelengths as well as the CO_2 laser energy. Kozodoy et al.[41] have studied the transmission of HGWs illuminated by Er:YAG (2.94 μm) and by Er:YSGG (2.79 μm) pulsed lasers. As mentioned in conjunction with the delivery of Er:YAG laser by fluoride glass fibers in Sec. 9.1.1, the output beam quality of the 3-μm solid state lasers is normally not single mode and, therefore, the loss of hollow guides is greater than for a low-order mode launch. The power transmission data for different bore size HGWs is shown in Fig. 9.16. The Er:YAG laser used by Kozodoy et al. had a repetition rate of 10 Hz and a pulsewidth of about 180 μs. A maximum pulse energy of 800 mJ (8 W average power) was transmitted through the 1000-μm-bore guide. Watanabe et al.[42] were able to deliver similar average powers of 2.94-μm light through their ZnS/Ag metallic hollow guides (see Sec. 7.3.2 for the structure of these guides). Their Er:YAG laser had a maximum pulse energy of 1.6 J at a 5-Hz repetition rate. This represented an average 8 W of power at the out-

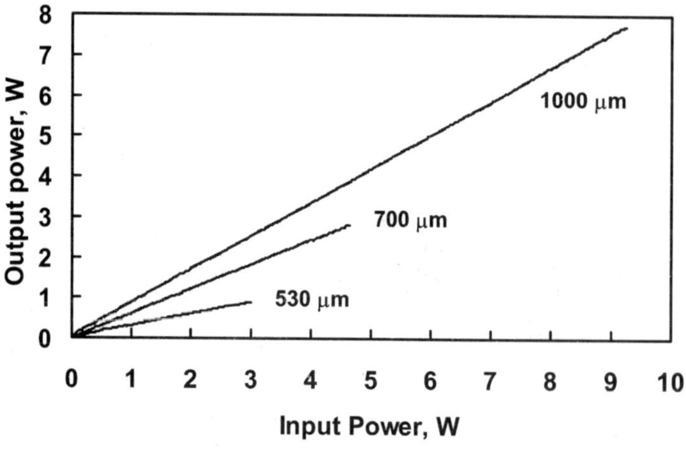

Figure 9.16 Average Er:YAG laser power delivery by HGWs. The input Er:YAG laser beam was multimode which is typical for these lasers when delivering this high average power.[41] [Reprinted with permission.]

put of the 800-μm-bore guide. Matsuura and Miyagi[43] used their ZnS/Ag coated metallic hollow guides to deliver Er:YAG, CO, and CO_2 laser power through the same guide. They were able to accomplish this through a judicious choice of the thickness of their ZnS coating. With the proper ZnS thickness they were able to achieve low loss at all three laser wavelengths; i.e., the minimum in the interference spectrum occurred at the three laser wavelengths. More recently, Matsuura et al.[44] used their small, 250-μm-bore HGWs composed of COP/Ag coatings (see Sec. 7.3.4) for the transmission of 108 mJ (1.08 W average power) of Er:YAG laser energy.

Several other lasers of interest for transmission through hollow guides are the free-electron laser (FEL), the Nd:YAG laser operating at 1.06 μm[45] and at the harmonic frequency of 0.532 μm, and short-pulse Ti:sapphire lasers. Matsuura et al.[46] used 530-, 700-, and 1000-μm-bore HGWs coated with AgI/Ag films to transmit 40- and 80-mJ pulses from a FEL operating at 6.45 μm. No damage to the guides was observed at these input energies. Matsuura et al.[47] studied the use of COP/Ag-coated HGWs to propagate Q-switched Nd:YAG laser pulses. In their setup, they evacuated the core of the guide to suppress air breakdown and, thus, increase the LID threshold. Using a 1000-μm-bore guide 1 m in length, they were able to transmit 158 mJ/pulse (20.1 J/cm^2) at 1.06 μm. This vacuum-core guide was then used successfully to ablate myocardium tissue. The same waveguides were also used by Matsuura et al. to deliver Q-switched second harmonic Nd:YAG laser energy at 532 nm. Instead of evacuating the waveguide they purged the core with He gas. This eliminated the air breakdown and was simpler in construction than the vacuum-core guide. The maximum output energy transmitted at 532 nm using their Q-switched Nd:YAG laser was 75 mJ for the 1000-μm bore guide. Finally, it is of interest to consider the transmission of ultrashort pulses through hollow guides as one would expect that the pulse dispersion through these guides would be significantly less than it would be for solid-core fibers. Matsuura et al.[48] studied pulse dispersion in HGWs using a femtosecond Ti:sapphire laser operating at 775 nm. The pulsewidth of the laser source was 196 fs and the COP/Ag guides had bore sizes of 700 and 1,000 μm. There was no damage to the guides using 700-μJ pulses at a rep rate of 1 kHz. Matsuura et al. found that the pulse broadening was 17 fs for a 1-m-long waveguide.

9.3 Power applications

Laser power applications for IR fiber optics involve the delivery of high powers ($>$100 W) for industrial cutting, welding, soldering, etc., applications and the delivery of low- or mid-level ($<$100 W) laser radiation for laser surgery. The need for a truly flexible, reasonably low loss fiber analogous to MM silica fibers for transmitting laser wavelengths greater than 2 μm has been a goal and a dream since the earliest days of IR fiber research at HRL in 1976, when there were visions of delivering tens of watts of CO_2 laser power for a myriad of laser surgical modalities. In 1977 your author was involved in the first joint agreement to transfer the HRL

IR fiber technology to a commercial product. An agreement was reached between HRL and American Hospital Supply to make a KRS-5 fiber optic probe to transmit only a few watts of 10.6-μm power for ophthalmic applications. This effort turned out to be much more challenging than originally anticipated, and a commercial IR fiber optic was never developed during this partnership. The reasons for this were the aging effects or the reduction of the fiber's transmission with time (see Sec. 6.1.4), the low laser damage threshold, the toxicity of the thallium halide fiber, and the fragile nature of these fibers in general. Subsequently, PC fibers have been perfected sufficiently to enable them to be adequately protected in a catheter and successfully used for low-power surgical specialties. At the time of the first KRS-5 fibers, however, the initial euphoria soon melted away once the difficulties of the surgical environment were fully appreciated.

9.3.1 Medical applications of IR fibers

One of the first successful IR fiber cables for CO_2 laser power delivery was made using KRS-5 fibers by a team of researchers at Matsushita Electric in Japan. Their unclad KRS-5 fiber had a diameter of 500 μm, and it was 1.5 m in length. The ends of the fiber were AR coated, and the fibers were constructed so that they would safely deliver from 20 to 40 W of laser power. Ishiwatari et al.[49] took great care in cabling their fiber so that the ends would not be damaged by laser power. Their first step was to place the KRS-5 fiber in a loose-fitting Teflon tube. The loose-tube concept minimizes losses to the wall of the tube while also allowing for expansion of the fiber in the jacket.[23] This plastic tube was surrounded by a metal sleeve for mechanical strength. The input connector, shown in Fig. 9.17(a), was rather complex. As may be seen from the sketch, the input end was left unsupported for a short distance. This was done to increase the input power level, as an end in a conventional fiber connector could be damaged by any vaporized connector material that might be deposited on the fiber end face. A ZnSe window covered the front face of the connector so that the fiber would be protected from any debris and as a

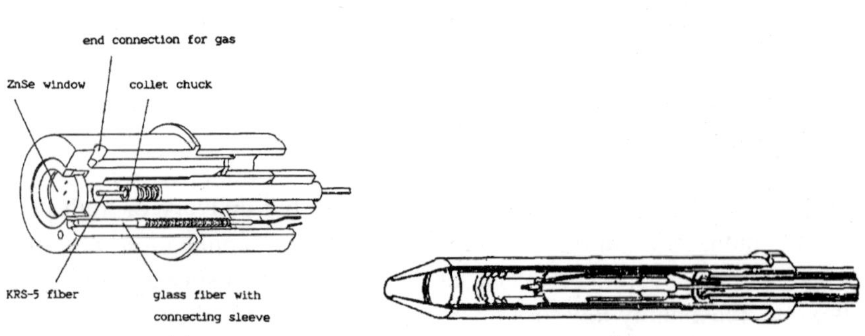

(a) (b)

Figure 9.17 KRS-5 fiber optic cable for CO_2 laser delivery showing special connector (a) and handpiece (b) with short focal length lens.[49] [Reprinted with permission.]

means of sealing the connector and cable to enable an N_2 gas fill. The connector also contains two silica fibers for transmission of the HeNe aiming beams and an input for the purge gas. The output end of the cable was terminated in a handpiece, shown in Fig. 9.17(b), which contained a ZnSe lens for focusing both the 10.6-μm and HeNe laser light onto tissue. The two silica fibers delivering the red aiming beams were set so that they crossed after the focusing lens in the same focal plane as the 10.6 μm light. This allowed the surgeon to visualize the focal point of the CO_2 laser beam by moving the probe to and from the tissue.

One of the key criteria for their probe was complete safety if the fiber should break or be damaged while transmitting laser power. To this end they made certain that the probe was totally sealed and that the bend radius was restricted by the metal jacket so that the fiber could not be bent too much. For further safety they built into the input end of the probe a fiber damage detector. This detector consisted of a ring detector that surrounded the fiber input end. For a normal fiber, virtually no light is reflected from the input end into the ring detector, but if a break occurs in the fiber significant energy is reflected into this outer-ring sensor to trigger a laser shutdown. The fiber/laser probe described by Ishiwatari et al. is really quite sophisticated. It incorporates many features designed to sense and to prevent failure of the fiber. Their probe is meant to be used in external procedures. That is, it is to replace the cumbersome and costly articulated arms typically found on surgical CO_2 lasers. A smaller catheter would need to be constructed if the fiber were to be introduced into the body for least-invasive procedures.

A very similar but less complex fiber probe was made by Takahashi and his team at Sumitomo Electric in Japan.[27,28] In their cable they used Ag-halide fibers packaged first in a Teflon tube that was then protected in a metal, interlocking cable for strength. The input end of the fiber was terminated in a special connector while the output radiation from the fiber was focused by a ZnSe lens at the end of the cable. Again, this probe was too large to be used inside the body, but it could replace articulated arms.

Artjushenko et al.[30,50] and his group at the General Physics Institute in Moscow cabled both Ag-halide and KRS-5 fibers for CO (5.25 μm) and CO_2 laser power delivery. One of their Ag-halide fiber probes is shown in Fig. 9.18 coupled to a small, 7-W CO_2 laser made by Asah Medico in Denmark. This laser was designed to be portable and with a sufficiently low power to be safely transmitted by these PC fibers.

Today it is more common to see hollow waveguides coupled to CO_2 lasers, as the PC fibers cannot reliably transmit sufficient power for many surgical applications. Usually the surgical CO_2 laser with a hollow-guide delivery system has a power less than 40 W, and the fiber permanently replaces the articulated arm. An example of a simple fiber/laser system employing a metallic silver-tube hollow waveguide (see Sec. 7.3.2) is the Sharplan surgical laser shown in Fig. 9.19. This laser has a small footprint and an output power of only 15 to 20 W. Nevertheless, from Fig. 9.19 it is clear that the delivery of power via a fiber rather than an articulated arm (not shown) is more appealing and also more cost effective from the

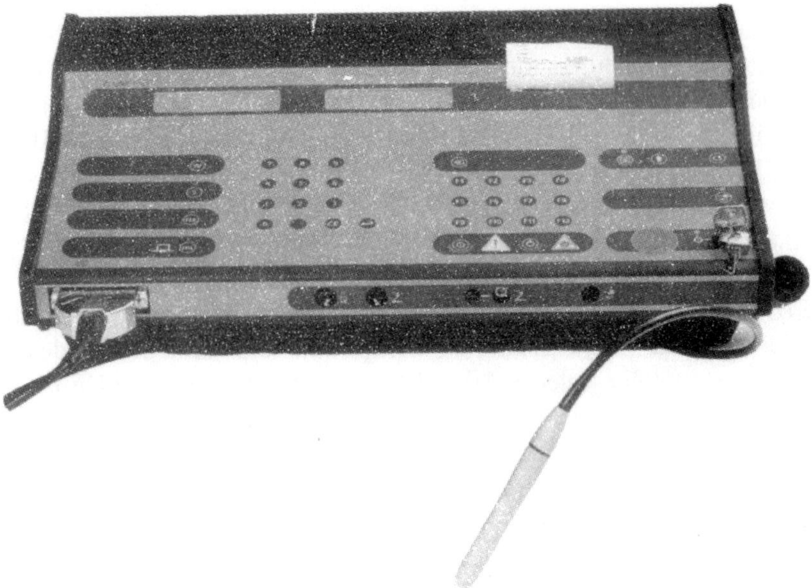

Figure 9.18 Low-power, portable CO_2 laser with Ag halide PC fiber delivery system.

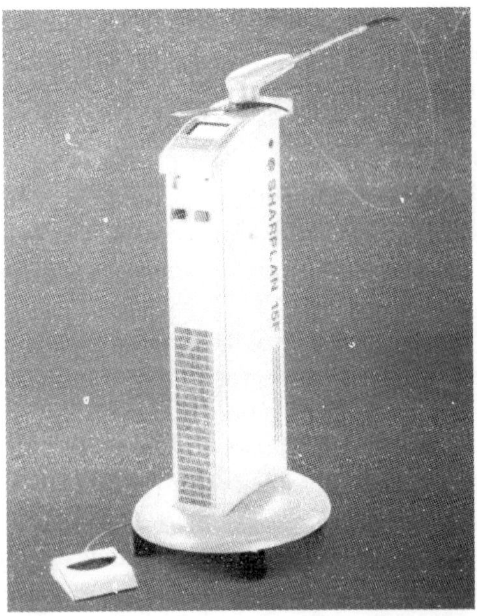

Figure 9.19 15-W surgical CO_2 laser with a hollow waveguide delivery system made by Sharplan Industries (now ESC Medical).

manufacturer's viewpoint. Another example is the HGW cable made by Clinicon shown in Fig. 9.20. The handpiece for their probe is rather novel. The 530-μm-bore HGW terminates at the end of a small, type IIA diamond blade. CO_2 laser

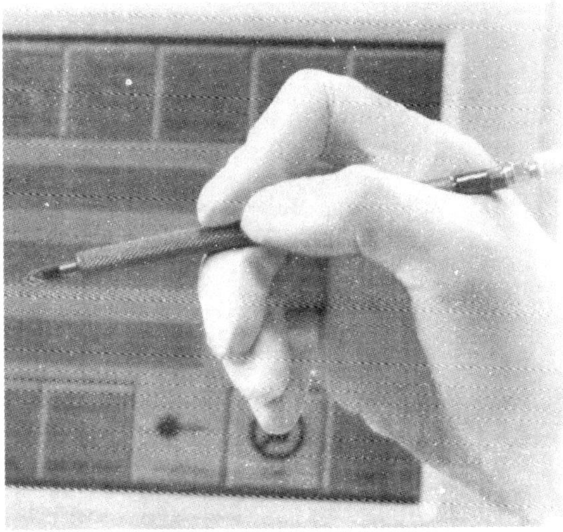

Figure 9.20 Light scalpel employing a diamond blade coupled to a low-power CO_2 laser via a HGW. The type IIA diamond blade transmits 10.6-μm radiation for both cutting and coagulating tissue [from Clinicon, Inc.].

radiation is transmitted by type IIA diamond and, in this case, the laser energy is used to cauterize and cut tissue as it exits the blade. The diamond blade itself is extremely sharp, so that it is the major cutting tool. The surgeon can turn the laser energy on when needed to cauterize bleeding tissue.

The Er:YAG laser operating at 2.94 μm is an excellent choice for both hard and soft tissue ablation. The 2.94-μm wavelength is highly absorbed in tissue because this wavelength is near the peak absorption of water, and water is the major component of most tissue. These lasers are therefore useful for making very clean cuts in tissue and for cutting and ablating hard tissue such as teeth and bone. One example of a commercial Er:YAG laser is that made by Hoya/Continuum. This laser, shown in Fig. 9.21, employs a fluoroaluminate fiber delivery system. This fluoride fiber (see Sec. 4.4.2.2) has a diameter of 450/500/600 μm and a loss at 2.94 μm of 0.1 dB/m. The laser itself delivers 30- to 350-mJ pulses up to 10 Hz. Again it can be seen from Fig. 9.21 that the footprint of the laser is small and that the fiber is protected so that it may not be bent less than the minimum-bending radius of 3 cm.

9.3.2 Miscellaneous power applications of IR fibers

Most applications for power delivery discussed so far relate to high-power industrial cutting and welding or lower-power ablation and cauterizing of tissue. There are other applications, however, that do not require much power and, thus, can be quite suitable for IR fibers. One of these is the use of IR fibers to transmit low levels of laser power for marking. Dekel et al.[51] studied the use of both Ag-halide and HGWs for the delivery of CO_2 laser power to mark plastic and paper. They needed only 1 W or less for marking. To rapidly articulate the fiber output end, they fixed

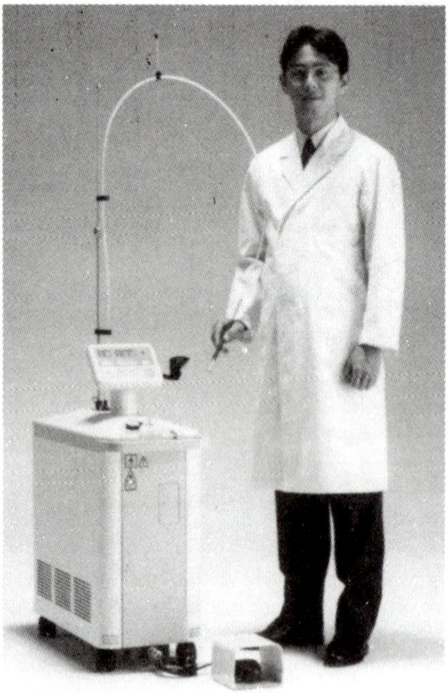

Figure 9.21 Medical Er:YAG laser using an AlF_3-based fluoride glass fiber delivery system. This laser/fiber device is made by Continuum/Hoya Corp.

small permanent magnets to the fiber end and then placed the fiber plus magnets in a bidirectional electromagnet. The current in the electromagnet was computer controlled to produce scanning velocities up to 5 mm/s over a scanning trajectory of 10 mm. Both fiber types seemed to work well in this application. Current laser marking systems usually do not employ a fiber optic; rather, the laser beam is scanned by movable mirrors. A fiber system could be of value in that the fiber can carry power to remote locations that are often hostile to moving mirrors. Harrington, in cooperation with Domino Laser Corp., looked at several methods of using HGWs to transmit low-power CO_2 laser radiation to remote locations for marking packages. The best method at the time involved the use of eight HGWs coupled to eight small CO_2 lasers. In this scheme the fiber output ends are not moved but left in a linear array, and the laser power to each fiber is computer controlled to produce the desired mark.

References

1. E. Garmire, T. McMahon, and M. Bass, "Flexible infrared waveguides for high-power transmission," *IEEE J. Quantum Electron.*, Vol. QE-16, pp. 23–32 (1980).
2. R. F. Bonner, L. G. Pervosti, M. B. Leon, K. Levin, and D. Tran, "New source for laser angioplasty: Er:YAG laser pulses transmitted through zirconium fluoride optical fiber catheters," Proc. SPIE, Vol. 906, pp. 288–293 (1988).

3. C. Whitehurst, M. R. Dickinson, A. Charlton, T. A. King, and P. France, "Transmission of 2.94 μm laser radiation by zirconium fluoride glass optical fibers," Proc. SPIE, Vol. 1048, pp. 141–144 (1989).
4. S. Wuthrich, W. Luthy, and H. P. Weber, "Optical damage thresholds at 2.94 μm in fluoride glass fibers," *Appl. Opt.*, Vol. 31, pp. 5833–5837 (1992).
5. K. Itoh, K. Miura, M. Masuda, M. Iwakura, and T. Yamagishi, "Low-loss fluorozirco-aluminate glass fiber," Proc. of 7th International Symposium on Halide Glass, pp. 2.7–2.12 (1991).
6. K. Itoh, K. Miura, M. Masuda, M. Iwakura, and T. Yamashita, "Low-loss fluorozirco-aluminate glass fiber," *J. Non-Cryst. Solids*, Vol. 167, pp. 112–116 (1994).
7. N. Anastasopoulou, C. Ziolek, A. A. Serafetinides, and H. Lubatschowski, "Q-switched Er:YAG radiation transmission through fluoride glass fibers and dielectric-coated metallic hollow waveguides," *Opt. Commun.*, Vol. 186, pp. 167–171 (2000).
8. T. Arai, M. Kikuchi, M. Saito, and M. Takizawa, "Power transmission capacity of As-S glass fiber on CO laser delivery," *J. Appl. Phys.*, Vol. 63, pp. 4359–4364 (1988).
9. S. Sato, S. Watanabe, and T. Fujioka, "High power, high intensity CO infrared laser transmission through As_2S_3 glass fibers," *Appl. Phys. Lett.*, Vol. 48, pp. 960–962 (1986).
10. S. Sato, K. Igarashi, M. Taniwaki, K. Tanimoto, and Y. Kikuchi, "Multihundred-watt CO laser power delivery through chalcogenide glass fibers," *Appl. Phys. Lett.*, Vol. 62, pp. 669–671 (1993).
11. J. Nishii, S. Morimoto, I. Inagawa, R. Iizuka, T. Yamashita, and T. Yamagishi, "Recent advances and trends in chalcogenide glass fiber technology: a review," *J. Non-Cryst. Solids*, Vol. 140, pp. 199–208 (1992).
12. J. Sanghera, L. B. Shaw, L. Busse, V. Nguyen, P. Pureza, B. Cole, B. B. Harbison, and I. Aggarwal, "Development and infrared applications of chalcogenide glass optical fibers," *Fiber and Integrated Optics*, Vol. 19, pp. 251–274 (2000).
13. T. Arai, M. Kikuchi, M. Saito, and M. Takizawa, "Simultaneous delivery of Nd:YAG and CO laser beams by a single As-S glass fiber," *Appl. Phys. Lett.*, Vol. 51, pp. 1887–1888 (1987).
14. K. Awazu, S. Ogino, A. Nagai, and T. Tomimasu, "Midinfrared free electron laser power delivery through a chalcogenide glass fiber," *Rev. Sci. Instrum.*, Vol. 68, pp. 4351–4352 (1997).
15. F. Smetktala, K. LeFoulgoc, L. LeNeindre, C. Blanchetiere, X. H. Zhang, and J. Lucas, "TeX-glass infrared optical fibers delivering medium power from a CO_2 laser," *Opt. Mater.*, Vol. 13, pp. 271–276 (1999).
16. S. Kobayashi, N. Shibata, S. Shibata, and T. Izawa, "Characteristics of optical fibers in infrared wavelength region," *Rev. Electrical Comm. Lab.*, Vol. 26, pp. 453–467 (1978).
17. D. Tran, U.S. Patent No. 5,274,728 (Issued 12-28-1993).

18. D. H. Jundt, M. M. Fejer, and R. L. Byer, "Characterization of single-crystal sapphire fibers for optical power delivery systems," *Appl. Phys. Lett.*, Vol. 55, pp. 2170–2172 (1989).
19. R. S. F. Chang, V. Phomsakha, and N. Djeu, "Recent advances in sapphire fibers," Proc. SPIE, Vol. 2396, pp. 48–53 (1995).
20. R. Nubling and J. A. Harrington, "Optical properties of single-crystal sapphire fibers," *Appl. Opt.*, Vol. 36, pp. 5934–5940 (1997).
21. R. Nubling and J. A. Harrington, "Single-crystal LHPG sapphire fibers for Er:YAG laser power delivery," *Appl. Opt.*, Vol. 37, pp. 4777–4781 (1998).
22. G. Clarke, D. Chadwick, R. Nubling, and J. A. Harrington, "Sapphire fibers for three micron delivery systems," Proc. SPIE, Vol. 2396, pp. 54–59 (1995).
23. D. A. Pinnow, A. L. Gentile, A. G. Standlee, A. J. Timper, and L. M. Hobrock, "Polycrystalline fiber optical waveguides for infrared transmission," *Appl. Phys. Lett.*, Vol. 33, pp. 28–29 (1978).
24. S. Sakuragi, Y. Kubo, K. Imagawa, H. Kotani, T. Morikawa, and J. Shimada, "KRS-5 optical fibers capable of tansmitting high-power CO_2 laser beam," *Opt. Lett.*, Vol. 6, pp. 629–631 (1981).
25. M. Ikedo, M. Watari, F. Tateishi, and H. Ishikawa, "Preparation and characteristics of the TlBr-TlI fiber for a high power CO_2 laser beam," *J. Appl. Phys.*, Vol. 60, pp. 3035–3039 (1986).
26. V. S. Alejnikov, V. G. Artjushenko, V. P. Belyaev, V. V. Vojtsekhovsky, E. M. Dianov, I. S. Lisitsky, L. N. Butvina, V. I. Masychev, T. N. Savenkova, and V. K. Sysoev, "Fibre-optic cable for CO and CO_2 laser power transmission," *Optics and Laser Technology*, Vol. 17, pp. 213–214 (1985).
27. K. Takahashi, N. Yoshida, and K. Yamauchi, "Silver halide infrared fiber," *Sumitomo Electric Tech. Rev.*, Vol. 26, pp. 371–378 (1987).
28. K. Takahashi, N. Yoshida, and M. Yokota, "Optical fibers for transmitting high-power CO_2 laser beam," *Sumitomo Electric Tech. Rev.*, Vol. 23, pp. 203–210 (1984).
29. F. Moser, D. Bunimovich, A. DeRowe, O. Eyal, A. German, Y. Gotshal, A. Levite, L. Nagli, A. Ravid, V. Scharf, S. Shalem, D. Shemesh, R. Simchi, I. Vasserman, and A. Katzir, "Medical applications of infrared transmitting silver halide fibers," *IEEE J. Selected Topics Q. Electron.*, Vol. 2, pp. 872–879 (1996).
30. S. S. Alimpiev, V. G. Artjushenko, L. N. Butvina, S. K. Vartapetov, E. M. Dianov, Y. G. Kolesnikov, V. I. Konov, A. O. Nabatov, S. M. Nikiforov, and M. M. Mirakjan, "Polycrystalline IR fibres for laser scalpels," *Int. J. Quantum Electron.*, Vol. 3, pp. 333–344 (1988).
31. R. Nubling and J. A. Harrington, "Hollow-waveguide delivery systems for high-power, industrial CO_2 lasers," *Appl. Opt.*, Vol. 34, pp. 372–380 (1996).
32. U. Kubo, Y. Hashishin, and K. Okada, "Optical beam guides for medical CO_2 and excimer lasers," Proc. SPIE, Vol. 906, pp. 214–219 (1988).
33. A. Hongo, K. Morosawa, T. Shiota, K. Suzuki, and S. Iwasaki, "Transmission of 1 kW-class CO_2 laser light through circular hollow waveguides for material processing," *Appl. Phys. Lett.*, Vol. 58, pp. 1582–1584 (1991).

34. A. Hongo, K. Morosawa, K. Matsumoto, T. Shiota, and T. Hashimoto, "Transmission of kilowatt-class CO_2 laser light through dielectric-coated metallic hollow waveguides for material processing," *Appl. Opt.*, Vol. 31, pp. 5114–5120 (1992).
35. R. K. Nubling and J. A. Harrington, "Hollow-waveguide delivery systems for high-power, industrial CO_2 lasers," *Appl. Opt.*, Vol. 34, pp. 372–380 (1996).
36. Y. Matsuura, T. Abel, J. Hirsch, and J. A. Harrington, "Small-bore hollow waveguide for delivery of near single-mode IR laser radiation," *Electron. Lett.*, Vol. 30, pp. 1688–1690 (1995).
37. J. Dai and J. A. Harrington, "High-peak-power, pulsed CO_2 laser light delivery by hollow glass waveguides," *Appl. Opt.*, Vol. 36, pp. 5072–5077 (1997).
38. J. Harrington, "A review of IR transmitting, hollow waveguides," *Fiber and Integrated Optics*, Vol. 19, pp. 211–227 (2000).
39. P. H. Cossmann, V. Romano, S. Sporri, H. J. Altermatt, N. Croitoru, M. Frenz, and H. P. Weber, "Plastic hollow waveguides-propeties and possibilities as a flexible radiation delivery system for CO_2-laser radiation," *Lasers in Surgery & Medicine*, Vol. 16, pp. 66–75 (1995).
40. R. George and J. A. Harrington, "Hollow plastic waveguides for sensor applications," Proc. SPIE, Vol. 4204, pp. 230–237 (2001).
41. R. L. Kozodoy, A. T. Pagkalinawan, and J. A. Harrington, "Small-bore hollow waveguides for delivery of 3-μm laser radiation," *Appl. Opt.*, Vol. 35, pp. 1077–1082 (1996).
42. T. Watanabe, M. Miyagi, S. Schrunder, T. Ertl, G. Muller, and L. Beerstecher, "Transmission of 1.6 J Er:YAG laser light by ZnS-coated silver hollow waveguides," *Optics and Laser Technology*, Vol. 27, pp. 389–391 (1995).
43. Y. Matsuura and M. Miyagi, "Er:YAG, CO, and CO_2 laser delivery by ZnS-coated Ag hollow waveguides," *Appl. Opt.*, Vol. 32, pp. 6598–6601 (1993).
44. Y. Matsuura, Y. W. Shi, Y. Abe, M. Yaegashi, G. Takada, S. Mohri, and M. Miyagi, "Infrared-laser delivery system based on polymer-coated hollow fibers," *Optics and Laser Technology*, Vol. 33, pp. 279–283 (2001).
45. H. Jelinkova, J. Sulc, P. Cerny, Y. W. Shi, Y. Matsuura, and M. Miyagi, "High-power Nd:YAG laser picosecond pulse delivery by a polymer-coated silver hollow-glass waveguide," *Opt. Lett.*, Vol. 24, pp. 957–959 (1999).
46. Y. Matsuura, K. Matsuura, and J. A. Harrington, "Power delivery of free electron laser light by hollow glass waveguides," *Appl. Opt.*, Vol. 35, pp. 5395–5397 (1996).
47. Y. Matsuura, G. Takada, T. Yamamoto, Y. W. Shi, and M. Miyagi, "Hollow fibers for delivery of harmonic pulses of Q-switched Nd:YAG lasers," *Appl. Opt.*, Vol. 41, pp. 442–445 (2002).
48. Y. Matsuura, M. Miyagi, K. Shihoyama, and M. Kawachi, "Delivery of femtosecond pulses by flexible hollow fibers," *J. Appl. Phys.*, Vol. 91, pp. 887–889 (2002).
49. H. Ishiwatari, M. Ikedo, and F. Tateishi, "An optical cable for a CO_2 laser scalpel," *J. Lightwave Tech.*, Vol. LT-4, pp. 1273–1279 (1986).

50. V. G. Artjushenko and E. M. Dianov, "Flexible cables for IR laser power delivery," Proc. SPIE, Vol. 1201, pp. 254–266 (1990).
51. B. Dekel, A. Inberg, N. Croitoru, S. Shalem, and A. Katzir, "Hollow glass waveguides and silver halide fibers as scanning elements for CO_2 laser marking systems," *Opt. Eng.*, Vol. 39, pp. 1384–1390 (2000).

Chapter 10

IR Fiber Amplifiers

10.0 Introduction

One of the most important technical advances in fiber optic telecommunication systems has been the optical fiber amplifier (OFA). The use of an all-optical fiber optic amplifier instead of a semiconductor optical amplifier has many inherent advantages. For example, the most popular erbium-doped fiber amplifiers (EDFAs) have a broad range of optical amplification in the 1550-nm window; these erbium-doped fibers are easy to incorporate directly into an all–fiber optic telecommunication system. Today the erbium-doped silica fibers are the most common rare-earth (RE)-doped fiber amplifiers, as they have high gain and low noise. EDFAs are quite versatile in that they may be used as preamplifiers, in-line and booster amplifiers, as well as sources. They can provide more than 20 dB of gain across a broad wavelength band, which enables them to be extremely useful as components in dense wavelength-division multiplexed (DWDM) systems. Additionally, they may be pumped with 800-, 980-, and 1480-nm sources and their quantum efficiency can be quite high.

There are many excellent references to OFAs, including several books on EDFAs.[1,2] In addition there has been a plethora of RE ions doped into silica glass, and their emission spectra and use as OFAs have been thoroughly studied. Partly as a consequence of the success of rare-earth active oxide glass fibers, it was natural to study the introduction of some of the same RE ions used so successfully in oxide glasses to dope the newer fluoride glasses. The motivation behind this concept was to take advantage of the lower phonon frequencies of the fluorides compared to those of oxide glasses. That is, an IR transparent glass like the HMFGs will have fewer phonons to contribute to nonradiative processes. Nonradiative transitions take energy away from the desired radiative or laser transitions and, therefore, lead to lower quantum efficiencies. It would seem, then, that merely choosing a host glass with a lower phonon frequency would lead to a much more efficient amplifier. Unfortunately, there are other considerations that have limited the use of the IR fiber hosts as a replacement for the oxide glass EDFAs. In fact, today the best application of the fluoride or other IR fiber hosts is for amplification in the 1300-nm telecom band. That is, erbium ions provide amplification in the 1550-nm band, so other ions are needed for the 1300-nm band. It is hoped that fluoride or chalcogenide glasses doped with ions like Pr^{3+} and Dy^{3+} may prove to be efficient OFAs for operation at the shorter telecom wavelength band.

After a brief review of OFAs in general, the emphasis in this chapter will turn to the RE doping in IR glass fibers and some more recent work on the use of the Ag-halide fibers as hosts for RE ions. One of the most intriguing hosts is the chalcogenide glasses, as they have even lower phonon frequencies than the fluoride glasses. However, it will be seen that one problem with the chalcogenide glasses is the low solubility of rare earths in these glass hosts. Nevertheless, doped IR fibers hold great potential as 1.3-μm amplifiers, as well as sources in the IR for chemical sensing. As a result, this is still an active field of investigation.

10.1 Brief review of optical fiber amplifiers

Fiber optic amplifiers operate on the same principle as a laser except that there is no external optical cavity as there is for a laser. The active lasing medium is a host ion, which in the case of the OFAs is a RE ion like Er^{3+} added to the host glass or crystal structure. Specifically, erbium ions in concentrations of about 1000 ppm-weight % are added to the core of an SM fiber to make an EDFA. A source of pump energy is provided by a diode laser. An energy-level diagram for Er^{3+} in silica is shown in Fig. 10.1. The pump laser light from either a 980- or 1480-nm source laser excites the Er ions to an excited state where they make a rapid, nonradiative decay to a long-lived metastable state, $^4I_{13/2}$. From there they undergo spontaneous or stimulated emission to the lower-energy $^4I_{15/2}$ state. The incoming signal, with wavelengths spanning the range from 1500 to greater than 1600 nm, will be amplified by stimulated emission. An example of how an EDFA can be used as an in-line amplifier is shown simply in Fig. 10.2. In this case, the incoming signal light at, say, 1550 nm is pumped with either a forward or backward diode

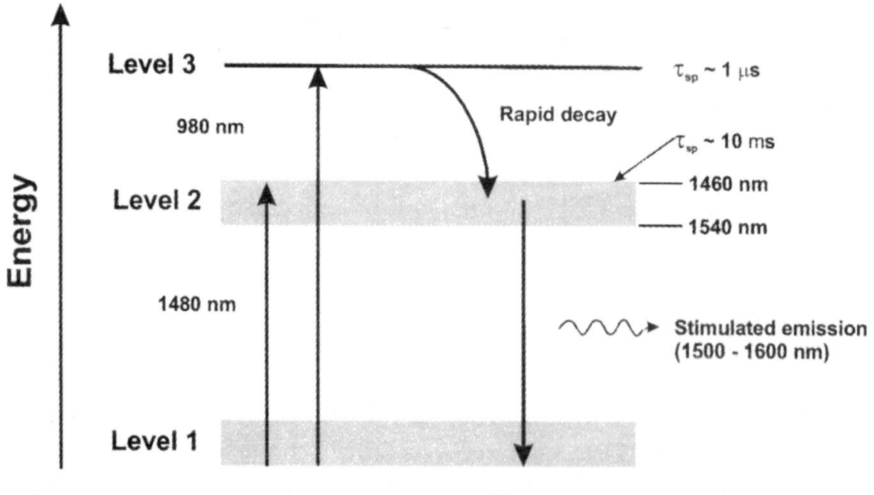

Figure 10.1 Three-level energy diagram for Er^{3+} in silica showing pump bands at 1480 nm and 980 nm and emission from metastable level 2 to ground state level 1.

laser. Isolators located before and after the EDFA prevent the pump signal from being transmitted along with the amplified signal.

A key to the operation of a broadband OFA is a wide emission band covering the important telecommunication bands. The stimulated emission cross section for Er^{3+} ions doped into several different glass hosts is shown in Fig. 10.3.[3] This data indicates the wavelength region for lasing or amplification by stimulated emission. From Fig. 10.3 it can be seen that the host fluoride glass and silica fibers have about the same cross section over approximately the same wavelength range. The emission band is broader for the telluride glass, and this is an advantage for increasing the number of channels in DWDM applications.

The most efficient lasing action occurs when depopulation of the metastable state leads to radiative transitions. However, there are competing transitions that can also depopulate the metastable state and, therefore, reduce the number of stim-

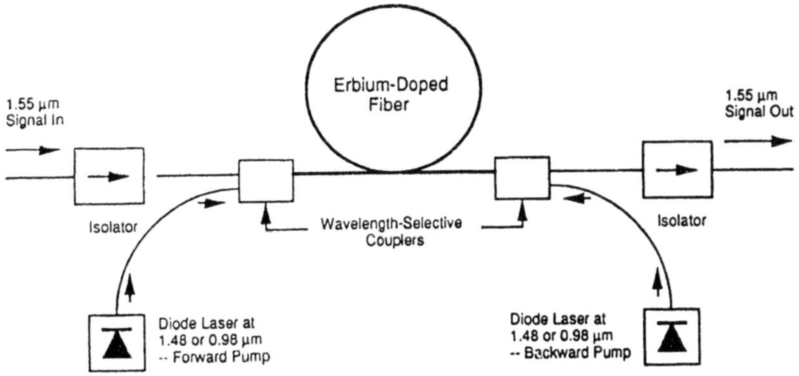

Figure 10.2 General telecommunications application of an EDFA configured for either forward or backward pumping. [Reproduced from J. Buck, "Optical Fiber Amplifiers," *Handbook of Optics, Vol. IV*, 2001, with permission from The McGraw-Hill Companies.]

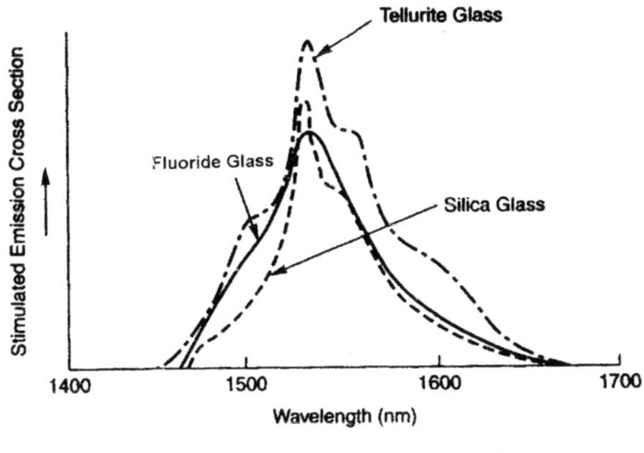

Figure 10.3 Simulated emission cross section for Er^{3+}-doped fibers for various glass hosts.[3] [Reprinted by permission of Pearson Education, Inc., Upper Saddle River, NJ.]

ulated photons emitted. These nonradiative transitions can dissipate energy as heat or phonons. Energy can also be lost from the metastable state as a result of excited state absorption (ESA) in which the pump light excites Er^{3+} ions from the metastable state to higher-energy states. Phonons with the highest frequencies are most likely to participate in the nonradiative process. This is because the dissipation of the ion's energy into only a few phonons with high energy is a low-order process and, therefore, more probable than higher-order processes involving a greater number of low-frequency phonons. Therefore, host glasses with a low-phonon-frequency spectrum, i.e., the IR glasses, will have relatively lower nonradiative relaxation rates compared to the oxide glasses. For example, the maximum phonon energy in silica glass is 1000 cm^{-1} versus 460 cm^{-1} for ZBLA glass. The nonradiative transition rate is greatly reduced for the ZBLA glass as a result of this difference in maximum phonon frequencies.

Another advantage of using IR fiber hosts for RE doping is that it is possible to obtain lasing or amplification at wavelengths longer than 1.55 μm, as well as at 1.3 μm where the EDFAs do not operate. For example, doping with Tm^{3+}, Er^{3+}, and Ho^{3+} ions in fluoride glass fibers results in laser operation at 2.3, 2.71, and 2.9 μm, respectively. While these are not important telecommunication wavelength bands, they are nevertheless potentially important for power delivery in microsurgical applications as well as convenient sources in the 2- to 3-μm region. When IR fibers are doped with Pr^{3+}, the fibers can be used to provide amplification at the important 1300-nm telecom band.

10.2 Fluoride glass optical fiber amplifiers

Fluoride glasses were seen to be potentially excellent hosts for RE doping as early as 1978, when these hosts were being viewed as solid-state laser materials.[4,5] One particularly attractive feature of the fluoride glass hosts is the large bandwidth of the emission spectra compared to the oxide glasses. In the excellent review article by Miniscalco,[1] he shows the normalized emission band for Er^{3+} in oxide and fluoride glass hosts. His data is reproduced in Fig. 10.4, where it can be clearly seen that the bandwidth is much larger for the heavier fluorohafnate compared to the silicate glasses. However, the emission spectra and bandwidths shown in Fig. 10.4 may be influenced by reabsorption, so instead it is more valuable to look at the gain bandwidth. The gain bandwidth refers to the bandwidth of the actual gain that can be expected in a high-gain amplifier. The data in Table 10.1, taken from Miniscalco, shows that the FWHM bandwidth at 1550 nm for which the gain remains above some value, say 20 dB or more, can be significantly greater for the fluoride compared to the silica glasses.

HMFG has been doped with a wide range of RE ions, including the more important Pr^{3+}, Er^{3+}, Ho^{3+}, Tm^{3+}, and Nd^{3+} ions. Spirit et al.[6] doped a 7.1-μm- core fluorozirconate fiber with 500 ppm Er^{3+} to amplify light in the 1550-nm band. A 1485-nm pump laser was used, and the length of the Er-doped fiber was 5 m. Using this fiber amplifier they were able to achieve a gain of 18 dB from 1530

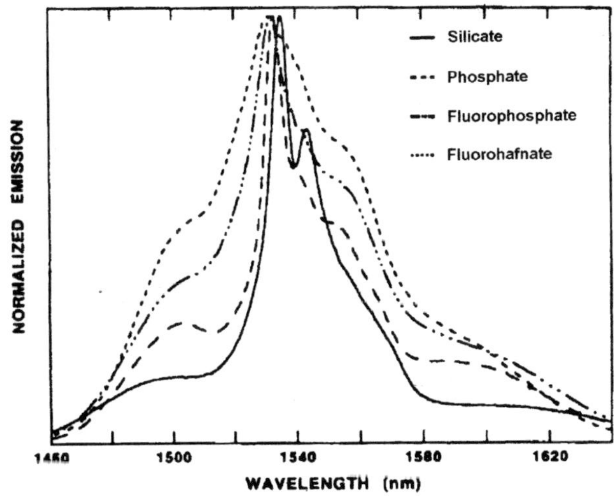

Figure 10.4 1500-nm emission spectra for different glass hosts doped with Er^{3+}. Note the variation in shape and linewidth for the various host glasses.[1] [Reprinted with permission.]

Table 10.1 1500-nm bandwidths for some Er^{3+}-doped glasses.

Glass	Gain Bandwidth, nm			
	FWHM, nm	20 dB peak	30 dB peak	40 dB peak
Al/P-doped silica	43.3	7.9	6.4	5.3
Fluorozirconate	63.3	16.4	11.7	9.0
Fluorohafnate	63.8	17.3	13.2	10.4

to 1565 nm. This range compares favorably with the 40-nm range from an Er-doped, Al co-doped silica fiber. The only advantage to the fluoride host is that the gain spectrum is flatter than it is for silica hosts. The efficiency for the Er-doped ZBLAN or silica fiber hosts is about the same.

While the results of Spirit et al. are encouraging, they are not appreciably better than those for Er-doped silica fibers. Furthermore, silica has much more favorable physical properties than fluoride glass, and, therefore, recent research on fluoride fiber amplifiers has turned from 1550 nm to the important, shorter- wavelength telecom band at 1300 nm where EDFAs do not operate. Probably the most important ion for fiber amplification at 1300 nm is Pr^{3+}. The energy level diagram for Pr^{3+} is shown in Fig. 10.5.[5] The important transition for amplification is $^1G_4 \rightarrow {}^3H_5$, but the $^1G_4$ metastable state is largely depopulated by nonradiative (phonon) transitions; therefore, the quantum efficiency of this transition is only a few percent. While this system is not as efficient as the EDFAs, it nevertheless is a promising 1300-nm OFA. Ainslie et al.[5] have doped ZBLAN fibers with 1000 ppm Pr^{3+} ions and studied their performance at 1300 nm. The gain bandwidth for a 24-m-long fiber longitudinally illuminated with a 1.047-μm 300-mW pump source is given in Fig. 10.6.[5] The peak of the gain curve is at 1300 nm and

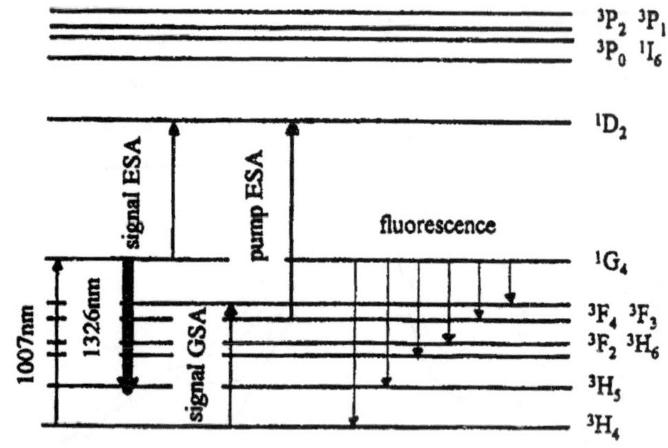

Figure 10.5 Energy level diagram for Pr^{3+} for emission into the 1300-nm bands. [Reprinted from Ref. 5, Copyright 1995, with permission from Elsevier.]

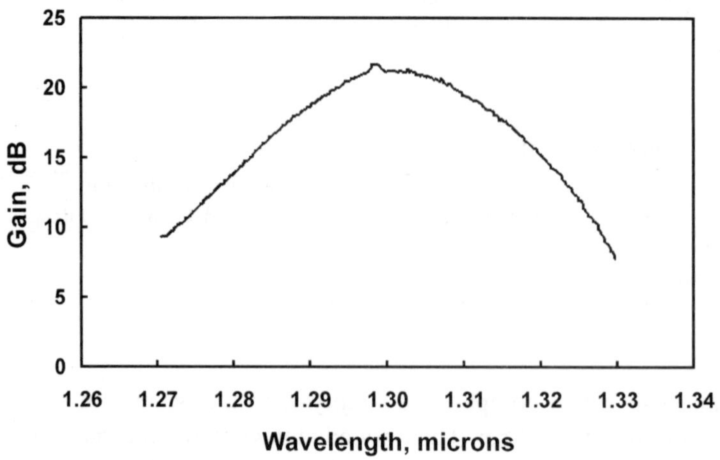

Figure 10.6 Small signal gain for Pr^{3+}-doped fluoride glass fiber amplifier with a 300-mW pump input power. [Reprinted from Ref. 5, Copyright 1995, with permission from Elsevier.]

the 3-dB bandwidth is 20 nm. This bandwidth is compatible with current telecom wavelengths.

Not all RE-doped fluoride fibers have applications as amplifiers for telecommunications systems. Fiber lasers were made from Nd^{3+}-doped HMFG in 1987 by Brierly and France.[7] These fiber lasers operated at 1.05 μm. A more efficient fiber laser was reported a year later by Brierly and France[8] when they used Er-doped fluorozirconate fibers. The Er-doped MM fiber, 75 cm in length, was pumped by 476.5-nm radiation from an Ar ion laser. The cw output from the $^4I_{11/2} \rightarrow {}^4I_{13/2}$ transition gives lasing action at 2.7 μm. Allen et al.[9] also achieved lasing action at the same wavelength for Er:ZBLAN fibers. In their case they used a much more convenient 792-nm laser diode to pump the fiber. The maximum output power was about 2 mW and the efficiency was 8%. Finally, a more efficient and powerful fluoride fiber laser was reported by Percival et al.[10] using a fluoride fiber co-doped with Ho^{3+} and Tm^{3+}. Lasing action takes place from a transition in Ho. The Tm ions are pumped by diode laser light between 820 and 830 nm and then the energy is transferred to the Ho ions. The maximum cw output power was 250 mW at 2.024 μm, with a maximum slope efficiency of 60%.

10.3 Chalcogenide glass optical fiber amplifiers

The chalcogenide glass fibers are also a potentially excellent host for RE ions, not only for the 1300- and 1550-nm telecommunication systems, but also for applications at the longer wavelengths. The chalcogenides have particularly low phonon frequencies and, therefore, they are potentially attractive hosts for laser transitions in the near- and middle-IR region.[11] That is, transitions that would be quenched by the high-frequency phonons in oxide and fluoride glasses are possible in these hosts. Some of the same RE ions have been doped into the chalcogenide glasses that have been used for the HMFGs. For example, Nd^{3+}-doped sulfide-glass fibers have lased at 1.08 μm, while both Nd^{3+}- and Pr^{3+}-doped sulfide fibers have been used as amplifiers at 1.08 and 1.34 μm, respectively. One of the exciting aspects for these host glasses is the possibility of lasing from transitions at long wavelengths. Pr^{3+}, for example, has emission bands at 1.3, 1.6, 2.9, 3.4, 4.5, 4.8, 4.9, and 7.2 μm.[11]

The discussion in Sec. 10.2 pertaining to OFAs operating at 1300 nm focused on Pr^{3+}-doped fluoride glass fibers. This fiber system, however, has a poor quantum efficiency (3% to 5%) and a small absorption cross section. If Pr ions are instead added to chalcogenides with their lower phonon frequencies, as low as 350 cm^{-1} for selenides compared to ~500 cm^{-1} for fluorides, then the efficiency is expected to be better. Additionally, Dy^{3+} ions are a good choice for 1300-nm applications, especially in selenide glasses, where efficiencies approaching 90% can be realized due to the low phonon energy of the selenide glass.

Cole et al.[12] have studied Pr^{3+} ions doped into GeAsGaSe glass fibers. Their work on Pr^{3+} doping de-emphasized the 1300-nm band in favor of the longer-wavelength emission bands in the 3- to 5-μm region. Pr^{3+} has two strong absorption bands at 1.5 and 2.0 μm for pumping into this region. The energy-level diagram for Pr^{3+} ions is given in Fig. 10.7(a) and the mid-IR emission from a GeAsGaSe MM fiber is shown in Fig. 10.7(b).[12] In Fig. 10.7(b) it can be seen that there is a reduction in the emission at ~4.1 and ~4.5 μm due to the strong absorption of S−H and Se−H impurities, respectively. These absorption bands are also shown in Figs. 5.10 and 5.11, for the sulfide and selenide glasses. Clearly, a reduction in these hydrogen-bonded impurities would be necessary for any active device operating most efficiently across this broad emission region.

Dy^{3+}, Pr^{3+}, and Tb^{3+} ions have been studied in chalcogenide host glasses for OFAs at 1300 nm.[13–18] Adding Pr^{3+} to GaLaS glass, for example, increases the quantum efficiency from a few percent in fluoride glass to about 30% in GaLaS glass.[19] Recently, however, there has been more emphasis placed on Dy^{3+} ions in

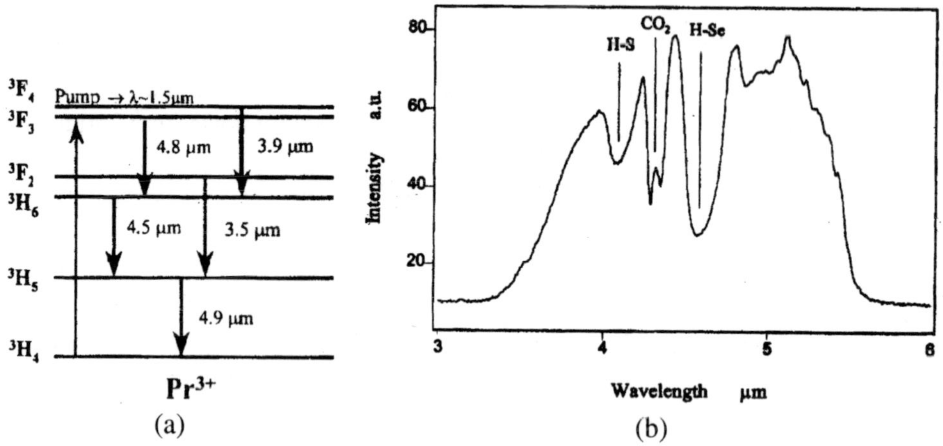

Figure 10.7 (a) Energy-level diagram for Pr^{3+} in chalcogenide glass host. (b) Emission spectrum for Pr^{3+} in selenide glass fiber. [Reprinted from Ref. 12, Copyright 1999, with permission from Elsevier.]

both the selenide and LaS glasses, as these glasses have low phonon frequencies.[20] Specifically, the total quantum efficiency at 1300 nm for Dy^{3+} in either GaLaS or selenide glasses is at least 30%, or considerably higher than Pr^{3+} in fluoride glasses. Also, because of the large cross sections, i.e., the high absorption for these glasses, the length of the fiber needed is only about 1/10 of that needed for Pr-doped fluoride glass fibers. The energy-level diagram for Dy^{3+} is shown in Fig. 10.8(a) along with the fluorescence of Dy^{3+} in GaLaS glass pumped with 0.907-μm radiation in Fig. 10.8(b). It is interesting to note two things from the data in Fig. 10.8. First, the $^6H_{9/2} \rightarrow {}^6H_{11/2}$ transition leading to 1300-nm emission would be entirely quenched for Dy^{3+} in fluoride glass hosts. Second, the emission peak shown in Fig. 10.8(b) is centered nearer the 1310-nm zero dispersion wavelength for standard telecommunications fibers than the same peak in Pr-doped fluoride or oxide glasses (compare to the data for Pr shown in Fig. 10.6). Dy^{3+} has also been doped into GeAsGaSe fibers by Cole et al.[12] These researchers added 400 ppm Dy^{3+} into a core-only fiber. The measured losses were a respectable 0.8 dB/m at 6 μm and 1.5 dB/m at 1.5 μm. When a 4-μm-core SM fiber was drawn for the same composition, the losses were found to increase to 3 dB/m at 1.55 μm. These fibers have applications both at 1.3 μm and in the 3- to 5-μm band. Longer wavelength emission from 7 to 9 μm has been demonstrated in Tb-doped GeAsGaSe and GeAsGaTe glasses.[16] This represents the longest wavelength emission seen from any RE-doped glass or crystal system. The measured lifetime of the excited state was 12 μs and 80 μs for the selenide and telluride glasses, respectively. Recently, Churbanov et al.[21] studied As-Se and As-S-Se glasses doped with Tb^{3+}, Dy^{3+}, and Pr^{3+}. Their As-Se fibers had a loss of 1.5 dB/m in the 6- to 9-μm range.

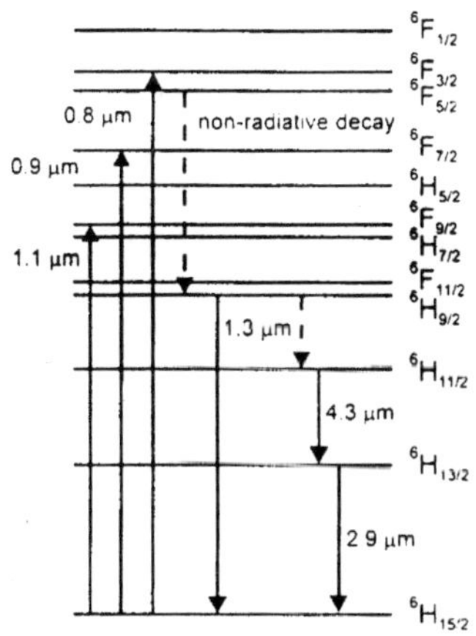

(a)

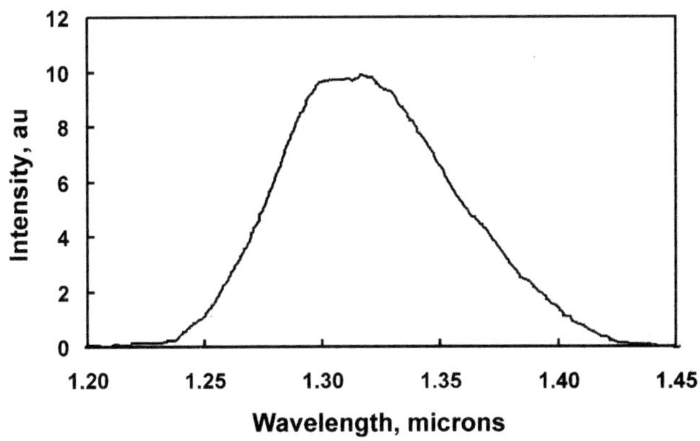

(b)

Figure 10.8 (a) Energy-level diagram for Dy^{3+}-doped gallium lanthanum sulfide (GLS) glass and (b) corresponding fluorescence spectrum of Dy^{3+}-doped GLS glass pumped at 0.907 μm.[20] [Reprinted with permission.]

10.4 Polycrystalline optical fiber amplifiers

Most of the studies to date on active IR fibers have concentrated on RE doping of glass fiber hosts. In fact, many of the studies on the RE-doped fluorides stem from the mid-1970s, when these materials were first developed. Very recently there has been some interest and research done on doping Ag-halides for possible use as OFAs. This effort has been largely led by Prof. Katzir and his group at Tel Aviv University. They have been able to grow mixed Ag-halide crystals with a host of RE and transition-metal elements, but most work has been done on Er^{3+}, Pr^{3+}, and Nd^{3+} doped into AgBrCl crystals.

After performing a series of measurements on RE dopants in the Ag halides, Oron et al.[22] focused on Nd^{3+} as the system to model for use as an OFA. They chose two configurations for their calculations and ultimately for fabrication into a PC fiber. The first was a single-clad 20-μm-core fiber and the second a double-clad 20-μm-core, 100-μm first clad, surrounded by a second cladding. The double-clad approach was chosen because this structure has been found to be a convenient method of increasing the pumping power without damage to the small core of an SM fiber. That is, higher powers are pumped into the larger-diameter first cladding, and as this power propagates down the fiber it transfers to the smaller core. Their results for the double-clad Ag-halide fiber are shown in Fig. 10.9. A sketch of a typical double-clad geometry is given in Fig. 10.9(a), and the results of their measurements using a 10-W pump beam for the Nd^{3+}-doped (Nd^{3+} concentration of 10^{25} m^{-3}) double-clad fiber are given in Fig. 10.9(b). The PC fiber had a core diameter of 20 μm and a first-cladding diameter of 200 μm. The data in Fig. 10.9(b) shows the amplification of 1.06-μm light from an input of 0.1 W to a maximum of 4 W near the end of a 1.2-m-long fiber. In general, these results at 1.06 μm do not show the strong amplification normally seen in OFAs. One reason for this is the rather strong absorption by the undoped fiber at these short wavelengths. As may be seen from Fig. 6.10, the short-wavelength losses are rather high below about 3 μm. More work is underway to investigate other RE ions that have transitions in the 5-μm region, where these Ag-halide fibers have considerably lower loss.[23–25]

Finally, there is a new and potentially interesting application for the Ag-halide fibers as fiber Bragg gratings. Uman et al.[26] have illuminated Ag halide crystals with 353-nm light using a phase mask to produce a periodic structure on the surface of the crystal. The photolytic process leaves silver strips on the surface that are then etched off, leaving grooves as deep as 1.1 μm. They have been able to make a grating with a 100-μm period on the Ag-halide crystal using this technique.

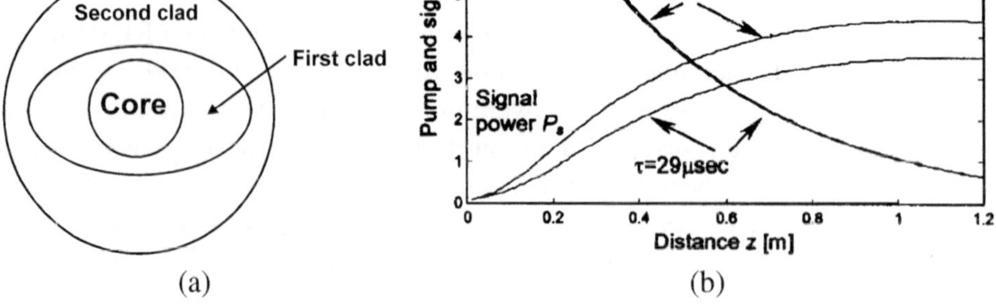

(a) (b)

Figure 10.9 (a) Configuration of double-clad Ag halide PC fiber for efficient pumping of light into the core and (b) pump attenuation and signal amplification for Nd^{3+}-doped Ag halide double clad fiber.[22]

References

1. W. J. Miniscalco, "Erbium-doped glasses for Fiber Amplifiers at 1500 nm," *J. Lightwave Tech.*, Vol. 9, pp. 234–250 (1991).
2. P. C. Becker, N. A. Olsson, and J. R. Simpson, *Erbium-Doped Fiber Amplifiers: Fundamentals and Technology*, Academic Press, San Diego, CA (1999).
3. J. Hecht, *Understanding Fiber Optics*, Prentice Hall, Upper Saddle River, NJ (2002).
4. M. Poulain, M. Chanthanasinh, and J. Lucas, "New fluoride glasses," *Mat. Res. Bull.*, Vol. 12, pp. 151–156 (1977).
5. B. J. Ainslie, S. T. Davey, D. Szebesta, J. R. Williams, M. W. Moore, T. Whitley, and R. Wyatt, "A review of fluoride fibres for optical amplification," *J. Non-Cryst. Solids*, Vol. 184, pp. 225–228 (1995).
6. D. M. Spirit, G. R. Walker, P. W. France, S. F. Carter, and D. Szebesta, "Characterization of diode-pumped erbium-doped fluorzirconate fibre optical amplifier," *Electron. Lett.*, Vol. 26, pp. 1218–1220 (1990).
7. M. C. Brierley and P. France, "Neodymium-doped fluorozirconate fibre laser," *Electron. Lett.*, Vol. 23, pp. 815–817 (1987).
8. M. C. Brierley and P. France, "Continuous wave lasing at 2.7 microns in an erbium-doped fluorozirconate fibre," *Electron. Lett.*, Vol. 24, pp. 935–937 (1988).
9. R. Allen, L. Esterowitz, and R. J. Ginther, "Diode-pumped single-mode fluorozirconate fiber laser from the $^4I_{11/2}$ to the $^4I_{13/2}$ transition in erbium," *Appl. Phys. Lett.*, Vol. 56, pp. 1635–1637 (1990).
10. R. M. Percival, D. Szebesta, S. T. Davey, N. A. Swain, and T. A. King, "Thulium sensitized holmium-doped cw fluoride fibre laser of high efficiency," *Electron. Lett.*, Vol. 28, pp. 2231–2232 (1992).
11. J. Sanghera, L. B. Shaw, L. Busse, V. Nguyen, P. Pureza, B. Cole, B. B. Harbison, and I. Aggarwal, "Development and infrared applications of chalcogenide glass optical fibers," *Fiber and Integrated Optics*, Vol. 19, pp. 251–274 (2000).
12. B. Cole, B. L. Shaw, P. Pureza, R. Mossadegh, J. Sanghera, and I. Aggarwal, "Rare-earth doped selenide glasses and fibers for active applications in the near and mid-IR," *J. Non-Cryst. Solids*, Vol. 256/257, pp. 253–259 (1999).
13. R. C. Schimmel, A. J. Faber, H. de Waardt, R. G. C. Beerkens, and G. D. Khoe, "Development of germanium gallium sulphide glass fibres for the 1.31 micron praseodymium-doped fibre amplifier," *J. Non-Cryst. Solids*, Vol. 284, pp. 188–192 (2001).
14. T. Schweizer, F. Goutaland, E. Martins, D. W. Hewak, and W. S. Brocklesby, "Site-selective spectroscopy in dysprosium-doped chalcogenide glasses for 1.3 micron optical-fiber amplifiers," *J. Opt. Soc. Am. B*, Vol. 18, pp. 1436–1442 (2001).
15. J. Heo and Y. B. Shin, "Absorption and mid-infrared emission spectroscopy of Dy^{3+} in Ge-As (or Ga)-S glasses," *J. Non-Cryst. Solids*, Vol. 196, pp. 162–167 (1996).

16. L. B. Shaw, B. Cole, J. S. Thielen, J. Sanghera, and I. Aggarwal, "Mid-wave IR and long-wave IR laser potential of rare-earth doped chalcogenide glass fiber," *IEEE J. Quantum Electron.*, Vol. 48, pp. 1127–1137 (2001).
17. J. Heo, "Rare-earth doped chalcogenide glasses for fiber-optic amplifiers," *J. Non-Cryst. Solids*, Vol. 326&327, pp. 410–415 (2003).
18. D. Lezal, J. Pedlikova, J. Zavadil, P. Kostka, and M. Poulain, "Preparation and characterization of sulfide, selenide, and telluride glasses," *J. Non-Cryst. Solids*, Vol. 326&327, pp. 47–52 (2003).
19. D. W. Hewak, J. A. Medeiros Neto, B. N. Samson, K. P. Jedrzejewski, and D. N. Payne, "Quantum efficiency of praseodymiuim doped Ga:La:S glass for 1.3 micron optical fibre amplifiers," *IEEE Photonics Technology Lett.*, Vol. 6, pp. 605–608 (1994).
20. D. W. Hewak, B. N. Samson, J. A. Medeiros Neto, R. I. Laming, and D. N. Payne, "Emission at 1.3 microns from dysprosium-doped Ga:La:S glass," *Electron. Lett.*, Vol. 30, pp. 968–970 (1994).
21. M. F. Churbanov, I. V. Scripachev, V. S. Shiryaev, V. G. Plotnichenko, S. V. Smetanin, E. B. Kryukova, Y. N. Pyrkov, and B. I. Galagan, "Chalcogenide glasses doped with Tb, Dy, and Pr ions," *J. Non-Cryst. Solids*, Vol. 326&327, pp. 301–305 (2003).
22. R. Oron, D. Bunimovich, L. Nagli, A. Katzir, and A. A. Hardy, "Design considerations for rare earth doped silver halide fiber amplifiers," *Opt. Eng.*, Vol. 40, pp. 1516–1520 (2001).
23. D. Bunimovich, L. Nagli, and A. Katzir, "Luminescence properties of praseodymium- and erbium-doped silver bromide crystals," *Appl. Opt.*, Vol. 36, pp. 7708–7711 (1997).
24. D. Bunimovich, L. Nagli, and A. Katzir, "The visible and infrared luminescence of activated silver bromide crystals," *Opt. Mater.*, Vol. 8, pp. 21–29 (1997).
25. L. Nagli, A. German, and A. Katzir, "Spectroscopic studies of Pr^{3+} ions in silver halide crystals," *Appl. Opt.*, Vol. 39, pp. 5070–5075 (2000).
26. I. Uman, L. Nagli, Z. Barkay, F. Moser, and A. Katzir, "Photolytic fabrication of phase gratings on silver halide crystals," *Appl. Opt.*, Vol. 41, pp. 4552–4556 (2002).

Chapter 11

New Concepts in IR Fiber Optics

11.0 Introduction

One of the most novel and unusual new fiber optics is the photonic crystal or photonic bandgap (PBG) fiber. Photonic crystal fibers (PCFs) were first formed in pure fused silica in 1995 by Russell[1,2] and his group at the University of Bath. PCFs are composed of a series of regularly spaced holes oriented along the fiber's optic axis. A schematic representation of a PCF is shown in Fig. 11.1(a). The air holes, with a refractive index of nearly 1, are surrounded by a honeycomb silica matrix. The defect site for the structure in Fig. 11.1(a) is the central air hole; but this defect could also be a solid core, which would be somewhat analogous to the solid core of a conventional SM fiber. Figure 11.1(b) shows a microphotograph of a silica PCF with its honeycomb structure of regularly spaced holes surrounding an air core.[3] These microstructured PCFs are today receiving enormous interest, as they have some very attractive properties that differ from those for solid-core/clad fibers.

The basic concept for photonic crystals was developed by Eli Yablonovitch in 1987.[4] Yablonovitch, then at Bell Laboratories, wanted to make a structure that had a forbidden region, or gap, in the photon bands analogous to the electronic bandgap in semiconductors. He reasoned that if he could construct a material that had a photonic bandgap, then it should be possible to tailor the optical properties of the material much like the electronic properties of semiconducting materials. Specifically, it would be possible to fabricate higher-capacity optical fibers, nanoscale lasers, and photonic integrated circuits. Just as the electrical properties of semiconductors are controlled by the lattice structure, the electrons and holes, and the dopants, photonic crystals are governed by the spacing of an array of holes in a matrix with a higher refractive index. The spacing of the holes is on the order of the wavelength of light. The fact that the holes form a periodic, hexagonal, close-packed array surrounding a larger central hole, as shown in Fig. 11.1(b), leads to a structure whose photonic energy spectrum has a gap for certain wavelength regions. That is, this PCF will have a series of wavelength regions for which light is not transmitted. All light in the bandgap will be reflected. For the 2D PCF shown in Fig. 11.1, this means that light traveling in the central hollow core and whose frequencies lie within the bandgap will be reflected. Therefore, the large central de-

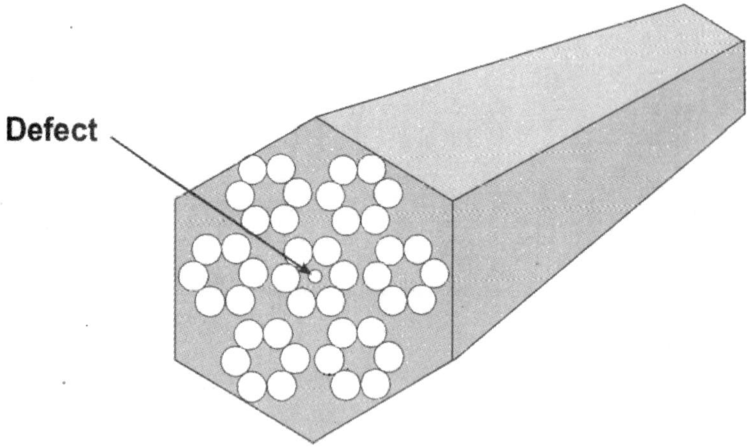

(a)

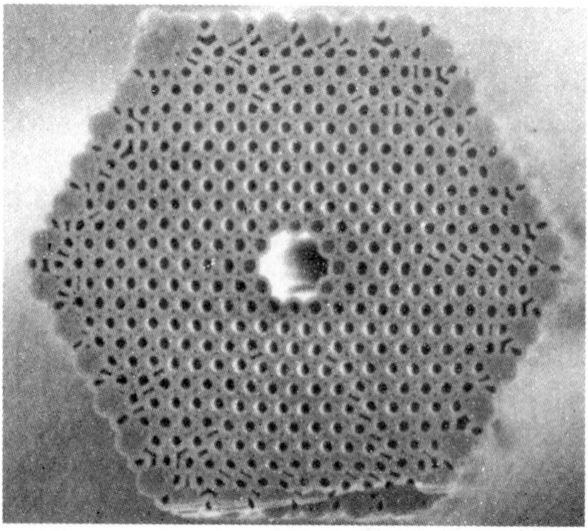

(b)

Figure 11.1 (a) Schematic representation of a photonic crystal fiber. The white circles are holes, and the central hole is a defect site in the photonic crystal at which light can be trapped by the photonic bandgap effect.[6] (b) Actual photonic bandgap fiber composed of lattice of holes surrounding an air-core defect.[3] [Reprinted with permission. Copyright 1999 American Association for the Advancement of Science.]

fect can be thought of as the fiber core surrounded by a photonic bandgap structure as its cladding.

PCFs like those shown in Fig. 11.1(b) have been extensively studied for a variety of applications. Some of the advantages of PCFs are their ability to handle higher laser powers, their very small or very large mode areas, and their ability to tailor their dispersive properties over a large spectral range. A disadvantage is that the PCFs with a hole as the central defect have higher loss than solid-core SM fibers or those PCFs with a solid core as the central defect. These holey silica fibers, as well as similar holey structures in a plastic fiber, however, do not transmit IR light.

Is there an IR fiber counterpart to the holey photonic bandgap silica or plastic fibers that transmit IR wavelengths beyond 2 μm? This question was answered in 1998 by Fink et al.[6] at MIT. This group of researchers proposed a bold, new idea for a hollow waveguide that would have low loss and also be omnidirectional. What Fink et al. showed was that if a hollow waveguide was fabricated with alternating layers of high- and low-index dielectric films deposited on the inner surface, then this waveguide would not only reflect light over a range of wavelengths in the photonic bandgap but also have the property that the losses would not depend on bending. The idea that the intensity of reflected light is independent of the angle of incidence is a remarkable discovery in view of Fresnel's equations, which are normally used to calculate the reflection coefficient for different incidence angles and polarization states. From the Fresnel equations for the coefficient of reflection, Eqs. (2.1a) and (2.1b), it may be seen that the reflectivity is strongly dependent on incident angle. The work of Fink et al. is in sharp contrast to this standard result. What Fink et al. show is that under the right conditions it is possible to make a reflector whose reflection is omnidirectional, or independent of incident angle and polarization. These concepts, discussed in detail in the next section, can be used to form a new type of hollow guide, which Fink et al. call an omnidirectonal waveguide, over a wide range of wavelengths. Additionally, the same principles apply to reflectors in general and, therefore, may be used to make a new class of mirrors and externally reflecting fiber optics.

11.1 Photonic bandgap hollow IR waveguides

The hollow waveguide structure that gives high reflectivity within select infrared wavelength bands is composed of alternating layers of high (n_H) and low (n_L) refractive index dielectric materials. The circular cross section of how such a waveguide would look is sketched in Fig. 11.2(a). In Fig. 11.2(b) a more detailed view, taken from Fink et al., is shown of the alternating layers in planar form. In general, the structure shown in Fig. 11.2 with a one-dimensional periodicity of concentric rings of alternating high/low index layers is called a Bragg fiber. Bragg fibers were first discussed in 1978 by Yeh et al.[7] The omnidirectional guide may be regarded as a special class of Bragg fibers for which the index contrast, i.e., the ratio of n_H/n_L, is large.

11.1.1 All-oxide glass Bragg fibers

First consider the simple case of a Bragg fiber formed in an all-oxide glass system for which n_H/n_L is not large. This structure would not have a photonic bandgap, and, therefore, it is not omnidirectional. As an example, assume that an oxide glass Bragg fiber preform would be fabricated using conventional MCVD methods; the same methods now used to make solid-core MM or SM fibers. The alternating layers of high/low oxide glasses would be deposited from two oxide glass materials,

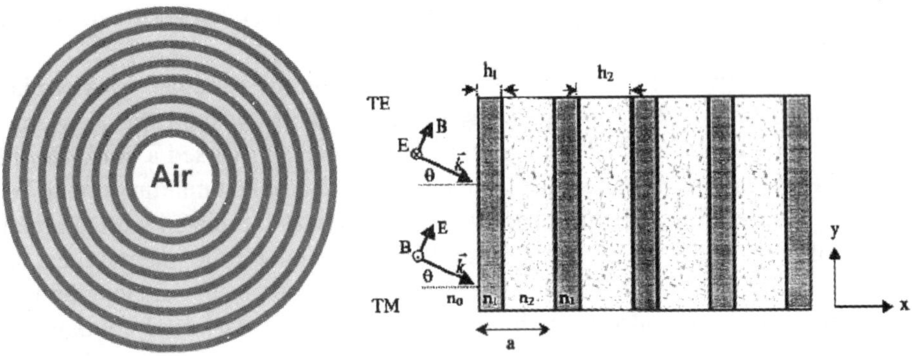

(a) (b)

Figure 11.2 (a) Structure of hollow Bragg fiber waveguide showing alternating layers of high (dark) and low (light gray) refractive index dielectric films. (b) Schematic of multilayer dielectric films with indices n_1 and n_2 and thicknesses h_1 and h_2, respectively. The incident wavevector $\vec{k}$ is indicated along with the E and B electric and magnetic fields, respectively. [Reprinted with permission from Ref. 6. Copyright 1998 American Association for the Advancement of Science.]

but with the condition that the index contrast should be as large as practically possible. The problem that exists at the outset is that it is extremely difficult to form a preform made of two oxide glasses that have high index contrast. As will be seen later for the omnidirectional guides, a good rule of thumb for the index contrast ratio necessary to achieve a relatively broad photonic bandgap is on the order of 2. This is virtually impossible to achieve if one restricts the choice of high- and low-index materials only to oxide glasses. Nevertheless, such a Bragg fiber can be constructed using oxide glasses even though the index contrast is rather small.

The author and his group at Rutgers University, along with Morse (then at Brown University), looked at some of the practical issues involved in making an oxide glass Bragg fiber. Morse suggested using silica doped with tantalum, n_H, and niobium, n_L, as the two oxide glasses to be used to make the preform and fiber. These two materials are thermally compatible and they can be deposited using the aerosol MCVD technology developed by Morse and his group.[8] The index difference for this combination is $\Delta n = n_H - n_L \cong 0.06$. Using the theory of Miyagi and Kawakami,[9] Harrington calculated the loss at 1.06 μm for a straight, 100-μm-bore Bragg fiber made from these materials propagating the lowest-order TE_{01} mode. The TE_{01} mode is the lowest-loss mode for this all-dielectric structure. By contrast, the HE_{11} mode is 1.5 times more lossy than the TE_{01} mode. The calculated losses are shown in Fig. 11.3 for this waveguide as a function of the number of dielectric layers (not pairs). This data illustrates a very important point—the number of layers needed to achieve a low loss is substantial. To reach a loss of 10^{-3} dB/m or 1 dB/km would require at least 60 layers! As will be seen in Sec. 11.1.2 on omnidirectional or photonic bandgap waveguides, many fewer layers are needed to achieve the same or even lower loss when the index contrast is high. The primary reason for this is the lack of index contrast for this Bragg fiber in comparison to

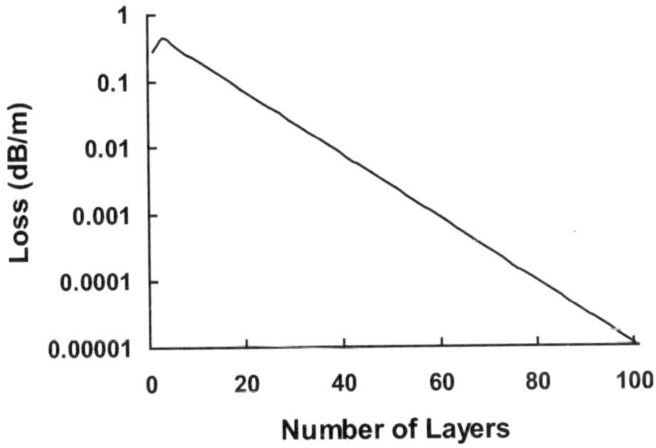

Figure 11.3 Calculated loss as a function of the number of alternating high/low dielectric layers for an all-oxide-glass, hollow Bragg fiber assuming an index difference of $\Delta n = 0.06$. The calculation is for the TE_{01} mode of a hollow guide operating at 1.06 μm with a bore size of 100 μm.

waveguides composed of dielectric layers with a much higher index contrast. To the author's knowledge no one has actually produced such an all-oxide glass Bragg fiber. One reason for this is the advent of the new omnidirectional fibers fabricated using high-index-contrast dielectric films.

11.1.2 Omnnidirectional hollow waveguides: MIT

The omnidirectional fiber proposed by Fink et al.[6,10,11] is based on two dielectric materials with a very large index contrast. The two materials chosen for their first omnidirectional or PBG hollow guide were tellurium ($n_H = 4.6$) and polystyrene ($n_L = 1.6$), which gives an index contrast ratio of 2.88. The ratio of the films' thickness, h_2(polystyrene)/h_1(tellurium) is 1.6/0.8 [see Fig. 11.2(b)]. The photonic band structure in Fig. 11.4 was calculated for these two materials for both TE, electric field perpendicular to the plane in Fig. 11.2(b), and TM, electric field parallel to the plane in Fig. 11.2(b), polarized light.[6] Figure 11.4 is plotted in terms of the wavevector k_y, which is parallel to the coating layers as shown in Fig. 11.2(b). The frequency of the photons is in units of c/a, where a is the thickness of one pair of high/low layers, i.e., $a = h_1 + h_2$. For light propagating in the core of a hollow guide the regions of interest in Fig. 11.4 are the regions above and contained within the light lines defined by $\omega = ck_y/n_0$, where n_0 is the refractive index of the air core equal to 1. For normal incidence (light perpendicular to the dielectric layers) $k_y = 0$. Most of the light in hollow guides, however, propagates at near-grazing incidence. For this case, light is incident at nearly 90°, $k_x = 0$, and the frequencies are those lying next to the light lines. Omnidirectional reflectivity or the photonic bandgap is the region within the light lines for which there is no transmitting state in the structure. In Fig. 11.4 this corresponds to the gray regions. From Fig. 11.4, one higher-order photonic bandgap can be seen above the fundamental bandgap region.

The omnidirectional range, defined by the frequency difference between the top and bottom of the gray bands in Fig. 11.4, and the midpoint frequency depend on the index contrast ratio, n_H/n_L, and the thickness of the films. For the example of tellurium/polystyrene (PS) films above, Fink et al. determined that the omnidirectional range was 5.6 μm centered at a wavelength of 12.4 μm. Their first experiments confirming omnidirectional behavior over this range were made using a planar structure composed of the nine-layer stack: Te/PS/Te/PS/Te/PS/Te/PS/Te. The spectral response of their reflector for TE and TM light incident from 0° up to 80° showed clearly high reflectivity throughout the calculated PBG region. In a subsequent paper Fink et al.[12] made a fairly large bore hollow guide using the same Te/PS film structure. The sample was prepared by coating the outside of a silica tube with an outer diameter of 1.92 mm with successive layers of Te and PS. The silica tube was then overcoated with a thick silicone rubber and then the silica tube was etched away, leaving the rubber tube with the inner layers of Te/PS films. High reflectivity for the tube was verified for both the straight 10-cm length and also when the tube was bent 90° with a 1-cm radius.

In a latter but related work, Temelkuran et al.[13] substituted polyethylene (PE) for polystyrene as the low-index material but retained Te as the high-index material. The photonic band structure was very similar to that for Te/PS, except that the results for their nine-layer Te/PE coatings exhibited two omnidirectional regions. The fundamental band went from 8.3 to 12.5 μm and the next higher order band

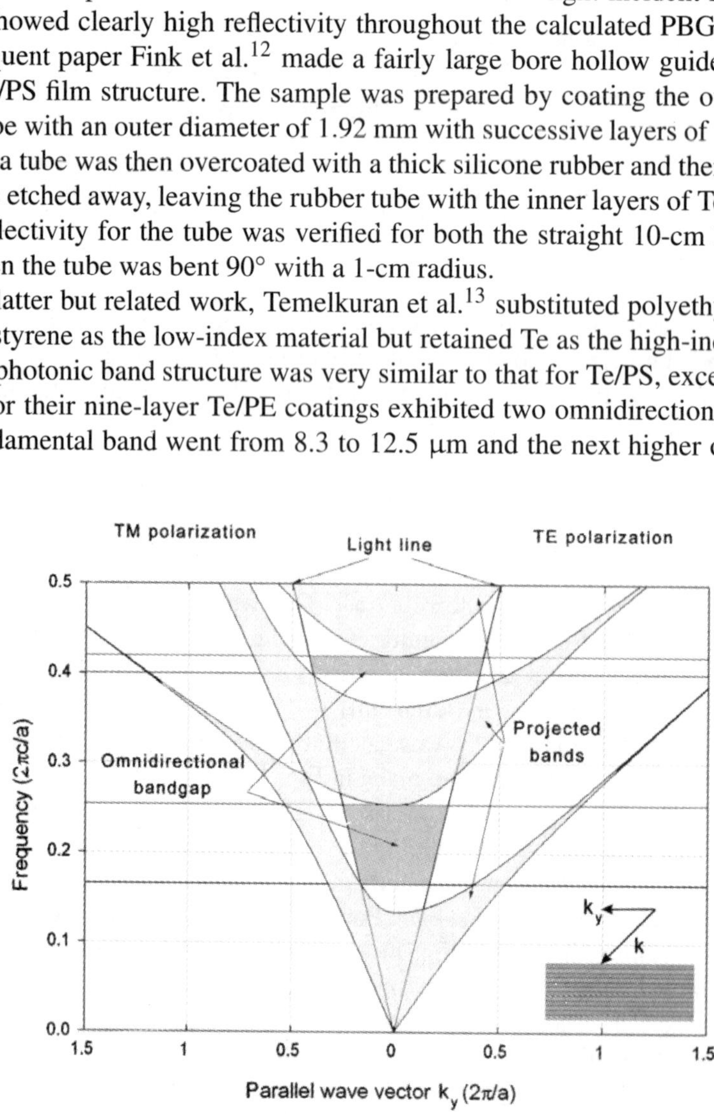

Figure 11.4 Photonic band diagram for a one-dimensional photonic crystal fabricated from alternating layers of high (2.8) and low (1.55) index materials. The gray regions within the projected bands represent omnidirectional or photonic bandgaps where high reflectivity will occur. [Reprinted from Ref. 6 with permission. Copyright 1998 American Association for the Advancement of Science.]

from 4.55 to 5.5 μm. Finally, Deopura et al.[14] demonstrated omnidirectional behavior in the visible region using a flat reflector composed of 19 alternating layers of tin sulfide ($n = 2.6$) and silica ($n = 1.46$).

One of the most exciting new waveguides to come out of the MIT work is a new structure composed of alternating layers of high-index chalcogenide glass and a polymer. The chalcogenide glass chosen was As_2Se_3, $n_H = {\sim}2.8$, and the low-index material was polyether sulphone (PES), $n_L = 1.5$. Temelkuran et al.[15] made a hollow preform using these two materials in a most unique manner. A 25- to 50-μm-thick sheet of PES was coated with As_2Se_3 using vapor deposition. The As_2Se_3/PES sheet was then rolled into a fiber preform for fiber drawing. A cross section of the drawn fiber is shown in Fig. 11.5. The innermost portion of the waveguide has a series of concentric layers clearly visible in the magnified microphotograph. These layers are surrounded by PES for support. As an example, one waveguide comprised PES films with a thickness of 900 nm and As_2Se_3 films with a thickness of 270 nm. The center wavelength was at 3.55 μm, yet they were able to observe up to the fourth-order bandgap, which appeared at 0.775 μm. Temelkuran et al. drew a variety of different bore sizes from their As_2Se_3/PES preforms. Some of their largest bore sizes varied between 700 and 750 μm. The

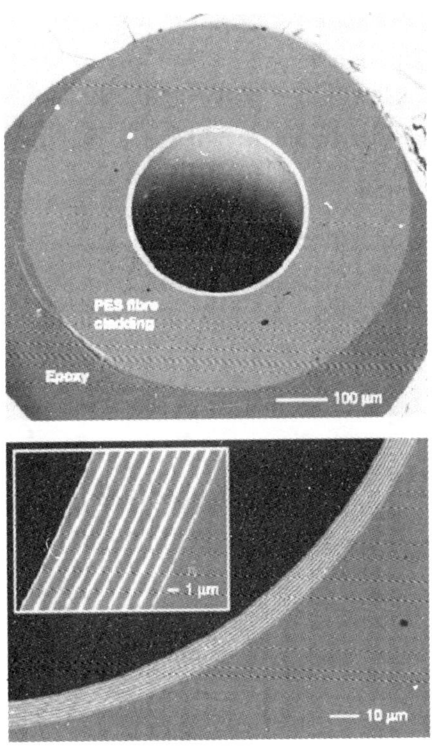

Figure 11.5 Cross section of multilayer 1-D PBG hollow waveguide drawn by MIT group. The high-index As_2Se_3 (white) and low-index PES (polyether sulphone) have film thicknesses that give the first-order photonic bandgap in the 10-μm region.[15] [Reproduced with permission.]

spectral response of one of these guides with a center frequency near 10 μm is shown in Fig. 11.6. The spectrum in Fig. 11.6 shows the fundamental PBG around 10 μm and the second-order gap at ∼5 μm. These are the regions of high transmission. The insert in Fig. 11.6 shows their results for loss measurements made at 10.6 μm using a CO_2 laser. They obtain a loss of 0.95 dB/m from their cutback method. While this loss is not as low as losses for the Ag/AgI HGWs, it is impressive, especially when one considers that the absorption in the two materials used to make their polymer/chalcogenide PBG guide is significantly higher than AgI or the other dielectrics used to make the Ag-based HGWs. Temelkuran et al. point out that the loss at 10.6 μm for their As_2Se_3 glass is at least 7 dB/m and for PES greater than 10^5 dB/m. Their results show that it is possible to combine high-loss materials in a multilayer stack to make a highly reflective PBG hollow guide. The reason for this is that the fields penetrate very little into the high-index-contrast dielectric layers. A high index contrast between the two materials is also the reason that one does not need many layers to form a good omnidirectional guide. This is in sharp contrast to the Bragg fibers discussed in Sec. 11.1.1, where it was seen that 50 to 100 layers are necessary to make a low-loss hollow waveguide. Finally, the authors did note that there was a loss on bending their guide. When the guides were bent to 90° the bending losses were found to be below 1.5 dB/m. In principle, one would expect no bending loss for an omnidirectional guide. The researchers feel that if they were to use more layers and a lower-loss polymer, the bending losses would be significantly reduced. Also, one problem with their method is that the preform is made by rolling, so that there is always an inner seam. If they were able to make a guide with concentric rings without a seam, then they expect that the losses would be lower yet.

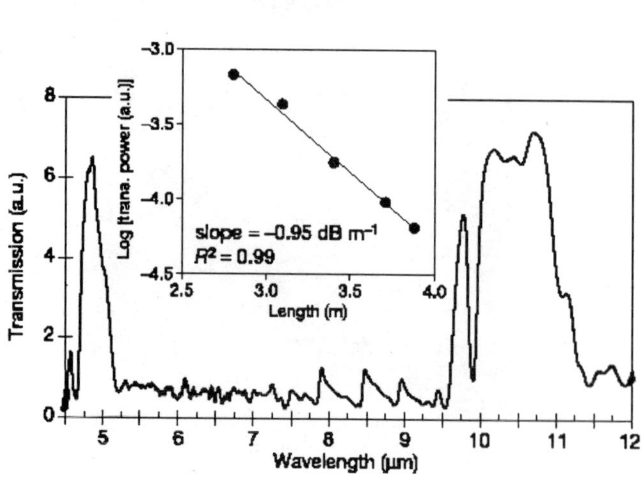

Figure 11.6 Transmission spectrum of waveguide shown in Fig. 11.5. Note the high transmission in the 10-μm region as well as the second-order bandgap ∼5 μm. The loss for a 700-μm-bore guide measured with a CO_2 laser, shown in the insert, is slightly less than 1 dB/m.[15] [Reproduced with permission.]

11.1.3 Other approaches to making PBG hollow waveguides

The fabrication of precision, multilayer thin-film stacks inside small-bore waveguides is clearly a challenging task. The approach taken by MIT to form PBG waveguides by rolling a sheet of plastic coated with chalcogenide glass is potentially an excellent approach for the fabrication of a preform that can be drawn into fiber. It is a distinct advantage of an all-glass and/or polymer system that a preform can be made and then drawn into long lengths of fiber. In this respect, methods for directly coating small-bore tubing using liquid or vapor phase processes are less efficient and more costly than if a preform can be made for fiber drawing. Nevertheless, the crystalline and polymeric materials used for liquid phase coating have proven to be very low loss and, therefore, they should prove to be very useful in making PBG waveguides if the layers can be deposited uniformly over relatively long lengths of tubing.

An unusual approach to making a PBG preform has recently been taken by Gibson and Harrington.[16] They use some of the same glassy materials used by the MIT group, but their approach to making the preform is entirely different. What Gibson and Harrington have done is to use a hot extrusion technique to press a stack of high- and low-index plates into a hollow preform. Specifically, they used As_2Se_3 glass for the high-index material, and for the low-index material they selected either polysulphone (PSU) or a low-temperature lead borosilicate or fluoride glass. The overall extrusion concept is diagrammed in Fig. 11.7. In Fig. 11.7(a) a stack of high- and low-index plates are positioned for insertion into the hot extrusion die shown in Fig. 11.7(b). There are several ways to extrude a tube, but

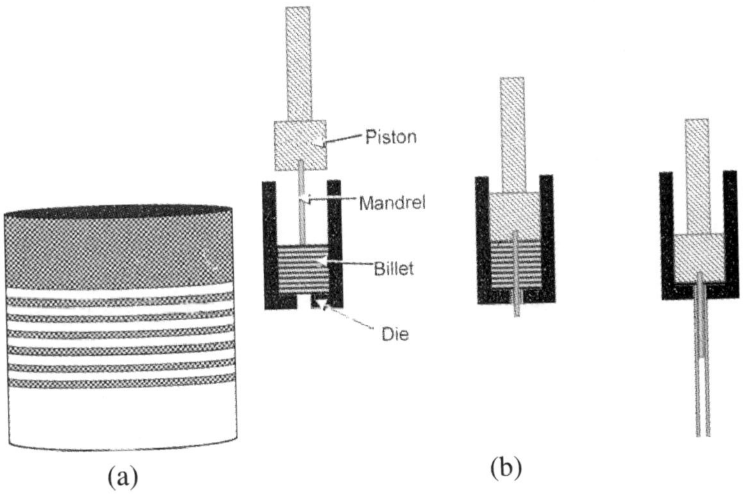

Figure 11.7 (a) Stack-of-plates billet for the extrusion of multilayer photonic bandgap hollow waveguide. The gray areas are high-index chalcogenide glass, and the light areas are low-index polymers like polysulphone or low melting oxide or fluoride glass. (b) Moving mandrel extrusion die/piston/billet assembly for extrusion of stack-of-plates billet into hollow PBG fiber.[16]

the most direct is to press a mandrel through the stack of plates as the material is extruded through a steel or tungsten carbide die. As_2Se_3 glass and PSU have compatible thermal properties so it is expected that they can be co-extruded with minimal mixing. Seddon et al.[17] and Itoh et al.[18] have extruded core/clad preforms of chalcogenide and fluoride glasses, respectively, using a stack of two plates to form the core and clad regions. To make a PBG hollow guide, however, it is necessary to extrude more than two plates, and this becomes more difficult. Examples of Gibson and Harrington's extrusion of more than 20 plates into a preform are shown in Fig. 11.8. This particular extrusion had a solid core, but Fig. 11.8 does indeed show that it is possible to extrude plates into a preform with cylindrical layers of alternating high- and low-index materials. But, it can be seen from the data that the cylindrical layers are not at all uniform in thickness nor is the circularity good. This group is working on perfecting their extrusion technique and also extruding hollow preforms. In principle this could be a good way to make a PBG preform with truly concentric circular layers, or even to directly extrude a hollow PBG fiber.

The approach used by Fink et al. and coworkers at MIT for making a PBG hollow guide is to use a series of high/low-index materials to form an all-dielectic stack inside a glass or plastic tube. The lowest-loss mode for these structures is the TE_{01} mode. Recently, Matsuura et al.[19] have altered this approach slightly by adding a metallic layer of Ag as part of the reflecting coating stack. Using a metal-

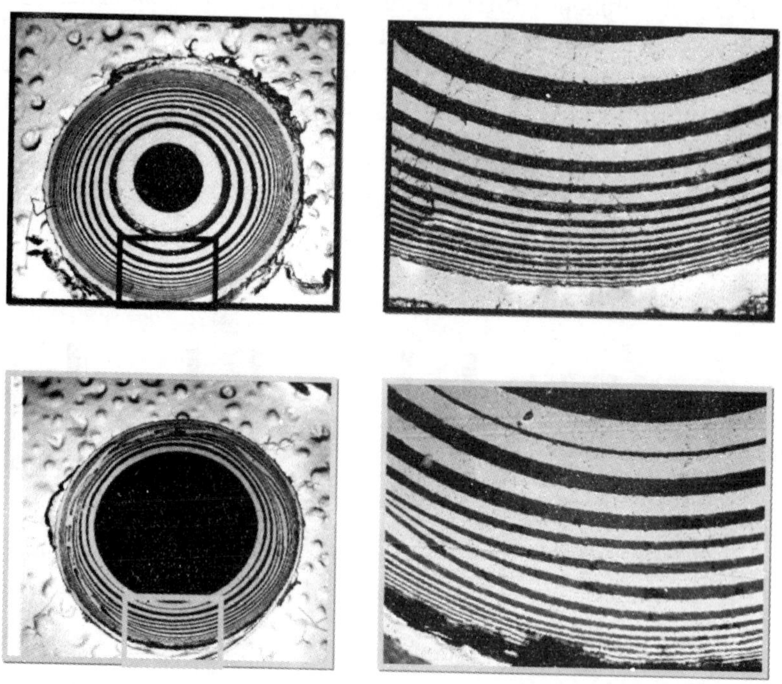

Figure 11.8 SEM micrographs of transverse sections taken 25 cm (top) and 33 cm (bottom) from the beginning of the co-extruded As_2Se_3–polysulfone rod. Images on the right are enlarged sections of the images on the left. (Dark regions: As_2Se_3; light regions: polysulfone.)[16]

lic layer will suppress the leaky losses. Specifically, Matsuura et al. point out that if a metallic layer is used as the first (nearest the tube wall) layer, then fewer dielectric layers are needed to achieve a low loss compared to the all-dielectric waveguides. As an example, they propose the structure $Ag/SiO_2/Ta_2O_5/SiO_2 \cdots$ deposited inside a silica tube. They have calculated that the loss for a 530-μm-bore waveguide coated with only five films of n_L (SiO_2) and n_H(Ta_2O_5) plus Ag will give a loss of 10^{-3} dB/m at 1.06 μm. In contrast, it would take 37 layers of these same two materials without the Ag coating to achieve the same loss. To obtain the lowest loss with only a few dielectric coatings it was also necessary to make the last dielectric layer in the stack (next to the air core) have a non–quarter-wavelength thickness. That is, all other dielectric layers would have quarter-wavelength thicknesses but the last one. The results of their calculations for these two dielectric materials in an all-dielectric PBG hollow waveguide as well as for a PBG hollow waveguide employing an Ag film with the last dielectric optimized for lowest loss are shown in Fig. 11.9. This data shows how an all-dielectric structure with 37 layers equals the loss for a Ag plus five-layer waveguide. Matsuura et al. also note that the bandwidth is wider for their guide versus the 37-layer guide. In some initial experiments these researchers coated a 530-μm-bore silica tube with $Ag/SiO_2/Ta_2O_5/SiO_2$. The coatings were applied using liquid-phase chemistry methods that included a sol-gel process for the Ta_2O_5 films. The last SiO_2 layer had a non–quarter-wavelength thickness of 0.17 μm at 1.06 μm. While the spectral losses for their waveguides were much higher than those shown in Fig. 11.9, the spectra did exhibit the requisite interference bands at 0.7, 0.9, and 1.4 μm. One reason for the higher loss is that the losses in Fig. 11.9 are for the lowest-loss HE_{11} mode only and their data was taken using a small f/# launch from their spectrometer. This resulted in launching higher-order modes and, therefore, higher losses.

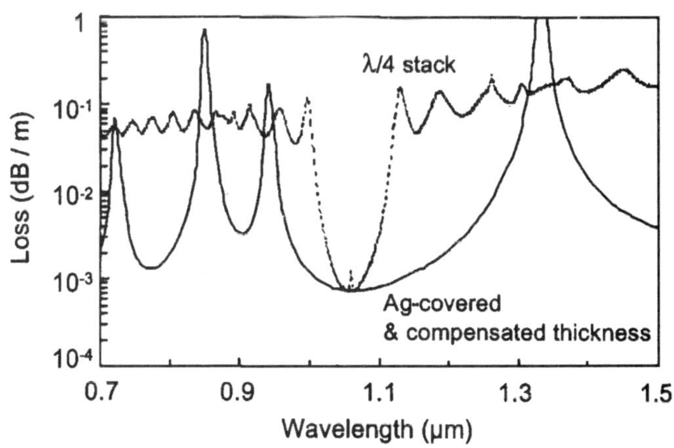

Figure 11.9 Calculated loss spectra for the HE_{11} mode in an all-dielectric structure having 37 dielectric layers (λ/4 stack) compared to using only five dielectric layers over an innermost Ag layer.

11.2 Other novel approaches to making waveguides

An unusual approach has been proposed by Matsuura et al.[20] for a hollow waveguide based on a leaky tube concept. They propose that if a glass tube is drawn with a very thin and uniform wall, then this tube will exhibit interference bands resulting from reflections within the thin glass wall. That is, there will be multiple interference bands resulting from light reflecting off the inner and outer glass walls, just as there would be from any thin film or thin transparent plate. Furthermore, if the outside of their thin-walled glass tube is coated with Ag, then the interference peaks will shift as a result of phase shifts occurring at the metal/glass boundary. To demonstrate this effect, Matsuura et al. drew Pyrex glass tubing with a wall thickness of 9.92 μm and an inner bore of 280 μm. Figure 11.10 gives their calculated and measured results for this glass tube waveguide designed to have minimal loss at 2.94 μm. The interference bands can clearly be seen in Fig. 11.10. Notice that there are many interference peaks as a result of the large spectral range and the thickness of the tubing wall. A wall thickness of 9.92 μm is substantially thicker than the thickness of the AgI or polymer films (<1 μm) used to coat Ag in the more conventional HGWs. The investigators are taking advantage of the minima in the interference spectrum to position a minimum at some convenient IR laser wavelength. In this case it is the Er:YAG laser wavelength at 2.94 μm. The higher experimental losses are attributed to the relatively high value of *k* for Pyrex beyond 3 μm. They point out that if a more transparent glass tube were used, then the losses would be less. For example, a good choice for longer wavelengths would be a chalcogenide glass.

Researchers at the Naval Research Laboratory (NRL) have recently made a chalcogenide glass photonic crystal fiber using $As_{39}Se_{61}$ glass. The cross section of their fiber is shown in Fig. 11.11(a). This photonic crystal fiber was made by

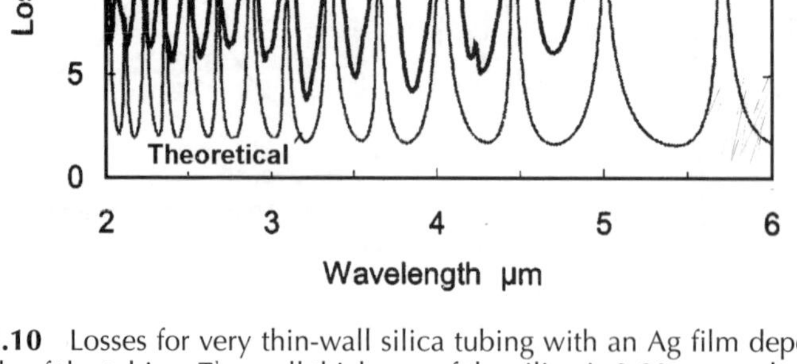

Figure 11.10 Losses for very thin-wall silica tubing with an Ag film deposited on the outside of the tubing. The wall thickness of the silica is 9.92 μm and, therefore, the silica acts as a thin film with multiple interference effects as seen in the data.[20] [Reprinted with permission.]

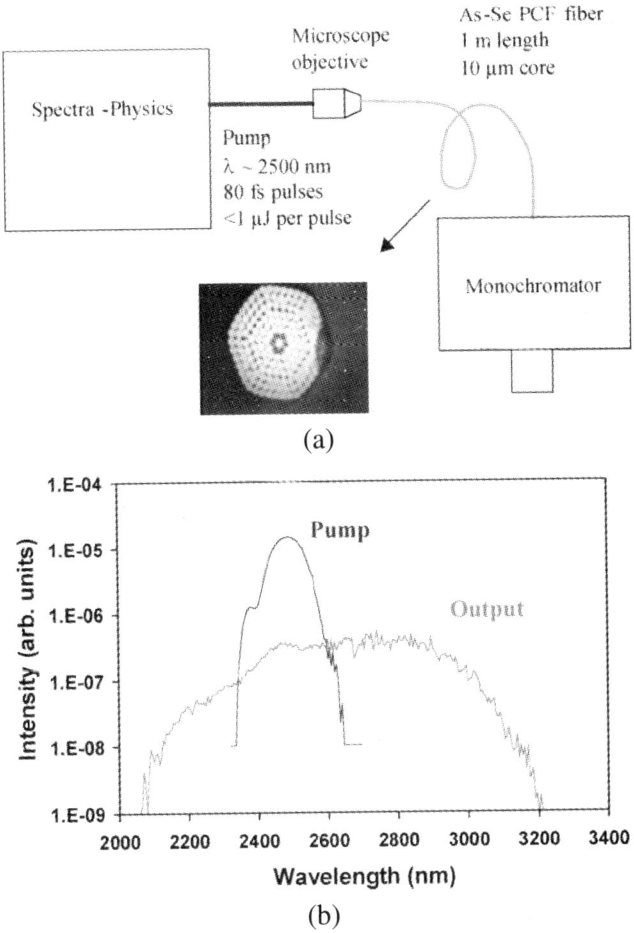

Figure 11.11 (a) Cross section of early $As_{39}Se_{61}$ photonic crystal fiber with a diameter of 130 μm and a solid core of about 10 μm. (b) Generation of super-continuum light on pumping with Ti:sapphire laser producing 80 fs pulses at 2.5 μm.

stacking tubes and canes to make a preform and then drawing the preform to produce the 130-μm-diameter fiber with a solid defect core of diameter approximately 10 μm shown in Fig. 11.11(a). The loss for this fiber was about 5 dB/m at 1.5 μm. Using a 1-m length of this fiber, J. Sanghera, L. B. Shaw, and I. D. Aggarwal[21] at NRL were able to excite the fiber by pumping with a Ti:sapphire femtosecond laser tuned to 2.5 μm, delivering a pulse with an 80 fs pulsewidth and an energy of less than 1 μJ per pulse. Figure 11.11(b) shows the super-continuum generation observed between 2 and 3 μm. Their work demonstrates the great potential for novel mid-IR sources.

References

1. J. C. Knight, T. A. Birks, P. Russell, and D. M. Atkin, "All-silica single-mode optical fiber with photonic crystal cladding," *Opt. Lett.*, p. 1549 (1996).

2. J. C. Knight, T. A. Birks, P. Russell, and D. M. Atkin, "All-silica single-mode optical fiber with photonic crystal cladding: errata," *Opt. Lett.*, Vol. 22, pp. 484–485 (1997).
3. R. F. Cregan, B. J. Mangan, J. C. Knight, T. A. Birks, P. Russell, P. J. Roberts, and D. C. Allan, "Single-mode photonic and gap guidance of light in air," *Science*, Vol. 285, pp. 1537–1539 (1999).
4. E. Yablonovitch, "Inhibited spontaneous in solid-state physics and electronics," *Phy. Rev. Lett.*, Vol. 58, p. 2062 (1987).
5. S. Barkou, J. Broeng, and A. Bjarklev, "Silica-air crystal fiber design that permits waveguiding by a true photonic bandgap effect," *Opt. Lett.*, Vol. 24, pp. 46–48 (1999).
6. Y. Fink, J. N. Winn, S. Fan, C. Chen, J. Michel, J. Joannopoulos, and E. Thomas, "A dielectric omnidirectional reflector," *Science*, Vol. 282, pp. 1679–1682 (1998).
7. P. Yeh, A. Yariv, and E. Marom, "Theory of Bragg fiber," *J. Opt. Soc. Am. A*, Vol. 68, pp. 1196–1201 (1978).
8. T. F. Morse, A. Kilian, L. Reinhart, and J. R. Cipolla, "Aerosol transport for optical fiber core doping: a new technique for glass formation," *J. Aerosol Science*, Vol. 22, pp. 657–666 (1991).
9. M. Miyagi and S. Kawakami, "Design theory of dielectric-coated circular metallic waveguides for infrared transmission," *J. Lightwave Technol.*, Vol. LT-2, pp. 116–126 (1984).
10. Y. Fink, D. J. Ripin, S. H. Fan, C. P. Chen, J. D. Joannopoulos, and E. L. Thomas, "Guiding optical light in air using an all-dielectric structure," *J. Lightwave Technol.*, Vol. 17, pp. 2039–2041 (1999).
11. B. Temelkuran, E. L. Thomas, J. D. Joannopoulos, and Y. Fink, "Low-loss infrared dielectric material system for broadband dual-range omnidirectional reflectivity," *Opt. Lett.*, Vol. 26, pp. 1370–1372 (2001).
12. Y. Fink, D. J. Ripin, S. H. Fan, C. P. Chen, J. D. Joannopoulos, and E. L. Thomas, "Guiding optical light in air using an all-dielectric structure," *J. Lightwave Technol.*, Vol. 17, pp. 2039–2041 (1999).
13. B. Temelkuran, E. L. Thomas, J. D. Joannopoulos, and Y. Fink, "Low-loss infrared dielectric material system for broadband dual-range omnidirectional reflectivity," *Opt. Lett.*, Vol. 26, pp. 1370–1372 (2001).
14. M. Deopura, C. K. Ullal, B. Temelkuran, and Y. Fink, "Dielectric omnidirectional visible reflector," *Opt. Lett.*, Vol. 26, pp. 1197–1199 (2001).
15. B. Temelkuran, S. D. Hart, G. Benoit, J. D. Joannopoulos, and Y. Fink, "Wavelength-scalable hollow optical fibres with large photonic bandgaps for CO_2 laser transmission," *Nature*, Vol. 420, pp. 650–653 (2002).
16. D. Gibson and J. A. Harrington, "Extrusion method for fabrication of hollow waveguides with 1D photonic bandgap structure," Proc. SPIE, Vol. 4616, pp. 91–96 (2002).
17. D. Furniss and A. B. Seddon, "Towards monomode proportioned fibreoptic preforms by extrusion," *J. Non-Cryst. Solids*, Vol. 256, pp. 232–236 (1999).

18. K. Itoh, K. Miura, M. Masuda, M. Iwakura, and T. Yamashita, "Low-loss fluorozirco-aluminate glass fiber," *J. Non-Cryst. Solids*, Vol. 167, pp. 112–116 (1994).
19. T. Katagiri, Y. Matsuura, and M. Miyagi, "Metal-covered photonic bandgap multilayer for infrared hollow waveguides," *Appl. Opt.*, Vol. 41, pp. 7603–7606 (2002).
20. Y. Matsuura, R. Kasahara, T. Katagiri, and M. Miyagi, "Hollow infrared fibers fabricated by glass-drawing technique," *Opt. Exp.*, Vol. 10, pp. 488–492 (2002).
21. J. Sanghera (private communication).

Index

James A. Harrington is currently Professor of Ceramic and Materials Engineering at Rutgers University. He has more than 30 years of research experience in the area of optical properties of solids. Since 1977, he has worked on all aspects of infrared fibers including fabrication, characterization, and applications. He is generally recognized as one of the world's leading experts in this continually evolving field.

Prior to joining the Fiber Optic Materials Research Program at Rutgers University in 1989, he was Director of Infrared Fiber Operations for Heraeus LaserSonics, where he led the research and development of infrared fiber optic delivery systems for CO_2 laser surgery. He was the inventor of both the hollow sapphire and hollow glass waveguides, which today are being used as CO_2 laser delivery systems in a variety of surgical applications. Before joining Heraeus LaserSonics, he was the Program Manager for IR fiber optics at Hughes Research Laboratories in Malibu, CA.

His current research interests include the development of a new class of ultralow-loss photonic bandgap HGWs, and the fabrication of HGW coherent image bundles for infrared imaging. Other research activities include the development of diffusing-tip silica fibers for use in photodynamic therapy (PDT) and prostate surgery, thin film coatings for hollow waveguides, IR fiber delivery systems for FEL medical lasers, and IR fibers for use in CO_2 laser-threat warning receivers.